Power Systems

Da Ruan · Paolo F. Fantoni (Eds.), Power Plant Surveillance and Diagnostics

Springer

Berlin
Heidelberg
New York
Barcelona
Hong Kong
London
Milan
Paris
Tokyo

Da Ruan · Paolo F. Fantoni (Eds.)

Power Plant Surveillance and Diagnostics

Applied Research with Artificial Intelligence

With 173 Figures
and 30 Tables

Springer

Dr. Da Ruan
Belgian Nuclear Research Centre
(SCK·CEN)
Boeretang 200
B- 2400 Mol
Belgium
e-mail: druan@sckcen.be

Dr. Paolo F. Fantoni
Institute for Energy (IFE)
P.O. Box 173
N-1751 Halden
Norway
e-mail: paolo.fantoni@hrp.no

ISBN 3-540-43247-7 Springer-Verlag Berlin Heidelberg NewYork

Library of Congress Cataloging-in-Publication Data

Power Plant Surveillance and Diagnostics: Applied Research with Artificial Intelligence; with 173 Figures and 30 Tables / Da Ruan; Paolo F. Fantoni (Eds.). – Berlin; Heidelberg; New York; Barcelona; Hong Kong; London; Milan; Paris; Tokyo: Springer, 2002
(Engineering online library)
(Power Systems)
ISBN 3-540-43247-7

Springer-Verlag is a company in the BertelsmannSpringer publishing group
http://www.springer.de

Printed in Germany

Typesetting: Camera ready by authors
Cover-Design: de´blik, Berlin
Printed on acid free paper SPIN: 10867802 62/3020/kk - 5 4 3 2 1 0

Foreword

In little more than a decade, the availability of nuclear power plants in the United States and most European and Pacific Rim countries has increased dramatically—from around 60% to about 90%. Such improvement is equivalent to adding approximately 30,000 MWe of new generating capacity in this period in the United States and perhaps an equal amount elsewhere. Some of this improvement has been due to longer fuel cycles—increases from 12 months to 18 or 24 months, but the major gains have been due to on-line maintenance and improved operation and planning by the utilities, accomplished primarily by training of plant operators and support personnel. However, there are emerging indications that the benefits of training are asymptotically approaching the limit beyond which it cannot improve availability and that plant personnel errors may also be approaching a comparable lower limit. Indeed, further improvement appears to be dependent upon the effective introduction and use of plant operational support systems.

The Halden Reactor Project (HRP) historically has focused on plant operational support system, i.e., such tools as **COPMA** (**C**omputerized **OP**eration **MA**nuals), **PEANO** (**P**rocess **E**valuation and **A**nalysis by **N**eural **O**perators), **TEMPO** (**T**h**E**r**M**al **P**erformance and **O**ptimization), **PLASMA** (**PL**Ant **S**afety **M**onitoring and **A**ssessment system), and others. Most of these tools fall into the category of "On-Line Monitoring of Instrument Channel Performance Systems" as described in Electric Power Research Institute (EPRI) Topical Report TR-104965 for which the United States Nuclear Regulatory Commission has issued a Safety Evaluation Report. While TR-104965 deals primarily with the Multivariate State Estimation Technique (MSET) developed by Argonne National Laboratory, the issues dealt with in the resultant NRC Safety Evaluation Report apply directly to the whole array of monitoring and diagnostic systems being developed by the HRP programs and others around the world.

Sponsorship of the recent Workshop on Power Plant Surveillance and Diagnostics (September 3-4, 2001) is one of many steps taken by HRP to present the attributes and capabilities of the above and other related plant operation support systems. Some 24 registered participants from 12 countries (six associated with OECD HRP) plus 16 HRP participants presented 20 technical papers over 1-½ days. The last half-day was spent visiting HRP, discussing PEANO and TEMPO, and demonstrating the Virtual Reality activities. Six additional papers were contributed for inclusion in this edited book (equivalent to a Workshop Proceedings) or in a special issue of *Intelligent Systems for Process Monitoring and Diagnosis*.

Topics presented ranged from detailed technical presentations on neuro-fuzzy logic-based surveillance and diagnostic systems to regulatory aspects of implementing such systems in licensed power plants. The results presented were impressive in a qualitative sense, but assessment of measurement errors of the systems described was not included in most papers. The potential adverse consequences of some systems being "ill posed" was discussed descriptively but not addressed analytically. These two topics represent areas where additional work is needed and which future HRP workshops might address.

One dark cloud hung over the conference and its participants. In a presentation the previous week at the IFE Summer School in Halden, Dominique Pirus of Electricité de France (EDF) said that EDF had initiated a moratorium on the installation of new plant operational support systems in their control rooms. The explanation for such drastic action was that the proliferation of such systems was introducing undue complexity in the instrumentation and processes systems of the plants and that the operation of such systems were diverting the operators from their other important duties. While this prohibition does not extend to engineering and maintenance systems, it is a "wake-up" call to those of us working in this field. The challenge to all the Workshop participants is to develop simpler, less intrusive and more useful and integrated operational support systems for the plant operators. They simply do not want multiple, independent plant operational support systems, each with its own special features and each requiring specialized training to operate. Equally important is for the plant operational support system developers to secure the "buy-in" and support of the plant operators of the systems that are installed, or they will not be used.

Overall, the Workshop was a great success, and HRP should be congratulated on its decision to sponsor this Workshop at this time. The success of this conference was due in large part to the tireless efforts of Da Ruan and Paolo Fantoni, and their efforts should be acknowledged and appreciated.

Robert E. Uhrig
University of Tennessee and
Oak Ridge National Laboratory
September 2001

To Greet and Maria Chiara

Preface

Condition monitoring and early fault detection, as an important issue in *Power Plant Surveillance and Diagnostics*, have gained momentum as increasingly competitive markets in the energy sector have forced utilities and vendors to look for better efficiency, at the system and component level, at a lower price. At the same time, the Nuclear Regulatory Commission (NRC) as from the Safety Evaluation Report (SER) dated July 24, 2000 has provided generic approval of on-line monitoring for calibration reduction of process instrumentation.

Many utilities in the energy sector are moving toward different maintenance strategies, based on condition monitoring, in order to minimise outage time and to increase the capacity factor. Still, better methods and systems for supporting plant personnel in assessing plant conditions during normal operation modes are needed to optimise plant operation and minimise maintenance costs. Early detection and recognition of faults on both system and component level is an important topic in this respect. Detection of faulty sensors, small leaks and component degradation is important to prevent thermal losses in the turbine cycle. Accurate and reliable determination of thermal power is important to maximise power output as thermal power is determining the safety limits.

Research on selected aspects of computer-assisted plant surveillance and diagnosis is an ongoing activity at the OECD Halden Project of the Institute for Energy Technology (IFE). The practical examples, which can be seen in some details in the book, are: (1) signal validation and on-line calibration monitoring with empirical models (PEANO), (2) thermal performance monitoring and optimisation, data reconciliation, and fault detection with physical models (TEMPO), and (3) event and fault detection and recognition with empirical models (ALADDIN).

In order to discuss the most recent developments and the on-going research related to all areas pertaining plant surveillance and diagnosis, as well as to identify the successful applications of already well settled methodological tools of the field, an OECD Halden workshop on *Power Plant Surveillance and Diagnostics* was successfully held on September 3-4, 2001 in Halden, Norway. The following topics of interest were discussed and presented at the workshop: (1) analysis of important issues relating to specification, development and use of systems for computer-assisted plant surveillance and diagnosis, (2) empirical and analytical methods for on-line calibration monitoring and data reconciliation, (3) noise analysis methods for early fault detection, condition monitoring, leak detection and loose part monitoring, (4) predictive maintenance and condition monitoring techniques, and (5) empirical and analytical methods for fault detection and recognition.

This book presents modern approaches and advanced applications for *Power Plant Surveillance and Diagnostics*. The root of this book is naturally traced back to the abovementioned workshop. Fourteen chapters are extended and selected from the 20 presentations at the workshop, and in addition, eight chapters are invited from leading researchers of the field in the world written exclusively for this book. All chapters have passed a peer-reviewing procedure, which represents about 10 countries' research progress related to *Power Plant Surveillance and Diagnostics*.

This book, as its title suggests, consists of modern approaches and advanced applications for *Power Plant Surveillance and Diagnostics*. The chapters of the book are generally divided into the following categories: (1) survey and introduction, including regulatory treatment of plant operational support systems, (2) application of current surveillance and diagnostic methodologies, (3) application of new or combined surveillance and diagnostics methodologies, and (4) identification and approaches to ill-posed issues.

This volume highlights the advantages of modern approaches and advanced applications for *Power Plant Surveillance and Diagnostics*, which can be viewed as complementary to traditional approaches; as a result, new or combined surveillance and diagnostics such as soft computing techniques provide a powerful tool for solving intricate problems pertaining in *Power Plant Surveillance and Diagnostics*. Each chapter of this book is self-contained and also indicates the future research direction on this topic of modern approaches and advanced applications for *Power Plant Surveillance and Diagnostics*.

The 23 chapters of the book are co-authored by 50 leading experts in *Power Plant Surveillance and Diagnostics* from 10 countries in the world. Special thanks are due to all contributors for their kind co-operation in helping to prepare this book; to Professor Uhrig for his stimulating role as the workshop chair as well as for his Foreword for this book; to Grete Bjerkely for her help related to all practical arrangements of the workshop; to Carl-Olof and Linn Fält for their reformatting of almost all chapters in this book; to Greet Ruan for her proof-reading of part of the manuscript, and to the OECD Halden Reactor Project for its support.

Da Ruan and Paolo F. Fantoni
Halden, December 2001

Contents

1 Modern Approaches and Advanced Applications for Plant Surveillance and Diagnostics: An Overview

Da Ruan[*]

The Belgian Nuclear Research Centre (SCK•CEN)
Boeretang 200, 2400 Mol, Belgium
[*]Visiting scientist at the OECD Halden Reactor Project (April 2001-September 2002)

druan@sckcen.be

1 Introduction

The goal of this introductory chapter is to briefly summarise all chapters in this book and to communicate to a wide audience by relating modern approaches and advanced applications for power plant surveillance and diagnostics.

In the context of nuclear engineering applications, modern approaches are usually refereed to as new information processing methods and/or new artificial intelligence techniques. Those terms are now commonly called as soft computing techniques or computational intelligent systems.

Soft computing (SC) has recently become a hot topic on conferences and in literature. As a consequence, the interest in SC has increased and has stimulated the development of new methods in the fields of fuzzy logic, neural networks, genetic algorithms, and others. Since the publication of Zadeh's paper on soft data analysis (Zadeh, 1981), the concept of soft computing has undergone many changes. In its latest incarnation, soft computing may be defined informally according to Zadeh as follows (Zadeh, 1997).

Soft computing (SC) is an association of computing methodologies centering on fuzzy logic (FL), neurocomputing (NC), genetic computing (GC), and probabilistic computing (PC). The methodologies comprising soft computing are for the most part complementary and synergistic rather than competitive.

Following Zadeh's statement (Zadeh, 1997), soft computing differs from conventional (hard) computing in that, unlike hard computing, it is tolerant of imprecision, uncertainty and partial truth. In effect, the role model for soft computing is the human mind. The guiding principle of soft computing is: exploit the tolerance for imprecision, uncertainty and partial truth to achieve tractability, robustness and low solution cost.

Recent developments in soft computing techniques, including fuzzy logic, neural networks, and genetic algorithms, have really blossomed. Many of these remarkable methodologies and their utility are fully appreciated by theoreticians and practitioners alike. Interestingly, one cannot really judge each methodology on its face value or its outside appearance. The fact of the matter is that each has, in its own way, very deep hidden power and high ideal philosophy. The power of each methodology as a design tool is limited only by the designer's imagination. Two features, in particular, stand out quite visibly: (1) many of them are biologically inspired, and (2) they are all capable of solving non-linear problems. One reason all of these methodologies are popular is precisely the two features just mentioned; they can solve many problems that were difficulty (Ruan, 1997).

In this overview paper, we briefly describe some of the technology in the soft computing areas that has been developed for application to nuclear power plants (NPPs). We review the available literature on applications of soft computing techniques in NPPs, and provide up-to-date coverage of new references on the subject from the international FLINS series conferences. We further outline some current activities related to power plant surveillance and diagnostics at the OECD Halden Project. Following Foreword by Uhrig and Preface by the editors, we summarise each chapter of the book to highlight the key issues related to the recent progress and developments in the field of power plant surveillance and diagnostics. And finally, we discuss briefly on the issue of soft computing techniques from the viewpoint of nuclear engineering applications.

2 Soft Computing Applications in NPPs

Nuclear engineering is one of the areas with a large potential for applications of soft computing, in which, however, the development is still in its infancy (Ruan, 1995, 1996). Most nuclear engineers today are involved in the development of nuclear power installations, either stationary power plants for the generation of electricity or plants for the propulsion of mobile systems. The Chernobyl accident and its cross-border consequences have reminded us that nuclear safety remains a short-term priority, both at home and abroad, as nuclear technology has not reached the same maturity in all countries. The need for on-line reactor operator decision support systems has become evident after the Three Mile Island accident in 1979. Since then, considerable attention has been paid by the engineering, scientific, economic, political communities and by society at large to prevent these types of event by using state-of-the-art artificial intelligence techniques.

Nuclear engineering is different from many other engineering areas in the following respects: (1) an experimentation is only possible to a certain degree which does not provide a safe basis for stochastic models or the validity of the law of large numbers; (2) the consequences of misjudgement, wrong decisions or bad design can be such horrendous that one is hesitant to accept even a tiny probability for such an event to happen; (3) the waste disposal is one of the most difficult problems yet unsolved; (4) the abuse of nuclear material for criminal purposes can have unprecedented consequences, and (5) the public opinion with respect to nuclear energy and related issues is much more controversial and influential than in other engineering disciplines (Ruan *et al.*, 1998).

Soft computing are new computing techniques that are tolerant to imprecision, uncertainty and partial truths. Applications of these techniques in NPPs present a tremendous challenge due to its strict nuclear safety regulation. The fields of NPPs and soft computing have nevertheless matured considerably during the last decade. NPPs can be supported by soft computing in many ways (Uhrig and Tsoukalas, 1999). We will list some applications of soft computing in NPPs in the following subsections.

2.1 Fuzzy Logic and NPPs

Fuzzy logic (FL) (Zadeh, 1965; Cox, 1994; Klir and Yuan, 1995) is designed to handle imprecise linguistic concepts such *as small, big, low, high, young*, or *old*. Systems based on FL exhibit an inherent flexibility and have proven to be successful in a variety of industrial control and pattern-recognition tasks ranging from handwriting recognition to traffic control. Central to the flexibility that fuzzy logic provides is the notion of fuzzy sets. Fuzzy sets are the basic concept supporting fuzzy theory. The main research fields in fuzzy theory are fuzzy sets, fuzzy logic, and fuzzy measure. Fuzzy reasoning or approximate reasoning is an application of fuzzy logic to knowledge processing. Fuzzy control is an application of fuzzy reasoning to control. One of the main strengths of fuzzy logic compared with other schemes to deal with imprecise data is that their knowledge bases, which are in a rule format, are easy to examine and understand. This rule format also makes it easy to update and maintain the knowledge base. As for the limitations of fuzzy logic, the main shortcoming is that the membership functions and rules have to be specified manually. Determining membership functions can be a time-consuming, trial-and-error process. Moreover the elicitation of rules from human experts can be an expensive, error-prone procedure.

We list here part of references from the literature concerning FL applications to NPPs: A fuzzy-logic approach to HTR NPP model control (Bubak and Moscinski, 1983), to a rule-based system for process control (Bernard, 1988), to a nuclear reactor control (Guth, 1989; Ramaswamy *et al.*, 1993; Heger *et al.*, 1995; Kim *et al.*, 1997; Lin and Yang, 1998; Ruan and van der Wal, 1998), to nuclear reactor system dynamics (Matsuoka, 1991), to a PWR-type NPP (Akin and Altin, 1991; Hah and Lee, 1994), to a site selection for nuclear plants (Gencay, 1991), to control of steam generator water-level in PWR (Kuan *et al.*, 92; Na *et al.*, 1995; Cho and No, 1996), and to the redundant sensor validation (Holbert *et al.*, 1994).

2.2 Neural Networks and NPPs

Neural networks (NNs) or artificial neural networks (ANNs) (Aleksander and Morton, 1990; Beale and Jackson, 1990; Hertz *et al.*, 1991) are computing devices inspired by the function of nerve cells in the brain. They are composed of many parallel, interconnected computing units. Each of these performs a few simple operations and communicates the results to its neighbouring units. In contrast to conventional computer programs where step-by-step instructions are provided to perform a particular task, NNs can learn to perform tasks by a process of training on many different examples. NNs provide a relatively easy way to model and forecast non-linear systems. This gives them an advantage over many current statistical methods that are primarily linear. They are also very effective in learning patterns in data that are noisy, incomplete and which may even contain contradictory examples. A main limitation of NNs is that they lack explanation capabilities. They do not provide users with details of how they reason with data to arrive at particular conclusions.

We list here part of references from the literature concerning NNs applications to NPPs: Potential application of NNs to the operation of NPPs (Uhrig, 1991; Reifman and Vitela, 1994), to the power prediction in NPPs (Roh *et al.*, 1991), to analyse NPP performance (Guo and Uhrig, 1992), to the non-linear identification of process dynamics (Parlos *et al.*, 1992), to the sensor validation and plant monitoring (Upadhyaya and Eryurek, 1992), to the PWR core parameter prediction (Kim *et al.*, 1993; Ikonomopoulos *et al.*, 1993), to the local power distributions in BWR fuel bundles (Tanabe *et al.*, 1993), to safety control of nuclear power operations (Jouse and Williams, 1993), to abnormal event identification in NPPs (Ohga and Seki, 1993), to a connectionist expect system for transient, to detecting faults in a NPP (Basu and Bartlett, 1994; Marseguerra and Zio, 1994; Marseguerra *et al.*, 1996), to an error prediction for a NPP fault-diagnostic advisor (Kim *et al.*, 1994), to noise analysis method for monitoring the moderator temperature coefficient of PWR (Thomas and Adams, 1994), to the monitoring feedwater flow rate and component thermal performance of PWR (Kavaklioglu and Upadhyaya, 1994), to incipient fault detection and identification in process systems (Parlos *et al.*, 1994), to a reactivity surveillance in a nuclear reactor (Arul, 1994), to self-tuning control of a nuclear reactor (Park and Cho, 1995), to control rod localisation (Garis and Pazsit, 1999), to full range signal validation of PWR nuclear power plant data (Fantoni, 2000), and to transient classification (Roverso, 2000).

2.3 Genetic Algorithms and NPPs

Genetic Algorithms (GAs) (Holland, 1975; Goldberg, 1989; Davis, 1991) are efficient problem-solving mechanisms that are inspired by the mechanisms of biological evolution. GAs have produced very good solutions for complex optimisation problems that have large numbers of parameters. The main idea of a GA is to start with a *population* of solutions to a problem, and then attempt to produce new generations of solutions that are better than the previous ones. Because of their inherent parallel structure, GAs have proven to be very effective at efficiently searching very large data sets. This search process also has another advantage in being highly suitable for parallel computer implementations. They have had particular success in large optimisation problems. GAs can learn complex relationships in incomplete data sets and they can be used as `data mining' tools to discover previously unknown patterns. They can adapt to changes in their operating environment and they can provide explanations of the decisions they produce in a format that humans can understand. A limitation of GAs is that the setting of parameters such as the crossover and mutation rates is problem dependent and is a time-consuming trial and error process. Further, the performance of a GA is affected by the representation scheme that is employed. The selection of an appropriate scheme may require extensive experimentation.

A genetic algorithm is a powerful search technique that simulates natural evolution in order to fit a population of computational structures to the solution of an optimisation problem. This technique presents several advantages over classical ones such as linear programming based techniques, often used in nuclear engineering optimisation problems. However, GAs demand some extra computational cost. Nowadays, due to the fast computers available, the use of GAs has increased and its practical application has become a reality. Nuclear power plant operation often involves very important human decisions, such as actions to be taken after a nuclear accident/transient, or finding the best core reload pattern, a complex combinatorial optimisation problem which requires expert knowledge. Due to the complexity involved in the decisions to be taken, computerised systems have been intensely explored in order to aid the operator. Following hardware advances, soft computing has been improved and, nowadays, intelligent technologies, such as GAs, NNs and FL, are being used to support operator decisions. In nuclear engineering there are many difficult optimisation problems related to nuclear reactor design. Genetic algorithm is a suitable technique to face such kind of problems. Although applications of GAs in nuclear engineering are still rare, GAs applied to the nuclear power plant operation and to nuclear reactor design optimisation are excellent examples (Pereira *et al.*, 1999; Schirru *et al.*, 1999).

3 FLINS International Forum and Its Related Activities

FLINS, an acronym for **F**uzzy **L**ogic and **I**ntelligent Technologies in **N**uclear **S**cience, has been launched in line with the Belgian Nuclear Research Centre (SCK•CEN)'s objective to give young talented people the opportunity to carry out future-oriented research. FLINS was initially built within one of the postdoctoral research projects at SCK•CEN in 1994. The main task for FLINS for the coming years is to solve many intricate problems pertaining to the nuclear environment by using modern technologies as additional tools and to bridge a gap between novel technologies and the industrial nuclear world.

The significant influence of soft computing in nuclear engineering (Ruan, 1995) was illustrated by the response to FLINS'94, the first international FLINS workshop on *Fuzzy Logic and Intelligent Technologies in Nuclear Science* (Mol, Belgium, September 14-16, 1994) (Ruan *et al.*, 1994). A successful application of fuzzy logic control to the feed-water control system of the 165 MWe Fugen Advanced Thermal Reactor (ATR) (1993) has enabled operators to control the steam drum water level more effectively than with a conventional proportional-integral (PI) control system (Iijima *et al.*, 1995). The Korea Atomic Energy Research Institute (Jung *et al.*, 1995) piloted a real-time self-tuning fuzzy controller for a steam generator with a scaling factor adjustment. This improves the performance of the water-level controller; a compact nuclear simulator itself simulates the controller. Of more than 70 papers submitted from 13 countries, 56 were accepted for the presentations at FLINS'94 workshop (see Appendix A).

FLINS'96, the second international FLINS workshop *on Intelligent Systems and Soft Computing for Nuclear Science and Industry*, was successfully held in Mol, Belgium, on September 25-27, 1996 (Ruan *et al.*, 1996). Of more than 80 papers submitted from 20 countries, 52 were accepted for publication in the proceedings of FLINS'96. In particular, contribution (Ruan, 1998) concentrates on the applications of soft computing and intelligent systems in NPPs (see Appendix B).

FLINS'98, the third international FLINS workshop (Antwerp, Belgium, September 14-16, 1998) aimed again to bring together scientists, researchers, and engineers from academia and industry and to introduce the principles of intelligent systems and soft computing such as fuzzy logic, neural networks, genetic algorithms and any combinations of these methodologies, to the field of knowledge-based expert systems and apply these techniques to complex problem-solving techniques within nuclear industry and related research fields (Ruan *et al.*, 1998). The FLINS'98 proceedings, in total 55 papers selected out of 80, submitted from 23 countries, concentrates on the applications of *Fuzzy Logic and Intelligent Technologies in Nuclear Science and Industry*, and covers several topics such as nuclear energy, nuclear safety assessment, radioactive waste management, nuclear measurements, nuclear safeguards, nuclear reactor operation, reactor controller design, fuel reload pattern design, signal validation, NPPs, telerobotics, and teleoperation in nuclear applications (see Appendix C).

Following FLINS'94, '96, and '98, the fourth international FLINS conference on *Intelligent Techniques and Soft Computing in Nuclear Science and Engineering* was held in Bruges, Belgium, August 28-30, 2000. In this framework, the special issue of FLINS'94 was published in *Fuzzy Sets and Systems* 74, number 1, August 25, 1995. The special issue of FLINS'96 was published in *International Journal of Intelligent Systems* 13, Numbers 2/3, 1998. The special issue of FLINS'98 was published in *International Journal of General Systems* 29, Numbers 1/2, 2000. And the special issue of FLINS 2000 will be published in *International Journal of Information Sciences* soon.

From the beginning, the organisers of FLINS conferences have aimed not to restrict the contributions to fuzzy logic, but instead, to encourage the submission of papers using other intelligent techniques, treating other facets of the information available. We strongly believe that this synergetic strategy will substantially enrich and fine-tune the existing models for our industrial applications in general and in nuclear science and engineering in particular. FLINS 2000, once again organised by SCK•CEN and continuously co-sponsored by the Fund for Scientific Research-Flanders (FWO), offered a unique international forum to present and discuss techniques that are new and useful for nuclear science and engineering and to launch international co-operations.

At FLINS 2000, more than 70 attendees from over two-dozen countries attended the conference. A total of about 55 speakers lectured on *Mathematical Tools and New Developments*, *Intelligent Engineering Applications*, *and Nuclear Engineering Applications*. Most lectures are published in the conference proceedings in which a total of 65 papers are selected out of over 90 submissions from 28 countries. The following research topics were presented and discussed: fuzzy set theory, rough set theory, fuzzy control, fuzzy modelling, decision support systems, man-machine interface, risk analysis, image processing, robotics, knowledge-based engineering, expert systems, process control integration, diagnosis, measurements and interpretation, and nuclear engineering applications of soft computing such as nuclear energy, nuclear safety assessment, radioactive waste management, nuclear measurements, nuclear safeguards, nuclear reactor operation, reactor controller design, fuel reload pattern design, signal validation, nuclear power plants, and optimisations in nuclear applications (Appendix D).

The fifth international FLINS conference on *Computational Intelligent Systems for Applied Research* is planned on September 16-18, 2002 in Ghent, Belgium. A special track on *Applications of Intelligent Real-Time Systems for Nuclear Engineering* is foreseen at FLINS 2002.

Among the FLINS projects at SCK•CEN, specific prototyping of fuzzy logic control (FLC) of SCK·CEN's BR1 research reactor has been chosen as FLINS' first priority. This R&D project for controlling BR1's power level started in 1995 and aimed at investigating the added value of FLC for nuclear reactors. In this framework, the availability of BR1 greatly simplifies the effort to validate the model description used. This allows us to concentrate on the optimal implementation of the overall control. The details of this project can be seen from the references (Ruan and van der Wal, 1998; Ruan, 2000). Some comments related to the use of SC techniques may be drawn from the project.

In the experience of the author it is wise to rely on certified, proven industrial hardware instead of developing *ad hoc* software when developing a controller for use in an environment that presents a potential high risk if control breaks down.

Generally, fuzzy adaptive control includes three aspects: (1) regulated scale factors; (2) regulated fuzzy control rules; (3) both regulated scale factors and rules. As for (3), it is not used widely as it is difficult to modify scale factors and control rules simultaneously with good and stable results. As for (1), it can indeed improve quality of fuzzy control, however, it is not optimal.

Normally, the methods of getting rules for fuzzy controllers can be broadly divided into two types, sourceable and non-sourceable. In the sourceable method the rules are obtained from some information source, such as human experience or historical input-output data. This type of experience is widely used by fuzzy controller designers. The problems of using a database of experience are that it is time-consuming and subjective to some degree. In order to overcome these problems, particularly avoiding the subjectivity, historical input-output data may be used. In order to obtain rules from those data, many methods are used, of which one example is neural networks (NNs). One problem with the sourceable method is that it depends strictly on the quality of the source, which will be transformed into rules. In case the source is noisy, i.e., contains erroneous information, then the rules will be noisy too. Another problem with the sourceable method is that it is usually non-adaptive, *i.e.*, all the rules are fixed, and therefore it cannot perform well in a dynamically varying environment.

Non-sourceable methods are source-free and they produce and choose rules according to a performance measurement of the controller, such as genetic algorithms (GAs), which mostly also generate membership functions and scale factors or self-organising controllers (SOC). GA's can find integrated optimal parameters but the method is computationally costly, and, furthermore, it is almost impossible apply it in a real complex system without a simulation model. The SOC is the only method that has the following advantages: objective, adaptive, less computationally expensive, more error tolerant, and simple. The general principle of the SOC is that the controller monitors its own performance and adjusts its control rules to improve performance for time-varying and unknown systems or events. The problem with SOC is how to do the performance measurement. One known method is to design a performance measurement table, which looks a little like a fuzzy control rule table and to use it to assess the performance of the controller (rules). The table is used to regulate (modify) the fuzzy control rules. But in practice setting up such a performance measure table is hardly easier than setting up a good fuzzy-control-rule-table. Hence, the design of such a performance measurement table may be very difficult and it is often system-dependent.

Fuzzy adaptive control (FAC), sometimes called self-organizing fuzzy control, uses an intelligent algorithm, which is able to self-regulate fuzzy control rules in a real-time control process. Although usually a fuzzy controller without an adaptive function may sometimes work very well, an adaptive function is considered to be a requirement as an advanced aid in some systems with too many uncertainties so that the rules cannot be determined accurately at the design stage. On the other hand an adaptive fuzzy controller has a potential for erratic or erroneous behaviour when confronted with certain unexpected aspects of the system or with certain unusual situations. Fuzzy logic is conventionally the main part of a fuzzy controller and the fuzzy controller usually has a plurality of inputs, *e.g.,* two inputs and two outputs. Conventionally such a fuzzy controller replaces a classical controller, *e.g.*, a PID controller. A fuzzy controller may be good for a complex system with many uncertain conditions, but it is perhaps not so good for a complex system with many certain conditions, where an analytic controller (non-fuzzy) works very well. Fuzzy logic has been applied successfully to some ill-defined industrial processes, but its value in the operation of well-characterised systems has not yet been demonstrated clearly. For example, a fuzzy rule-based approach has not yet achieved general acceptance in a nuclear reactor control. One reason is the high level of safety requirements in safety critical systems. A safe design of a fuzzy controller for safety critical systems such as a nuclear reactor with very high safety requirements is considered technically very difficult. Such a controller preferably has inherent safety features, however, it is practically impossible to predict all possible accidents and unusual situations. Thus, a controller that is too rigidly designed, even with respect to its adaptive function, may fail in unusual situations. The performance of any controller that is used in safety critical systems is preferably determinable within safe limits. Organisations entrusted with approving the use of such controllers must be able to satisfy themselves that the controller will behave well and properly. One problem with genetic algorithms and neural networks is that their operation can be so complex that it is difficult to for a supervising organisation to rule out the possibility that some state of the controller exists whose behaviour may be sufficiently sub-optimal that dangerous conditions may develop.

4 Related Activities at the OECD Halden Project

Safe and reliable operation of nuclear power plants (NPPs) benefits from R&D advances and related technical solutions. The OECD Halden Reactor Project at the Institute for Energy (IFE), a joint undertaking of national organizations in 20 counties sponsoring a jointly financed programme under the auspices of the OECD Nuclear Energy Agency, is an international network fostering these advances with programmes devised to provide answers in a direct and effective manner. The Project's strong international profile and solid technical basis represent an asset for the nuclear community at a time in which maintaining centres of expertise at accessible cost becomes increasingly important.

In addition to the joint programme work, a number of organizations in the participating countries execute their own development work in collaboration with the Project. These bilateral arrangements constitute an important complement to the joint programme and normally address issues of commercial interest to a participant organization or group of organizations (Vitanza, 2001).

In the framework of Safety Man Technology-Organisation (MTO) at the OECD Halden Reactor Project, the safe operation of nuclear power plants, other energy and process plants, and transportation systems depends on operators receiving relevant information about the state of the process, interpreting the information correctly, and reacting effectively to the actual situation. Several recent accidents have resulted from fatal weaknesses in the interaction between people (operators), technology and organization. The concept Safety-MTO is based on the fact that major improvements are possible when the operator(s), technology and organization are viewed as a single system. Their interaction and ability to function properly within every area of importance to total system safety must be optimised. Such areas include design, management, operation, maintenance, the supervision of safety, and control (IFE, 2000).

Among the activities carried out at the MTO, are PEANO, ALADDIN, and TEMPO of particular interest for power plant surveillance and diagnostics.

A tool called PEANO has been developed by IFE. The aim of this work is to monitor whether processes or instruments change over time and to detect failures at an early stage. This helps prevent serious accidents and enhances the availability of equipment to maintain continuous production. The core methods of PEANO are based on neural networks and fuzzy logic models trained to recognize behaviour. Training data can either be obtained from a large number of observations and measurements, or generated from traditional simulations. The methods are especially well suited for processes that are difficulty to describe mathematically on the basis of first principles or which require extensive resources for the preparation and execution of models. PEANO provides functions for model training involving historical data and for subsequent online system applications. Considerable emphasis has been put on robustness, so the system is able to adapt to normal process changes. One important feature of PEANO is that the system can tell if the input data is outside the network's validation region (training data). The methods have been tested by Electricité de France (EdF) using comprehensive simulated data on different failures, and the results have been good. Practical tests have been carried out with the Halden Reactor. Recently, PEANO was tested with data from US plants, and has also been tested with data from a conventional power plant in Italy. The results are promising and these techniques seem to generate increased interest among industrial enterprises. In the last chapter of this book, Fantoni gives a detailed description of the progress and the current state of PEANO.

Similar methods are utilized in a system called ALADDIN. This system aims at detecting and classifying events to help operators rapidly identify any type of transient that might have occurred. Roverso deliveries the full details on ALADDIN in Chapter 22.

One important task for the operator during normal operation is to optimise plant performance with respect to economy and safety. The TEMPO (Thermal Performance Monitoring and Optimisation) system developed by the Halden Project helps the operations staff identify and correct process failures, which can reduce thermal efficiency at nuclear/conventional power plants (www.ife.no/tempo). Typical problem are fouling and leaks in heat exchangers, leaky valves, and degradations in pumps and compressors, as well as drift in sensors (especially flow meters). TEMPO is a versatile system, and plans have been made to adapt it to different types of plants with a potential for improving performance in different operational states. The possibility of combining TEMPO and PEANO will be considered with a view to enhancing the robustness of the overall systems.

In connection with the accumulated knowledge and expertise of PEANO and ALADDEN, a cooperation between Tecnatom s.a. in Madrid and the Halden Reactor Project in Halden is currently carried out aiming at enhancing the accuracy of flow measurements, which is to reduce as much as possible the measurement uncertainty. The principles of cross-correlation flow measurement are based on noise analysis. The preferred method in the literature for fluid velocity estimation by cross correlation analysis, is based on the computation of the cross correlation of an up-stream and a down-stream signal, and on the location of the maximum of the cross correlation function, which indirectly indicates the time-delay and the velocity of the fluid. A different approach is based on delaying the up-stream signal by a delay time τ, until the mean square difference between the two signals reaches a minimum. Yet another possibility is to analyse the phase shift of a range of frequencies of the two signals. To better estimate the feedwater flow, new techniques based on artificial neural networks (ANNs) and possibly with a combination of fuzzy logic applications are being investigated. The idea is to develop a "virtual flow meter" based on neural cross-correlation analysis of signals obtained from sensor pairs placed at spatially separated locations along the subject pipe. Inputs to the neural virtual flow meter will also include other plant measurements that have an influence on the velocity profile of the fluid in the pipe, e.g., temperature and pressure measurements. Feature extraction from PSD spectra is also considered as a promising approach for facilitating the neural cross-correlation analysis. The validation of the techniques developed within the project is carried out in an experimental loop facility of Tecnatom s.a. in Madrid. It is expected that the results obtained in the project would be used for the development of new type of sensors that provide better estimates of critical process parameters. The extrapolation of the results for making measurements in a real nuclear power plant will be demonstrated at a later stage in the project. Enhancement of feedwater flow measurement by means of computational intelligence techniques is expected in at least one of two possible scenarios: either by providing a higher accuracy than existing methods, or by providing an acceptable level of accuracy but at a much reduced cost. Either direction is expected to be beneficial to the market of the feedwater flow measurement.

5 Summary of Each Chapter of the Book

Chapter 2 on Regulatory Treatment by Uhrig

For almost two decades, various university and national laboratory research groups have been developing surveillance and diagnostic techniques for application to nuclear power plants. For more than a decade, many of these techniques included various types of artificial intelligence based systems to develop a databased model as a substitute for first principles models that were found deficient. The motivation for such techniques was to monitor the condition of signal channels with a view to reducing the number of calibrations required by the Nuclear Regulatory Commission (NRC). When the Electric Power Research Institute (EPRI) approached the NRC for approval of these systems, such use was not approved, primarily because of the uncertainty of the magnitude of the errors associated with the monitoring technique when used for sensor calibration. After eight years of interaction with the NRC, the EPRI submitted Topical Report #104965 "On-Line Monitoring of Instrument Channel Performance." The NRC replied with 14 conditions that had to be met for acceptance. Chapter 2 discusses the EPRI topical report and NRC's response.

Chapter 3 on Optimised Maintenance and Management by Hashemian

New power generation economics and politics in the 2000 decade, recent interest in nuclear power plant life extension, and challenges from other electricity production systems such as combined cycle gas plants are providing new opportunities and new challenges to the nuclear power industry. Chief among these is the challenge to continue to improve efficiency, reduce costs, and manage the aging of critical components. Chapter 3 presents a review of testing and maintenance technologies that have great potential to help the nuclear industry in meeting these challenges. These technologies range from new methods for on-line calibration and testing of process instrumentation to predictive maintenance of the reactor vessel and its internals.

Chapter 4 on Recent KFKI AEKI Activities by Kiss et al.

Application of noise diagnostics techniques for the regular surveillance of the four VVER-440/213 type reactors of the Paks NPP (Hungary) has long time traditions at KFKI AEKI. Chapter 4 is devoted to the description of those regular reactor noise diagnostics measurements that are routinely performed by KFKI AEKI at the Paks NPP in co-operation with experts from the power plant. Details of the in-core coolant velocity mapping and vibration monitoring of core internals (e.g., control rods) are presented, emphasising their importance in practical plant operation. Recent plans for noise analysis system upgrading and for introducing new on-line process analysis and signal validation services are given as well.

Chapter 5 on Adaptive Model-Based Control by Melin and Castillo

Adaptive control is a method of designing a controller with some adjustable parameters and an embedded mechanism for adjusting these parameters Adaptive controllers have been used mainly to improve the controller's performance online. Normally, an adaptive controller is designed based on one of the available techniques. Each technique is originally designed for a specific class of dynamic system. The controller is then adjusted as data are collected during run time to extend its effectiveness to control a large class of dynamic systems. Chapter 5 proposes an extension to the principle of adaptive control using soft computing techniques First, the concept of adaptive fuzzy control is described. A neuro-fuzzy approach is used to learn the parameters of the fuzzy system using real data from the plant The proposed adaptive control was tested with a specific non-linear plant, to evaluate the performance of this approach. A stepping motor was used as the non-linear plant in this study. The simulation results clearly show the advantage of using soft computing techniques to improve the concept of adaptive control.

Chapter 6 on Bayesian Networks by Carlsson and Wennersten

Since the beginning of industrialism, the complexity of industrial processes has been continuously increasing. In modern plants during normal operation this complexity is handled by the computerised control system. However, some process conditions outside the design of the control system still have to be managed by the human operator. These conditions can be due to equipment malfunction, unknown inputs and disturbances. Due to the complexity and the limited direct process interaction that operators experience, they often have difficulties making the right decisions in these abnormal situations. To resolve this task, various computerised decision support tools have been suggested. These decision support tools have often been to aid the operator in searching for the cause of the observed process symptoms. A problem common to all fault diagnosis approaches is that they have to deal with uncertainty. Bayesian networks provide probability theory to deal with this uncertainty. In addition, it is also possible to train Bayesian networks on process data. Chapter 6 discusses the potential of Bayesian networks to resolve some of the current problems of the decision support tools as outlined above. The chapter is not a complete review on fault diagnosis with Bayesian networks, but should rather be considered as an introduction to the benefits of Bayesian methods.

Chapter 7 on Hidden Markov Models by Kwon

Transient identification in nuclear power plants (NPPs) means classifying the type of transients by interpreting the major plant variables and operating status of the equipment. The transient is defined as when a process proceeds to an abnormal state from a normal state. The term identification, as applied to an engineering system or process, means the classification of the cause, which brought an undesirable state or failure of the system. The identification can be done at several different levels, e.g., component, subsystem, function or event. For the proposed transient identification system, identification is made at the event level to determine which transient has occurred in NPPs. It is necessary to identify the type of transient through continuous monitoring of NPPs during its early stage to provide sufficient information to the human operators in order to assist for proper operator action selection to prevent a more severe situation or to mitigate the accident consequence. Chapter 7 provides a transient identification system for use in NPPs that demonstrates the spatial and temporal modelling and learning capabilities with hidden Markov models. To demonstrate these capabilities, simulated NPP parameters are collected and clustered to build hidden Markov models for transient identification purposes.

Chapter 8 on Expert System-Based Implementation by Kim et al.

When transitions are occurred in large and complex systems such as nuclear power plants or industrial process plants, it is often difficult to diagnose them. Various computer-based operator-aiding systems have been developed in order to help operators diagnose the cause of transitions of the plants. However, most of the operator support systems carried out only a function of alarm processing and the other only a function of a fault diagnosis. ADIOS (Alarm and Diagnosis-Integrated Operator Support System) has been developed to perform functions from alarm process to fault diagnosis. It has been composed of two modules: an alarm process module and a fault diagnosis module. The alarm process module can offer key alarms to operator using dynamic alarm filtering methods and the fault diagnostic module was able to diagnose the component failure caused by a sensor fault and a hardware fault. ADIOS has been built in an object-oriented AI environment of G2 expert system software tool. Chapter 8 discusses the architecture of the knowledge base of ADIOS, focusing on alarm and diagnostic processing, along with the inference scheme. Also discussed herein are the advantages of, and some issues in, developing an improved alarm and diagnostic system in an object-oriented environment, and also using a real-time AI tool, such as G2.

Chapter 9 on Detection of Incipient Signal by Nieman and Singer

A 161 megawatt co-generation plant operated by the US Steel Corporation is highly instrumented, providing real-time monitoring of turbine, extraction, generator, condenser and bearing functions, including pressure, valve position, temperature, amperage, voltage and displacement. An equipment condition monitoring (SmartSignal eCM™) system based on an empirical profile of normal operational conditions and utilizing pattern recognition techniques was used to monitor 60 signals on this plant at one-minute intervals. A study of plant data from a two month period shows that the Plant eCM™ fault detection and identification technology generates a high fidelity estimated value for each of the signals and enables early detection of significant, but subtle, variations in pressures, temperatures, flows, and energy level. Alerts annunciated by the system focus attention on groups of signals that can be early indications of equipment failure or process imbalance. Chapter 9 demonstrates the uses and effectiveness of equipment condition monitoring using an empirical process model based upon an advanced and highly robust method of pattern recognition.

Chapter 10 on On-Line Determination of the MTC by Demazière and Pázsit

The moderator temperature coefficient of reactivity (MTC) is an important safety parameter in pressurized water reactors (PWRs). It is shown that the main reason of the MTC underestimation by noise analysis that was noticed experimentally previously lies with the heterogeneous structure of the moderator temperature noise throughout the core. The resulting deviation of the reactor response from point-kinetics only accounts for a negligible part. Chapter 10 proposes a new MTC noise estimator. This estimator relies on the core average temperature noise that has to be evaluated with a proper core weighting function. The coolant temperature noise has to be measured in many points of the reactor so that the core average temperature noise could be approximated. Gamma-Thermometers (GTs) could be used for that purpose since it is demonstrated (both theoretically and via real measurements) that in the frequency range of interest for the MTC estimation they work as thermocouples.

Chapter 11 on Detecting Impacting of BWR Instrument Tubes by Arzhanov and Pázsit

Detection of impacting of detector tubes, also called instrument strings, has been a matter of interest both in Swedish and foreign BWRs. Although detection of the vibrations is relatively simple, the discovery and quantifying of the severity of impacting is far more complicated. There exists no single method that gives absolute results without the need for calibration or comparison with reference measurements. Most known methods that were frequently applied in the past require comparison with a reference, i.e., impacting-free state, such as the broadening of the peak, decreasing of the decay ratio, or distortion of the probability distribution function. Chapter 11 describes a calibration-free method to detect and quantify impacting of instrument tubes in BWR reactors. The method is tested on measurements, taken by GSE Power Systems AB at the Swedish BWR Oskarshamn-2 during three fuel cycles.

Chapter 12 on Advanced Core Noise Monitoring System by Mori et al.

For an efficient nuclear power plant operation, advanced core design is very important. However, advances in design may result in approaching unexplored areas where unforeseen events are increasingly common. Chapter 12 is to describe a BWR core noise monitoring system for addressing core anomaly problems in future advanced core operation. To monitor in-core status from a limited number of signals, Mori *et al.* introduce various new algorithms to compensate for a lack of information. They demonstrate that up-to-date signal processing technologies, such as independent component analysis, non-linear principal component regression, factor analysis and neural network, are effective tools for core monitoring. Also, the combination of the first principle or design model and the empirical fitting model is applied to estimate unobserved in-core state variables. Several concrete examples, such as stability monitoring, reactivity coefficient estimation, core flow estimation and in-core signal validation, are shown based on long-term monitoring experiences in real BWR plants. Through these examples, they demonstrate that the core noise monitoring system is a general platform for providing a variety of monitoring tools to meet the requirements of an advanced core operation strategy.

Chapter 13 on Diagnosis of Measuring Systems by Traichel et al.

For a continuous monitoring, control, and diagnosis of high transient technical processes in safety-related systems as Nuclear Technology it is necessary to provide the best knowledge about the actual state of process. Especially the supervision of water level within pressure vessels with water-steam mixture (pressuriser, steam generator, reactor pressure vessel) during operational and accidental transient processes is very important. For the determination of water level hydrostatic measuring systems are almost exclusively applied, because they stand the test over many decades in conventional and nuclear plants. Chapter 13 deals with the new part of fault detection of hydrostatic water level measurement by cluster analysis methods. The procedure of a cluster analysis is demonstrated. The results of fault diagnosis are demonstrated as for one Blow Down Experiment at the authors' pressuriser test facility.

Chapter 14 on Time Series Analysis by Castillo and Melin

Diagnostic systems are used to monitor the behaviour of a process and to identify certain pre-defined patterns that are associated with well-known problems. These problems, once identified, imply suggestions for specific solutions. Most diagnostic systems are in the form of a rule-based expert system: a set of rules is used to describe certain patterns. Observed data are collected and used to evaluate these rules. If the rules are logically satisfied, the pattern is identified, and the problem associated with that pattern is suggested. In general, the diagnostic systems are used for consultation rather than replacement of human expert. Most current plant monitoring systems only check a few variables against individual upper and lower limits, and start an audible alarm should each variable move out of its predefined range. Other more complicated systems normally involve more sensors that provide more data but still follow the same pattern of independently checking individual sets of data against some upper and lower limits. The warning alarm from these systems only carries a meaning that there is something wrong with the process in the plant. Chapter 14 proposes a new fuzzy-fractal approach for plant monitoring. The concept of the fractal dimension is used to measure the complexity of the time series of relevant variables for the process. A set of fuzzy rules is used to represent the knowledge for monitoring the process. In the fuzzy rules, the fractal dimension is used as a linguistic variable to help in recognizing specific patterns in the measured data. The fuzzy-fractal approach has been applied before in problems of financial time series prediction and for other types of problems, but now it is proposed to the monitoring of plants.

Chapter 15 on Failure Detection by Na et al.

In sensor monitoring applications, there are a large number of input signals related to an output and it would require a large amount of time to train a fuzzy neural network with all the input signals since the number of connection weights for neural networks and parameters for fuzzy inference would be extremely large. Also, by eliminating unimportant sensors and sensor parameters, the cost and time of collecting the data can be reduced. As the number of input features grows, the structure of the neuro-fuzzy system would become complex. Thus, it is essential to reduce the number of inputs to a neural network and to select optimum number of mutually independent inputs that are able to clearly define the input-output mapping. The process of selecting these important input variables is usually known as feature selection. Depending on the problem encountered, various criteria can be used to rank feature subsets. If the task is to predict the properties of the measurements like the sensor signal estimation, a criterion that evaluates the predictive ability of the selected features should be used. Many feature selection methods have been developed including principal component analysis (PCA), genetic algorithm (GA), and others. Chapter 15 describes an input feature selection method in which PCA, correlation analysis, and GA are combined to select important input features and this combined feature selection method will be called a PCG method. The failure detection algorithm will be applied to the failure detection of the steam generator water level; the steam generator steam flowrate, the pressurizer (PRZR) water level and the PRZR pressure sensors in pressurized water reactors and compared with other algorithms.

Chapter 16 on Diagnostic and Decision Making Tool by Boger

Fault diagnosis of a complex system, such as nuclear power plant (NPP), is an important part of the operator task. Artificial neural networks (ANN) model learns from the plant past behaviour and is easy to develop if enough data is available for training, be it from actual plant history or from full-scale simulator fault situations. The operator is reluctant to depend on a "black-box" advice, and any reasonable explanation is welcome. Two examples of using ANN modelling for fault diagnosis are presented. The first is based on NPP full-scale simulator time history data of three instrument or component faults. The second example is the diagnosis of a seemingly erratic excessive steam turbine bearing vibration. The ANN models were able to correctly classify the causes of the faults. In the first case a knowledge extraction technique from trained ANN models, Causal Index, was used. It was able to provide plausible explanations for the ANN faults classification. In the second case, Genetic Algorithm (GA) was used as an optimisation technique, with the ANN model providing the target function. It was able to identify the cause of the excessive vibration in a power plant turbine bearing. After reviewing the principal component analysis conjugate gradient (PCA-CG) algorithm for large-scale ANN modelling and the Causal Index (CI) method of analysing trained ANN, Chapter 16 proposes a novel use of GA-ANN combination for fault analysis.

Chapter 17 on Transient Identification by Mol et al.

In the last years, many different approaches based on neural network (NN) have been proposed for transient identification in nuclear power plants (NPPS). Some of them focus the dynamic identification using recurrent neural networks, however, they are not able to deal with unrecognised transients. Other kind of solution uses competitive learning in order to allow the "don't know" response. In this case, dynamic features are not well represented. Chapter 17 presents a new approach for neural network based transient identification which allows either dynamic identification or "don't know" response. Such approach uses two multilayer neural networks trained with the backpropagation algorithm. The first one is responsible for the dynamic identification. This NN uses, as input, a short set (in a movable time window) of recent measurements of each variable avoiding the necessity of using starting events. The other one is used to validate the instantaneous identification (from the first net) through the validation of each variable. This net is responsible for allowing the system to provide "don't know" response. To validate the method, an NPP transient identification problem comprising 15 postulated accidents, simulated for a pressurized water reactor, was proposed in the validation process it has been considered noising data in order to evaluate the method robustness. Obtained results reveal the ability of the method in dealing with both dynamic identification of transients and correct "don't know" response.

Chapter 18 on Planning Surveillance Test Policies by Lapa et al.

Nuclear power plant systems are comprised of both online and cold-standby components. Cold-standby components differ from the online ones, as they may be unavailable due to unrevealed failures. The usual procedure employed to reveal failures is to submit the components to surveillance tests. The surveillance tests policy may deal with two conflicting scenarios: the test frequency must be sufficiently high in order to minimize the occurrence of failures but, on the other hand, it must be low enough due to its influence on the component unavailability. Obtaining an optimum surveillance test policy at the system level, which involves consideration of many interdependent components, is a very difficult optimisation problem. Chapter 18 proposes the use of genetic algorithms for searching the optimum surveillance tests policy, due to its robustness and efficiency in complex problem solving. Here, the probabilistic model considers: a) wear-out effects on cold-standby components when they undergo surveillance tests; b) that when a failure is revealed during a surveillance test, corrective maintenance is performed and, c) that components are distinct (that is, each has distinct test parameters, such as: outage time, wear-out factors, etc.); and d) that tests are not necessarily periodic.

Chapter 19 on Possibilistic Approach for Transient Identification by Almeida et al.

Transient identification in industrial plants is often a very hard task and may involve a great amount of human cognition. Observing and analysing the behaviour of state variables (such as pressure, temperature, flux, etc.), the expert (commonly the operator) can diagnose malfunctions in the system. In some situations, however, the time needed for the expert to conclude its diagnosis may be considerably large Chapter 19 describes a possibilistic approach for transient identification based on the minimum centroids set method, proposed in previous work, optimised by genetic algorithm. The idea behind this method is to split the complex classification problem into small and simple ones, so that the performance in the classification can be increased. In order to accomplish that, a genetic algorithm is used to learn, from realistic simulated data, the optimised time partitions, which the robustness and correctness in the classification are maximized. The use of a possibilistic classification approach propitiates natural and consistent classification rules, leading naturally to a good heuristic to handle the "don't know" response, in case of unrecognised transient, which is fairly desirable in transient classification systems where safety is critical. Application of the proposed approach to a nuclear transient identification problem reveals good capability of the genetic algorithm in learning optimised possibilistic classification rules for efficient diagnosis including "don't know" response.

Chapter 20 on Ill-Posed Surveillance and Diagnostic Measurements by Gribok et al.

The accurate knowledge of the state of nuclear power plants is of vital importance for plant safety and efficiency optimisation. For the past fifteen years, The University of Tennessee's Nuclear Engineering (UTNE) Department has been applying both conventional and artificial intelligence based predictive techniques to assist in gathering and presenting this information to operators and engineers. Over the past five years, the focus has been on optimising the predictive models to assure their accuracy and repeatability using several categories of regularization techniques These techniques range from principal component regression and training with noise for autoassociative neural networks, to ridge regression and complexity regularization for linear models. Chapter 20 presents a historical overview of the use of regularization methods in surveillance and diagnostics at UTNE. The methods range from a heuristic approach used with non-linear models such as neural networks, to systematic approaches for linear models, and finally to informational approaches applicable to both linear and non-linear modelling techniques.

Chapter 21 on Early Detection and Diagnostics by Sepielli et al.

The operation of industrial plants is based on the readings of a set of sensors. Their reliable operation is essential as the output of sensors provides the only objective information from the process. The task of signal validation is to confirm whether the sensors are functioning properly. Chapter 21 presents the work done in collaboration between ENEA and the OECD Halden Reactor Project (HRP), which resulted in the application of HRP neuro-fuzzy tools- PEANO and ALADDIN to problems of gas plant control, either related to measurement calibration or to process instability. The first part of the chapter describes results and experiences obtained using the PEANO toolbox for validation of critical process sensors at ENEA's ICARO cogeneration plant. PEANO's neuro-fuzzy techniques are able to track in real-time the expected behaviour of a complex process, both in steady state and transient conditions. Fuzzy and possibilistic clustering algorithms are used to classify the operating region where the validation process has to be performed. The possibilistic approach used (rather than probabilistic), allows a "don't know" classification, which results in a fast detection of unforeseen plant conditions or outliers. Specialized Artificial Neural Networks (ANN) are used for the validation process, one for each fuzzy cluster in which the operating map has been divided. They work concurrently on each signal pattern presented to the system and the overall contribution is weighted according to the membership function of the pattern in each cluster. The second part of the chapter describes the work with the transient identification tool ALADDIN, which has been tested as one of the ANN techniques to be applied to the pattern recognition of flame instability statuses in industrial combustion chambers, in the framework of an Italian national research program on fossil fuel combustion. In particular, acoustic waves have been investigated as input signals for ALADDIN to recognize the flame behaviour in the combustion process.

Chapter 22 on ALADDIN by Roverso

Many industrial processes are characterized by long periods of steady-state operation, intercalated by occasional shorter periods of a more dynamic nature in correspondence of either normal events, such as minor disturbances, planned interruptions or transitions to different operation states, or abnormal events, such as major disturbances, actuator failures, instrumentation failures, etc. This second class of events represents a challenge, and possibly a threat, to the smooth, safe, and economical operation of the monitored process. The prompt detection and recognition of such an event is of the essence for the performance of the most effective and informed response to the challenge. Chapter 22 describes a set of techniques, developed as part of the internationally sponsored OECD Halden Reactor Project, which can be employed to design such a transient classification COSS. The ALADDIN methodology integrates these techniques into a unified model, which has undergone several preliminary tests, and which is now used as the basis for the implementation of a complete system dedicated to the recognition of control system anomalies in nuclear power plants.

Chapter 23 on PEANO by Fantoni

On-line monitoring evaluates instrument channel performance by assessing its consistency with other plant indications. On-line monitoring of instrument channels provides information about the condition of the monitored channels through accurate, more frequent monitoring of each channel's performance over time. PEANO is a system for on-line calibration monitoring developed in the years 1995-2001 at the Institutt for energiteknikk (IFE), Norway, which makes use of Artificial Intelligence techniques and Fuzzy Logic for its purpose. The system has been tested successfully in Europe in off-line tests with EDF (France), Tecnatom (Spain) and ENEA (Italy), and in US in co-operation with EPRI. PEANO is currently installed and used for on-line monitoring at the HBWR reactor in Halden. Chapter 23 presents the PEANO tool and its applications to on-line monitoring techniques for calibration reduction of process instrumentation in power plants.

6 Discussions on Soft Computing Techniques

At FLINS'96 there was a general discussion on Hard computing *vs.* Soft computing (Li *et al.*, 1998; Lin and Wildberger, 1994).

- Hard computing is precise and quantitative. Soft computing is inexact and qualitative.
- Hard computing seeks to achieve absolute optimisation. Soft computing accepts *good enough* results. It "satisfies" according to the term invented by Nobel laureate Herbert Simon to express typical human economic behaviour.
- Hard computing is embodied in programs that are carefully designed to exhibit unchanging behaviour regardless of the inputs they are given and the problems they are used to analyse. Soft computing may use training methods to evolve programs that are explicitly intended to adapt their behaviour to changing data and a changing environment.
- Hard computing treats data sets as *crisp* and their members as perfectly discernible. Soft computing accepts data that may be *fuzzy* or *rough*, and responds by producing information that also has these characteristics.

The methods of soft computing, implying as they do, the inclusion of adaptation or machine learning in previously invariable numerical algorithms, bring with them new difficulties in proving that the solutions provided are correct and reliable. The hard part of soft computing is to achieve quality assurance without recourse to the very methods that soft computing was invented to avoid.

The definition of "soft computing" adopted in (Li *et al.*, 1998): *Every computing process that purposely includes imprecision into the calculation on one or more levels and allows this imprecision either to change (decrease) the granularity of the problem, or to "soften" the goal of optimisation at some stage, is defined as to belong to the field of SC.* In the view of the author, SC is still unknown to many nuclear engineers. As a complementary tool to existing methods, intelligent systems and SC as a global approach to some ill defined and to more complex engineering problems in nuclear industry would be helpful. However, to accept SC related techniques in nuclear areas, as a standard tool remains a great challenge.

The nuclear power industry puts special demands on plant safety, surpassing all other industries in its safety culture. The regulatory environment in which NPPs operate reflects these needs, and also the demands of the public for high levels of assurance about safety and regulatory compliance. This culture is not one that encourages innovation in control systems and philosophy, yet nowhere are there greater potential benefits from high reliability systems, automated fault recognition and rationally supported decision-making. A demonstration of the use of intelligent control in an actual plant is a vital step in prototyping the next generation of nuclear power plants. These demonstrations must prove not only the ability to safely survive major disturbances, but also the ability to operate efficiently and reliably in normal operation and to recover smoothly from the minor events that will occur on a regular basis, without challenge to future operations (Ruan, 1998).

7 Fusion of Soft Computing and Hard Computing Techniques: Future Research Direction

Modern approaches and advanced applications for *Power Plant Surveillance and Diagnostics* have been proved to be an extremely challenging problem. The well-developed field of conventional hard computing (HC) offers efficient solutions to a wide variety of existing applications of *Power Plant Surveillance and Diagnostics*. However, the HC techniques often rely upon strong mathematical modelling, analysis, and synthesis. It does not seem that they can easily cope very well with uncertain and unpredictable environments all the time. On the other hand, the advent of soft computing (SC) techniques does provide us a powerful tool to solve demanding real-world problems related to some relevant topics of *Power Plant Surveillance and Diagnostics* as demonstrated in this book. Soft Computing is an emerging field that consists of complementary elements of fuzzy computing, neural computing, evolutionary computation, machine learning, probabilistic reasoning and many others. Thanks to their strong leaning and cognitive ability and good tolerance of uncertainty and impression, the SC techniques have shown their great potential to solve demanding problems in the field of *Power Plant Surveillance and Diagnostics*. The emerging SC and conventional HC should not be viewed as competing with each other but rather as complementary. From the existing literature and successful applications, it can be conclude that the fusion of the SC and HC techniques has provided and will continue to provide innovative solutions for modern approaches and advanced applications of *Power Plant Surveillance and Diagnostics*.

References

Akin HL, Altin V (1991) Rule-based fuzzy logic controller for a PWR-type nuclear power plant, *IEEE Transaction on Nuclear Science* **38** (2): 883-890

Aleksander I, Morton H (1990) *Introduction to Neural Computing*, North Oxford Press

Arul AJ (1994) Reactivity surveillance in a nuclear reactor by using a layered artificial neural network, *Nuclear Science and Engineering* **117:** 186-193

Basu A, Bartlett EB (1994) Detecting faults in a nuclear power plant by using dynamic node architecture artificial neural networks, *Nuclear Science and Engineering* **116:** 313-325

Beale R, Jackson T (1994) *Neural Computing-an Introduction*, Aam Hilger, Bristol

Bernard JA (1988) Use of a rule-based system for process control, *IEEE Control Systems Magazine* **8** (5): 3-13

Bubak M, Moscinski J (1983) A fuzzy-logic approach to HTR nuclear power plant model control, *Annals of Nuclear Energy* **10** (9): 467-471

Cho BH, No HC (1996) Design of stability-guaranteed fuzzy logic controller for nuclear steam generators, *IEEE Transaction on Nuclear Science* **43** (2): 716-730

Cox E (1994) *The Handbook of Fuzzy Systems*, Academic Press, New York

Cheon SW, Chang SH (1993) Application of neural networks to a connectionist expect system for transient identification in nuclear power plants, *Nuclear Technology* **102**:177-191

Davis L (1991) *Handbook of Genetic Algorithms*, Van Nostrand Reinhold, New York

Fantoni PF (2000) A neuro-fuzzy model applied to full range signal validation of PWR nuclear power plant data, *International Journal of General Systems* **29** (2): 305-320

Garis NS, Pazsit I (1999) Control rod localization with neural networks using a complex transfer function, *Progress in Nuclear Energy* **34** (1): 87-98

Goldberg DE (1989) *Genetic Algorithms in Search Optimization and Machine Learning*, Addison-Wesley

Gencay S (1991) Site selection for nuclear plants using fuzzy decision analysis, *Kerntechnik* **56** (5): 320-327

Guo Z, Uhrig R (1992) Use of artificial neural networks to analyze nuclear power plant performance, *Nuclear Technology* **99**: 36-42

Guth MAS (1989) Some uses and limitations of fuzzy logic in artificial intelligence reasoning for reactor control, *Nuclear Engineering and Design* **113**: 99-109

Hah YJ, Lee BW (1994) Fuzzy power control algorithm for a pressurized water reactor, *Nuclear Technology* **106**: 242-253

Heger AS, Alang-Rashid NK, Jamshidi M (1995) Application of fuzzy logic in nuclear reactor control. Part I: an assessment of state-of-the-art, *Nuclear Safety* **36** (1): 109-121

Hertz J, Krogh A, Palmer RG (1991) *Introduction to the Theory of Neural Computation*, Addision-Wesley Publishing Company

Holbert KE, Sharif Heger A, Alang-Rashid NK (1994) Redundant sensor validation by using fuzzy logic, *Nuclear Science and Engineering* **118**: 54-64

Holland JH (1975) *Adaptation in Natural and Artificial Systems*, Ann Arbor, University of Michigan

IFE (2000) Annual Repoprt

Iijima T, Nakajima Y., Nishiwaki Y (1995) Application of fuzzy logic control systems for reactor feed-water control, *Fuzzy Sets and Systems* **74** (1): 61-72

Jung CH, Ham CS, Lee KL (1995) A real time self tuning fuzzy controller for the steam generator through scaling factor adjustment for the steam generator of NPP, *Fuzzy Sets and Systems* **74** (1): 53-60

Jouse WC, Williams JG (1993) Safety control of nuclear power operations using self-programming neural networks, *Nuclear Science and Engineering* **114**: 42-54,

Ikonomopoulos A, Tsoukalas LH, Uhrig RE (1993) Integration of neural networks with fuzzy reasoning for measuring operational parameters in a nuclear reactor, *Nuclear Technology* **104**: 1-12

Kavaklioglu K, Upadhyaya BR (1994) Monitoring feedwater flow rate and component thermal performance of pressurized water reactors by means of artificial neural networks, *Nuclear Technology* **107**: 112-123

Kim HG, Chang SH, Lee BH (1993) Pressurized water reactor core parameter prediction using an artificial neural network, *Nuclear Science and Engineering* **113**: 70-76

Kim HK, Lee SH, Chang SH (1993b) Neural network model for estimating departure from nucleate boiling performance of a pressurized water reactor core, *Nuclear Technology* **101**: 111-122

Kim HG, Chang SH, Lee BH (1993c) Optimal fuel loading pattern design using an artificial neural network and a fuzzy rule-based system, *Nuclear Science and Engineering* **115**, 152-163

Kim K, Bartlett EB (1994) Error prediction for a nuclear power plant fault-diagnostic advisor using neural networks, *Nuclear Technology* **108**: 283-297

Kim DY, Seong PH (1997) Fuzzy gain scheduling of velocity PI controller with intelligent learning algorithm for reactor control, *Annals of Nuclear Energy* **24** (10): 819-827

Klir GJ, Yuan B (1995) *Fuzzy Sets and Fuzzy Logic*, Prentice-Hall Inc.

Kuan CC, Lin C, Hsu CC (1992) Fuzzy logic control of steam generator water level in pressurized water reactors, *Nuclear Technology* **100**: 125-134

Li X, Ruan D, van der Wal AJ (1998) Discussion on soft computing, *International Journal of Intelligent Systems* **13** (2/3), 287-300

Lin C, Yang DH (1998) Design of a fuzzy logic controller for water level control in an advanced boiling water reactor based on input-output data, *Nuclear Technology* **122**: 318-329

Lin TY, Wildberger AM (eds) (1994) *Soft Computing*, the Third International Workshop on Rough Sets and Soft Computing (RSSC94), California

Marseguerra M, Zio E (1994) Fault diagnosis via neural networks: the Boltzmann machine, *Nuclear Science and Engineering* **117**: 194-200

Marseguerra M, Ricotti ME, Zio E (1996) Neural network-based fault detection in a pressurized water reactor pressurizer, *Nuclear Science and Engineering* **124**: 339-348

Matsuoka H (1991) A simple fuzzy simulation model for nuclear reactor system dynamics, *Nuclear Technology* **94**: 228-241

Na N, Kwon K, Ham C, Bien Z (1995) A study on water level control of PWR steam generator at low power and the self-tuning of its fuzzy controller, *Fuzzy Sets and Systems* **74** (1): 43-51

Ohga Y, Seki, H (1993) Abnormal event identification in nuclear power plants using a neural network and knowledge processing, *Nuclear Technology* **101**: 159-167

Park MG, Cho NZ (1995) Self-tuning control of a nuclear reactor using a Gaussian function neural network, *Nuclear Technology* **110**: 285-293

Parlos AG, Atiya AF, Chong KT, Tsai WK (1992) Nonlinear identification of process dynamics using neural networks, *Nuclear Technology* **97**: 36-42

Parlos AG, Muthusami J, Atiya AF (1994) Incipient fault detection and identification in process systems using accelerated neural network learning, *Nuclear Technology* **105**: 145-161

Pereira CMNA, Schirru R, Martinez AS (2000) Genetic algorithms applied to nuclear reactor design optimization, in: Ruan D (ed) *Fuzzy Systems and Soft Computing in Nuclear Engineering*, Physica-Verlag, Heidelberg, pp 315-334

Reifman J, Vitela JE (1994) Accelerating learning of neural networks with conjugate gradients for nuclear power plant applications, *Nuclear Technology* **106**: 225-241

Ramaswamy P, Edwards RM, Lee KY (1993) An automatic tuning method of a fuzzy logic controller for nuclear reactors, *Transactions on Nuclear Science* **40** (4): 1253-1262

Roh MS, Cheon SW, Chang SH (1991) Power prediction in nuclear power plants using a Back-Propagation learning neural network, *Nuclear Technology* **94**: 270-278

Roverso D (2000) Soft computing tools for transient classification, *Information Sciences* **127** (3/4): 137-156

Ruan D (1995) Fuzzy logic in the nuclear research world, *Fuzzy Sets and Systems* **74** (1): 5-13

Ruan D (ed) (1996) *Fuzzy Logic Foundations and Industrial Applications*, Kluwer Academic Publishers, Boston

Ruan D (ed) (1997) *Intelligent Hybrid Systems: Fuzzy Logic, Neural Networks, and Genetic Algorithms*, Kluwer Academic Publishers, Boston

Ruan D, D'hondt P, Govaerts P, Kerre EE (eds) (1994) *Fuzzy Logic and Intelligent Technologies in Nuclear Science*, World Scientific, Singapore

Ruan D, D'hondt P, Govaerts P, Kerre EE (eds) (1996) *Intelligent Systems and Soft Computing for Nuclear Science and Industry*, World Scientific, Singapore

Ruan D, Aït Abderrahim H, D'hondt P, Kerre EE (eds) (1998) *Fuzzy Logic and Intelligent Technologies for Nuclear Science and Industry*, World Scientific, Singapore

Ruan D, Aït Abderrahim H, D'hondt P, Kerre EE (eds) (2000) *Intelligent Techniques and Soft Computing in Nuclear Science and Engineering*, World Scientific, Singapore

Ruan D (1998) Intelligent systems in nuclear applications, *Int. J. of Intelligent Systems* **13** (2/3): 115-125

Ruan D, van der Wal AJ (1998) Controlling the power output of a nuclear reactor with fuzzy logic, *Information Sciences* **110**: 151-177

Ruan D, Li X, Van den Eynde G (1999) Adaptive fuzzy control for a simulation of hydraulic analogy of a nuclear reactor, in: Ruan, D (ed.) *Fuzzy Systems and Soft Computing in Nuclear Engineering*, Physica-Verlag, Heidelberg, pp 65-82

Ruan D (2001) Implementation of adaptive fuzzy control for a real-time control demo-model, *Real-Time Systems* **21**: 219-239

Schirru R, Pereira CMNA, Martinez AS (2000) Genetic algorithms applied to nuclear power plant operation, in: Ruan D (ed) *Fuzzy Systems and Soft Computing in Nuclear Engineering*, Physica-Verlag, Heidelberg, pp 335-350

Tanabe A, Yamamoto T, Shinfuku K, Nakamae T (1993) Development of neural network for analysis of local power distributions in BWR fuel bundles, *Journal of Nuclear Science and Technology* **30** (8): 804-812

Thomas JR, Adams JT (1994) Noise analysis method for monitoring the moderator temperature coefficient of pressurized water reactors: neural network calibration, *Nuclear Technology* **107**: 236-240

Tsoukalas L, Modarres M (1993) Artificial intelligence, expert systems, and fuzzy logic in nuclear technology, *Transactions on America Nuclear Society* **68**: 133-139

Uhrig RE (1991) Potential application of neural networks to the operation of nuclear power plants, *Nuclear Safety* **32** (1): 68-79

Uhrig RE, Tsoukalas LH (1999) Soft computing techniques in nuclear engineering applications, *Progress in Nuclear Energy* **34** (1): 13-75

Upadhyaya BR, Eryurek E (1992) Application of neural networks for sensor validation and plant monitoring, *Nuclear Technology* **97:** 170-176

Vitanza C (2001) Overview of the OECD Halden Reactor Project, *Nuclear Engineering and Design* **207**: 207-221

Zadeh LA (1965) Fuzzy Sets, *Information and Control* **8** (3): 338-353

Zadeh LA (1981) Possibility theory and soft data analysis, In: Cobb L, Thrall RM (eds) *Mathematical Frontiers of the Social and Policy Science*, Westview Press, Boulder, Colorado, pp 69-129

Zadeh LA (1994) Fuzzy logic, neural networks, and soft computing, *Communications of the ACM* **37** (3): 77-83

Zadeh LA (1997) What is soft computing? *Soft Computing* **1**: p1

Appendix A: Selected References from FLINS'94

[A1] Abrosimov VK, Verbin ES (1994) Principles of design for nuclear reactor safety system on the basis of neural network, in: (Ruan et al., 1994), pp 216-218

[A2] Alang-Rashid NK, Sharif Heger A (1994) Self-tuning fuzzy logic nuclear reactor controller, in: (Ruan et al., 1994), pp, 285-289

[A3] Averkin AA (1994) Fuzzy logic acquisition and simulation modules for expert systems to assist operator's decision for nuclear power stations, in: (Ruan et al., 1994), pp 277-278

[A4] Chung HY, Bien Z (1994) Real-time diagnosis of incipient multiple faults with application for nuclear power plants in: (Ruan et al., 1994), pp 232-237

[A5] Gradetsky VG, Rizzotto G, Pagni A, Slesarev YV, Pospelov DA, Ul'yanov SV, Yamafuji K (1994) Principles of expert fuzzy controller design: AI mobile wall climbing robots for decontamination of nuclear power-station, in: (Ruan et al., 1994), pp 242-246

[A6] Lyapin B, Ryjov A (1994) A fuzzy linguistic interface for data bases in nuclear safety problems, in: (Ruan et al., 1994), pp 212-215

[A7] Orlovski SA, Integrated fuzzy cluster, choice, and knowledge acquisition technology in problem of nuclear safety, in: (Ruan et al., 1994), pp 271-273

[A8] Perfilyeva IG, Postnikov VV (1994) Two models for energy distribution control in the nuclear power plants, in: (Ruan et al., 1994), pp 274-276

[A9] Van den Durpel L, Ruan D (1994) Fuzzy model-based control of a nuclear reactor, in: (Ruan et al., 1994), pp 247-253

Appendix B: Selected References from FLINS'96

[B1] Carlsson C, Fullér R (1996) Problem-solving with multiple interdependent criteria: better solutions to complex problems in: (Ruan et al., 1996), pp 89-97

[B2] Chung HY, Park IS, Bien Z (1996) Improved method for incipient multiple fault diagnosis with application to nuclear power plant, in: (Ruan et al., 1996), pp 365-371

[B3] Galdoz EG, H.R. Fontanini HR, Tapia ER (1996) Integrating information in a real-time data visualization system on nuclear power plant, in: (Ruan et al., 1996), pp 395-400.

[B4] Guido PN, Oggianu S, Etchepareborda A, Fernandez O (1996) Expert environment for the development of nuclear power plants failure diagnosis systems, in: (Ruan et al., 1996), pp 395-400

[B5] Kozin IO, Petersen KE (1996) Obtaining reliability and safety assessments on the basis of non-probabilistic methods, in: (Ruan et al., 1996), pp 250-257

[B6] Liu X, Ruan D (1996) Application of fuzzy decision making to countermeasure strategies after a nuclear accident, in: (Ruan et al., 1996), pp 308-315

[B7] Liu Z, Ruan D (1996) Experiments of fuzzy logic control on a nuclear research reactor, in: (Ruan et al., 1996), pp 336-348

[B8] Modarres M (1996) Functional modelling for integration of human-software-hardware in complex physical systems, in: (Ruan et al., 1996), pp 189-204

[B9] Moon BS, Lee J (1996) A fuzzy control algorithm for a mobile robot to move pass obstacles, in: (Ruan et al., 1996), pp 329-335

[B10] Na NJ, Kim IS, Kim JT, Hwang IK, Lee DY, Ham CS (1996) AI-based alarm processing for a nuclear power plant, in: (Ruan et al., 1996), pp 388-394

[B11] Schildt GH (1996) A fuzzy controller for NPPs, in: (Ruan et al., 1996), pp 379-387

[B12] Schoonewelle H, van der Hagen THJJ, Hoogenboom JE (1996) Process monitoring by combining several signal-analysis results using fuzzy logic, in: (Ruan et al., 1996), pp 316-322
[B13] Sharif Heger A, Jamshidi M, Alang-Rashid NK (1996) Self-tuning fuzzy logic nuclear reactor controller, in: (Ruan et al., 1996), pp 349-358
[B14] Smets P (1996) Applying the transferable Belief model to diagnostic problems, in: (Ruan et al., 1996), pp, 285-292
[B15] Oh YG, Hong HP, Han SJ, Chun CS, Kim BK (1996) Fuzzy logic utilization for the diagnosis of metallic loose part impact in nuclear power plants, in: (Ruan et al., 1996), pp 372-378
[B16] Onisawa T (1996) System reliability analysis with natural language and expert's subjectivity, in: (Ruan et al., 1996), pp 243-249
[B17] Wang PP, Gu XL (1996) Intelligent engineering & technology for nuclear power plant operation, in: (Ruan et al., 1996), pp 359-364
[B18] Zimmermann HJ (1996) Fuzzy algorithmic and knowledge-based decision support in nuclear engineering, in: (Ruan et al., 1996), pp 1-8

Appendix C: Selected References from FLINS'98

[C1] Baek S, Park DS, Cho J, Lee YB (1998) A robot endeffector tracking system based on feedforward neural network, in: (Ruan et al., 1998), pp 324-331
[C2] Benedikt J, Nishiwaki Y (1998) A fuzzy clustering application to land use classification in satellite images, in: (Ruan et al., 1998), pp 368-375
[C3] Brendeford TS, Bye A, Hollnagel E, Hoffmann M (1998) FAME - a tool for function allocation in NPP control rooms, in: (Ruan et al., 1998), pp 444-450
[C4] Bjørlo TJ, Berg Ø (1998) Development of plant surveillance and operations systems at the OECD Halden reactor project, in: (Ruan et al., 1998), pp 435-443
[C5] Cai D, McTear MF, McClean SI (1998) Risk analysis in distributed databases using evidence theory, in: (Ruan et al., 1998), pp 179-186
[C6] Chien SI, Nam SW, Lee NH, Lee YB (1998) Stroke based gesture recognition using polhemus sensor for remote work monitoring, in: (Ruan et al., 1998), pp 310-317
[C7] Dahll G, Gran BA (1998) The use of Bayesian belief nets in safety assessment of software based systems, in: (Ruan et al., 1998), pp 187-194
[C8] Decreton M, De Geeter J, Colon E (1998) The challenges for telerobotics in a nuclear environment, in: (Ruan et al., 1998), pp 467-474
[C9] Erkan K, Bütün E, Can B (1998) Triga Mark-II reactor controller design using genetic algorithm, in: (Ruan et al., 1998), pp 400-407
[C10] Espinosa J, Vandewalle J (1998) Predictive control using fuzzy models applied to a steam generating unit, in: (Ruan et al., 1998), pp 151-160
[C11] Fantoni PF, Figedy S, Racz A, Papin B (1998) A neuro-fuzzy model applied to full range signal validation of PWR nuclear power plant data, in: (Ruan et al., 1998), pp 451-458
[C12] Gregoire E, Saïs L (1998) Practical inconsistency management for critical-tasks decision-support Systems, in: (Ruan et al., 1998), pp 384-391
[C13] Hardy LLE, Polanco EM, De La Vega DMG (1998) A small scale simulator of the pressurizer with expert system, in: (Ruan et al., 1998), pp 416-423

[C14] Kacprzyk J (1998) Multistage fuzzy control in some power engineering applications, in: (Ruan et al., 1998), pp 171-178

[C15] Kahraman C, Tolga E (1998) Nuclear energy and safety assessment, in: (Ruan et al., 1998), pp 345-352

[C16] Kästner W, Fenske A, Hampel R (1998) Improvement of the robustness of model-based measuring methods using fuzzy logic, in: (Ruan et al., 1998), pp 129-142

[C17] Li X, Ruan D (1998) Comparative study of fuzzy control, PID control, and advanced fuzzy control for simulating a nuclear reactor operation, in: (Ruan et al., 1998), pp 424-434

[C18] Moon BS, Lee YB (1998) A neural algorithm for moving control of mobile robots in an office hall, in: (Ruan et al., 1998), pp 318-323

[C19] Nissan E, Soper A, Zhao J, Knight B, Petridis M (1998) Fuel reload pattern design within a family of hybrid architectures, in: (Ruan et al., 1998), pp 408-415

[C20] Park HW, Yang HS, Park YP, Kim SH (1998) Position and vibration control of a flexible robot manipulator using fuzzy logic, in: (Ruan et al., 1998), pp, 302-309

[C21] Pilato V, Tola F, Martinez JM, Huver M (1998) Contribution of neural networks to nuclear measurements, in: (Ruan et al., 1998), pp 360-67

[C22] Roverso D, Fantoni PF (1998) ALADDIN: A neural classifier of fast transients for alarm filtering in nuclear power plants, in: (Ruan et al., 1998), pp 459-466

[C23] van der Wal AJ (1998) The importance of soft computing methods for military observation systems, in: (Ruan et al., 1998), pp 163-170

[C24] Zhang J, Schwert V, Knoll A (1998) Comprehensive fuzzy control of systems with complex sensor patterns, in: (Ruan et al., 1998), pp 286-293

[C25] Zhou C, Ruan D (1998) Integration of linguistic and numerical information for biped locomotion, in: (Ruan et al., 1998), pp 294-301

[C26] Zimmermann HJ (1998) Computational intelligence and nuclear engineering, in: (Ruan et al., 1998), pp 1-18

Appendix D: Selected References from FLINS 2000

[D1] Bien Z, Song WK (2000) Blend of soft computing techniques for effective human-machine interaction/interface in service robotic systems, in: (Ruan et al., 2000), pp 35-42

[D2] Castillo O, Melin P (2000) Intelligent adaptive model-based control of robotic dynamic systems with a new hybrid neuro-fuzzy-fractal approach, in: (Ruan et al., 2000), pp 351-358

[D3] Costa Nunes ME, Pereira CMNA, Frutuoso e Melo PF (2000) Optimization of the device of stages through genetic algorithms for non-Markovian systems reliabillity evaluation: an application to nuclear safety systems, in: (Ruan et al., 2000), pp 527-534

[D4] Fantoni PF, Hoffmann M, Nystad BH, De Oliveira MV (2000) Integration of sensor validation in modern control room alarm systems, in: (Ruan et al., 2000), pp 462-469

[D5] Guan JW, Bell DA (2000) Knowledge discovery for controlling nuclear power plants, in: (Ruan et al., 2000), pp 429-436

[D6] Gómez-Ortega J, Ramírez DR, Limón D, Camacho EF (2000) Predictive mobile robot navigation using soft computing techniques, in: (Ruan et al., 2000), pp 335-342

[D7] Hampel R, Chaker N, Wagenknecht M (2000) Knowledge representation using fuzzy logic based characteristics for safety related applications Part I: basic investigations; Part II: applications, in: (Ruan et al., 2000), pp 311-318, 319-326

[D8] Hampel R, Fleischer S, Dräger F, Maekawa T (2000) Water level measurement system for boiling water reactors using internal gamma radiation – neural network application, in: (Ruan et al., 2000), pp 504-511

[D9] Lapa CMF, Pereira CMNA, Frutuose e Melo PF (2000) Coupled emergency diesel generator - component coolant water system maintenance scheduling optimization by genetic algorithm, in: (Ruan et al., 2000), pp 519-526

[D10] Melin P, Castillo O (2000) Adaptive intelligent control of aircraft dynamic systems with a new hybrid neuro-fuzzy-fractal approach, in: (Ruan et al., 2000), pp 359-368

[D11] Na MG, Yang WS, Choi H (2000) A CANDU fuel pin power reconstruction using an adaptive fuzzy inference system, in: (Ruan et al., 2000), pp 437-444

[D12] Ortiz JJ, Requena I (2000) Optimization of fuel reload for a BWR using neural networks and genetic algorithms, in: (Ruan et al., 2000), pp 512-518

[D13] Oussalah M, De Schutter J (2000) Noise identification and its influence on Kalman filter divergence, in: (Ruan et al., 2000), pp 80-86

[D14] Park HS, Kim K, Kim GN, Shon JS, Bae KK (2000) A study on the incipient fault diagnosis of a radwaste solar evapolation processing system using a neural network, in: (Ruan et al., 2000), pp 488-495

[D15] Petruzela I (2000) Fuzzy system of automatic failure classification, in: (Ruan et al., 2000), pp 455-461

[D16] Roverso D (2000) Wavelet-based recurrent neural networks for transient classification, in: (Ruan et al., 2000), pp 496-503

[D17] Si Fodil M, Siarry P, Tyran JL (2000) Optimization of the number of fuzzy rules towards a better temperature control of nuclear reactors, in: (Ruan et al., 2000), pp 445-454

[D18] Šorf M, Janků L, Lhotská L, Eck V (2000) Application of expert system and machine learning approach to intelligent man-machine interface, in: (Ruan et al., 2000), pp 191-200

[D19] Zhou C, Kanniah J, Meng Q (2000) Intelligent robotic control using reinforcement learning agents with fuzzy evaluative feedback, in: (Ruan et al., 2000), pp 327-334

[D20] Zimmermann HJ (2000) Dynamic fuzzy data analysis and uncertainty modeling in engineering, in: (Ruan et al., 2000), pp 3-15

2 Regulatory Treatment of On-Line Surveillance and Diagnostic Systems

Robert E. Uhrig

The University of Tennessee

and

Oak Ridge National Laboratory, USA

ruhrig@utk.edu

1 Introduction

For almost two decades, various university and national laboratory research groups have been developing surveillance and diagnostic techniques for application to nuclear power plants. For more than a decade, many of these techniques included various types of artificial intelligence based systems to develop a data-based model as a substitute for first principles models that were found deficient. The motivation for such techniques was to monitor the condition of signal channels with a view to reducing the number of calibrations required by the Nuclear Regulatory Commission (NRC). When the Electric Power Research Institute (EPRI) approached the NRC for approval of these systems, such use was not approved, primarily because of the uncertainty of the magnitude of the errors associated with the monitoring technique when used for sensor calibration. After eight years of interaction with the NRC, the EPRI submitted Topical Report #104965 "On-Line Monitoring of Instrument Channel Performance." The NRC replied with 14 conditions that had to be met for acceptance. This paper discusses the EPRI topical report and NRC's response.

2 Regulatory Treatment of On-Line Surveillance and Diagnostic Systems

In 1998, the Electric Power Research Institute submitted Topical Report 104965 "On-Line Monitoring of Instrument Channel Performance" on behalf of several utilities. Of the several different types of surveillance and diagnostic systems being developed, two systems were the focus of the topical report: 1) the Instrumentation Calibration Monitoring Program (ICMP), an algorithm using a comparison of the outputs of redundant sensors developed by EPRI and 2) the Multivariate State Estimation Technique (MSET), a nonlinear regression system developed by Argonne National Laboratory. EPRI proposed a new generic approach for monitoring instrument calibrations during normal plant operation. These techniques used this on-line monitoring approach with a "calibrate-as-required" approach. EPRI proposed to allow commercial nuclear power plant licensees to use this on-line monitoring technique as a calibration assessment tool to extend calibration intervals required by the Technical Specifications for safety critical measurements.

The EPRI technical report demonstrated an on-line technique for obtaining real-time instrument performance data in a non-intrusive manner. It incorporated these data with field calibration results to verify whether or not the monitored instrument channel's performance was within acceptable limits.

The benefits claimed by EPRI for this approach were 1) it helped eliminated unnecessary field calibrations, 2) it reduced associated labor costs, 3) it limited personnel radiation exposures, and 4) it limited the potential for miscalibration of the instruments channels. The system was not to be connected to the plant instrumentation permanently, but rather, instrument data was to be collected in a batch mode and analyzed later to assess instrument calibration, performance and operability. The system was to be used as "measurement and test equipment" (M&TE) to monitor calibration and operational status of safety-related instruments.

The system monitors the steady-state output of each channel and evaluates the monitored value to determine whether the channel is operating outside its acceptable limits. During evaluation, the monitored value is compared to a calculated value of the process variable to assess deviations. The calculated value was defined as the "process parameter estimate." It represented the instantaneous value of the process at the monitored operating point. EPRI proposed to use this system on a quarterly basis. However, evaluation of the data could be performed continuously or at more frequent discrete intervals, manually or automatically.

3 Types of On-Line Surveillance and Diagnostic Systems Considered by the NRC

The ICMP algorithm developed by EPRI compares measurements from redundant channels to determine if one or more channels are operating beyond their specific limits. In contrast, MSET uses empirical type, statistically-based pattern recognition modules that interact and operate to provide the user with needed information by detecting, locating and identifying subtle changes that could lead to future problems well in advance of significant degradation. MSET is used for non-redundant sensors and to catch common mode failure of redundant sensors.

However, NRC did not limit the monitoring systems to ICMP and MSET. Rather, they considered on-line surveillance and diagnostic systems as a general class and based their requirements on the functional requirements for the system. Among the alternative systems included under the NRC regulations, provided it met their requirements, were:

(1) The artificial autoassociative neural network (AANN) system that was originally developed by the University of Tennessee. In an autoassociative neural network, the network is trained so that each output is equal to the corresponding input. The intercorrelation between the various inputs makes each output dependent upon all inputs associated with it. A "sequential probability ratio test" (SPRT) is used to identify deviations beyond prescribed limits that are based on "missed alarms" and "false alarms."

(2) PEANO is an improved version of the autoassociate neural network system above. It uses fuzzy classification of the input variables to establish multiple ranges of the input variables, thereby creating and training neural networks for each range to enhance the prediction accuracy.

(3) The Nonlinear partial Least Squares (NLPLS) System was also developed at the University of Tennessee and has been implemented at Tennessee Valley Authority's Kingston Fossil Plant. This system involves the linear transformation of the partial least squares methodology to provide a supervised reduction of the process system's data while preserving the information most relevant to the specific parameter(s) of the system to be modeled. By replacing the linear regression with a neural network, nonlinear relationships can be incorporated into the model.

4 Configuration of a Typical On-Line Monitoring and Diagnostic System

A typical off-line monitoring and diagnostic configuration consists of a separate off-line computer on which the system resides with communications hardware and software for automatic acquisition of data from the plant process computer and on-line monitoring software to archive, analyze and display the data interactively in graphs and reports.

A typical on-line monitoring system collects data from instrument channels and processes it on an off-line computer that has excellent graphical display capabilities for the individual instrument channel deviation from the process channel estimate as a function of time.

Regardless of the algorithm employed, the on-line monitoring technique evaluates the deviation of an instrument with reference to its process parameter estimate as determined by one of the monitoring systems. This determines whether 1) the performance exhibited by the instrument is acceptable, 2) the instrument must be scheduled for calibration, or 3) the instrument channel is inoperable.

5 Advantages of On-Line Monitoring and Diagnostic Systems

The advantages of on-line-monitoring claimed by EPRI in Topical Report 104965 are:

(1) Compared to the traditional calibration processes, on-line monitoring is non-intrusive, more frequent, and will result in a reduced number of field calibrations.

(2) On-line monitoring can be continuous and will identify calibration problems when they occur.

(3) Elimination of unnecessary field calibrations will reduce associated labor costs, personnel radiation exposure and potential for miscalibration.

(4) On-line monitoring allows evaluation of instrument performance under normal operating operation whereas normal calibration usually takes place during plant shutdown.

(5) By reducing the time spent on normal calibration, refueling outages will be shorter.

(6) Long term trends in instrument performance developed using on-line monitoring could be used for predictive maintenance scheduling.

6 NRC Response to EPRI's TR-104965

The current Technical Specifications require that all redundant safety-related instruments be calibrated once each fuel cycle (i.e., traditional calibration). The changes proposed by EPRI in response to the NRC's initial concerns include:

(1) Establish on-line monitoring as an acceptable procedure for assessing instrument performance and its calibration while the plant is in its normal operating mode.

(2) On-line monitoring be used as a basis for extending the calibration interval from once per fuel cycle for each safety sensor to once in a maximum of eight years by implementing the processes:

a. At least one redundant sensor will be calibrated each fuel cycle.

b. For "n" redundant sensors, all sensors will be calibrated every "n" fuel cycles.

c. Sensors that are identified as out of calibration by the on-line monitoring process will also be calibrated as necessary.

Hence, traditional calibration would be replaced by on-line calibration monitoring and a "calibrate as required" approach.

7 NRC Staff Initial Position on On-Line Monitoring

The justification for replacing traditional monitoring with on-line monitoring and calibration as needed is either:

(1) the proposed on-line monitoring technique can perform all the required designated functions better than or as good as the current traditional calibration with the same or better reliability, or

(2) if due to inherent deficiencies in the proposed technique, it cannot be demonstrated to be either better than or at least as good as the current practice, then the licensee should verify that the impact of the proposed technique on plant safety will be insignificant and that the advantages of using it will outweigh the deficiencies.

8 Deficiencies of On-Line Monitoring Systems Identified by the NRC

The NRC identified what they perceived as the deficiencies of the on-line monitoring system.

(1) They are not capable of monitoring instrument performance for its full range including "trip set point" (TSP).

(2) They do not have an accurate reference by which to compare the monitored value to a calculated reference (the process parameter estimate) that itself is less accurate compared to the simulated input used in traditional calibration.

(3) They do not provide accuracy traceable to standards.

(4) They do not allow frequent physical inspection of the instrument or allow technicians to observe instrument anomalies.

(5) Hence, because of these deficiencies, on-line monitoring may be unable to verify an instrument's performance adequately to establish its operability, thereby degrading the plant safety.

9 EPRI's and Utilities' Response to NRC's Concerns

EPRI responded to NRC's concerns with the following statements:

(1) For assessing the capability of on-line monitoring to perform functions performed by traditional calibration either better or as well, functions should be evaluated aggregately rather than one function at a time.

(2) While little more uncertainty is associated with the process parameter estimate than with a simulated reference input traceable to NIST, the accuracy of the process parameter estimate is sufficient for its proposed purpose which is to provide a reference value against which subsequent drift could be measured.

(3) The uncertainties associated with the process parameter estimate will be quantitatively bounded and accounted for in either the on-line monitoring acceptance criteria or the applicable set point and uncertainty calculations

(4) EPRI believes that, taken as a whole, the on-line monitoring technique is superior to traditional calibration and provides greater assurance of instrument operability throughout an operating cycle.

On-line monitoring would not result in a loss of function or a significant safety degradation because:

(1) The error introduced by the additional uncertainty in the process parameter estimate would be accounted for in determining the test acceptance criteria and/or the related setpoint calculations.

(2) At least one channel would be calibrated during each outage by a method traceable back to **NIST** Standards.

(3) Instrument performance would be monitored by on-line monitoring more frequently than by traditional means.

(4) Redundancy exists in instrument channels.

(5) V. C. SUMMERS Nuclear Station representatives stated that they had been using the proposed on-line technique (ICMP) for monitoring instrument performance for the last eight years and found that the benefits were overwhelming and outweighed the insignificant degradation in plant safety due to the deficiencies.

10 NRC Staff Requirements for On-Line Monitoring Systems

After the extensive exchange of views, the NRC issued the following 14 requirements for on-line monitoring systems:

(1) The submittal for implementation of the on-line monitoring technique must confirm that the impact of the deficiencies inherent in the on-line monitoring technique (inaccuracy in process parameter estimate single-point monitoring and untraceability of accuracy to standards) on plant safety be insignificant, and that all uncertainties associated with the process parameter estimate have been quantitatively bounded and accounted for either in the on-line monitoring acceptance criteria or in the applicable set point and uncertainty calculations.

(2) Unless the licensees can demonstrate otherwise, instrument channels monitoring processes that are always at the low or high end of an instrument's calibrated span during normal plant operation be excluded from on-line monitoring programs.

(3) The algorithm used for on-line monitoring is able to distinguish between the process variable drift (actual process going up or down) and the instrument drift and to compensate for uncertainties introduced by unstable processes, sensor locations, non-simultaneous measurements, and noisy signals. If the implemented algorithm and/or its associated software cannot meet these requirements, administrative controls, including the guidelines in Section 3 of the topical report for avoiding a penalty for non-simultaneous measurements could be implemented as an acceptable means to ensure that these requirements are met satisfactorily.

(4) For instruments that were not included in the EPRI drift study, the value of the allowance or penalty to compensate for single-point monitoring must be determined by using the instrument's historical calibration data and by analyzing the instrument performance over its range for all modes of operation, including startup, shutdown, and plant trips. If the required data for such a determination is not available, an evaluation demonstrating that the instrument's relevant performance specifications are as good as or better than those of a similar instrument included in the EPRI drift study, will permit a licensee to use the generic penalties for single point monitoring given in EPRI Topical Report 104965.

(5) Calculations for the acceptance criteria defining the proposed three zones of deviation ("acceptance," "needs calibration" and "inoperable") should be done in a manner consistent with the plant-specific safety-related instrumentation setpoint methodology so that using an on-line monitoring technique to monitor instrument performance and extend its calibration interval will not invalidate the setpoint calculation assumptions and the

safety analysis assumptions. If new or different uncertainties should require the recalculation of instrument trip setpoints, it should be demonstrated that relevant safety analyses are unaffected. The licensee should have a documented methodology for calculating acceptance criteria that are compatible with the practice described in regulatory Guide 1.105 and the methodology described in acceptable industry standards for trip setpoint and uncertainty calculations.

(6) For any algorithm used, the maximum acceptable value of deviation (MAVD) shall be such that accepting the deviation in the monitored value anywhere in the zone between parameter estimate (PE) and MAVD will provide high confidence (level of 95%/95%) that drift in the sensor-transmitter and/or any part of an instrument channel that is common to the on-line monitoring loop is less than or equal to the value used in the setpoint calculations for that instrument.

(7) The instrument shall meet all requirements of the above requirement for the acceptable band or acceptable region.

(8) For any algorithm used, the maximum value of the channel deviation beyond which the instrument is declared "inoperable" shall be listed in the technical specifications with a note indicating that this value is to be used for determining the channel operability only when the channel's performance is being monitored using an on-line monitoring technique. It could be called "allowable deviation value for on-line monitoring" (ADVOLM) or whatever name the licensee chooses. The ADVOLM shall be established by the instrument uncertainty analysis. The value of the ADVOLM shall be such to ensure:

   A. that when the deviation between the monitored value and its PE is less than or equal to the ADVOLM limit, the channel will meet the requirements of the current technical specifications and the assumptions of the setpoint calculations and safety analyses are satisfied; and

   B. that until the instrument channel is recalibrated (at most until the next refueling outage), actual drift in the sensor-transmitter and/or any part of an instrument channel that is common to the instrument channel and the on-line monitoring loop will be less than or equal to the value used in the setpoint calculations and other limits defined in 10-CFR50.36 as applicable to the plant-specific design for the monitored process variable are satisfied.

(9) Calculations defining alarm setpoint (if any), acceptable band, the band identifying the monitored instrument as needing to be calibrated earlier than its next scheduled calibration, the maximum value of deviation beyond which the instrument is declared "inoperable," and the criteria for determining the monitored channel to be an outlier," shall be performed to ensure that all safety analysis assumptions and assumptions of the associated

setpoint calculations are satisfied and the calculated limits for the monitored process variables specified by 10CFR50.36 are not violated.

(10) The plant-specific submittal shall confirm that the proposed on-line monitoring system will be consistent with the plant's licensing basis, and that there continues to be a coordinated defense-in-depth against instrument failure.

(11) Adequate isolation and independence, as required by Regulatory Guide 1.75, GDC 21, GDC 22, IEEE Std 279 or UEEE Std 603, and IEEE Std. 384 shall be maintained between the on-line monitoring devices and class 1 E instruments being monitored.

(12) A. QA requirements are delineated in 10CFR Part 50, Appendix B, shall be applicable to all engineering and design activities related to on-line monitoring, including design and implementation of the on-line system, calculations for determining process parameter estimates, all three zones of acceptance criteria (including the value of the ADVOLM), evaluation and trending of on-line monitoring results, activities (including drift assessments) for relaxing the current TS-required instrument calibration frequency from "once per refueling cycle" to "once per maximum period of 8 years," and drift assessments for calculating the allowance or penalty required to compensate for single-point monitoring.

B. The plant-specific QA requirements shall be applicable to the selected on-line monitoring methodology, its algorithm, and the associated software. In addition, software shall be verified and validated and meet all quality requirements in accordance with NRC guidance and acceptable industry standards.

(13) All equipment (except software) used for collection, electronic transmission, and analysis of plant data for on-line monitoring purposes shall meet the requirements of 10CFR Part 50, Appendix B, Criterion XII, "Control of Measuring and Test Equipment." Administrative procedures shall be in place to maintain configuration control for the on-line monitoring software and algorithm.

(14) Before declaring the on-line monitoring system operable for the first time, and just before each performance of the scheduled surveillance using an on-line monitoring technique, a full-features function test, using simulated input signals of known and traceable accuracy, should be conducted to verify that the algorithm and its software perform all required functions within acceptable limits of accuracy. All applicable features shall be tested.

11 NRC Conclusions on On-Line Monitoring Systems

(1) The staff concluded that the generic concept of an on-line monitoring technique as presented in the topical report, is acceptable for on-line tracking of instrument performance.

(2) The staff agrees with the topical report's conclusion that, on-line monitoring has several advantages, including timely detection of degraded instrumentation.

(3) The staff believes that on-line monitoring can provide information on the direction which instrument performance is heading and in that role, it can be useful in preventive maintenance activities.

(4) Although the proposed on-line monitoring technique compared to traditional calibration process will render results with less accuracy, the staff finds EPRI's conclusion acceptable that accuracy rendered by the process parameter estimate is sufficient to assess instrument operability.

(5) Compared to traditional calibration once per refueling outage, the on-line monitoring technique when taken as a whole, provides higher assurance of instrument operability throughout a plant operating cycle.

12 Current Status

Under the auspices of EPRI, an Instrument Monitoring and Calibration Users Group has been formed and is presently carrying out an aggressive program implementing on-line monitoring systems. They are gathering the data needed to satisfy the NRC requirements listed above. The University of Tennessee is under contract to carry out the error analysis of the on-line monitoring measurements, and another organization is carrying out the verification and validation of the software used for monitoring. Discussions between the utilities and the Nuclear Regulatory Commission are continuing with a view to implementing the on-line monitoring and calibration system at an early date.

References

Memorandum to Stephen Dembek from Evangelos C. Marinos, June 22, 2000 Subject: The Staff Review of EPRI Topical Report (TR) 104965, "On-Line Monitoring of Instrument Channel Performance," Final Report, November 1998, (TAC No. M93653)

EPRI Topical Report (TR) 104965, "On-Line Monitoring of Instrument Channel Performance," Final Report, November 1998, (TAC No. M93653)

3 Optimized Maintenance and Management of Ageing of Critical Equipment in Nuclear Power Plants

H.M. Hashemian

Analysis and Measurement Services Corporation

AMS 9111 Cross Park Drive, Building A-100
Knoxville, TN 37923, USA

hash@ams-corp.com

New power generation economics and politics in the 2000 decade, recent interest in nuclear power plant life extension, and challenges from other electricity production systems such as combined cycle gas plants are providing new opportunities and new challenges to the nuclear power industry. Chief among these is the challenge to continue to improve efficiency, reduce costs, and manage the aging of critical components. This paper presents a review of testing and maintenance technologies that have great potential to help the nuclear industry in meeting these challenges. These technologies range from new methods for on-line calibration and testing of process instrumentation to predictive maintenance of the reactor vessel and its internals.

1 Introduction

The nuclear power industry is enjoying a renaissance as evident by a number of new developments over the last five years. These new developments include: (1) announcement by nearly all nuclear power plants in the United States to implement 20 year life extension; (2) life extension already granted by the U.S. Nuclear Regulatory Commission (NRC) to a number of nuclear power plants for an additional 20 years of operation beyond their current 40 year license; (3) a strong desire to reduce dependence on oil imports; (4) increased interest in the role of the nuclear energy in protecting the environment; (5) improved efficiency and safety of current generation of plants; (6) success of advanced reactor designs; (7) progress toward disposal of high-level nuclear waste; and (8) improved legislative environment for increase use of nuclear energy. On the other hand, in light of deregulation/liberalization of the electric power industries around the world, pressure has increased on the nuclear power industry to become more competitive. The combination of these new developments has stimulated the nuclear industry to concentrate on reducing operations and maintenance (O&M) costs to become more competitive. For example, O&M costs in the USA are presently responsible for about 70 percent of the overall cost of nuclear energy generation, with fuel being responsible for the remaining 30 percent. Therefore, reducing O&M costs can help make the nuclear industry more competitive. This is not as easy in coal and gas power generation where more than 60 percent of the cost of electricity generation is spent for the purchase of the fuel.

In an effort to reduce O&M costs, while maintaining and improving safety, the nuclear industry has implemented risk-based maintenance, and is also depending on new technologies such as computer-aided testing, on-line calibration, and on-line maintenance to reduce manpower requirements and human errors, increase accuracy and reliability, and trend the maintenance data to account for ageing effects on the performance of the plant equipment. Table 1 lists some of the specific techniques that have been developed, validated, and used in nuclear power plants to meet the plant technical specification requirements, optimize maintenance, or manage the ageing of critical equipment. These techniques are reviewed individually in the following sections of this paper.

Table 1. Examples of technologies for on-line maintenance and management of ageing of critical plant equipment

1	On-line verification of calibration of process instrumentation channels
2	On-line detection of Venturi clogging
3	In situ response time testing of pressure transmitters
4	On-line detection of blockages and voids in pressure sensing lines
5	In situ calibration of primary coolant RTDs and core exit thermocouples
6	In situ response time testing of RTDs and thermocouples
7	New methods for remote testing of the attachment/adhesion/embedment of temperature sensors and strain gages to solid materials
8	In situ testing of cables and connectors
9	Predictive maintenance and management of ageing of reactor internals
10	On-line measurement of stability margins (decay ratios) in BWRs
11	Measurement of core flow and detection of core flow anomalies

Abbreviations:

RTD: Resistance Temperature Detector
LCSR: Loop Current Step Response Test
BWR: Boiling Water Reactor
CRDM: Control Rod Drive Mechanism

2 On-Line Calibration Verification

According to present practice, hundreds of instruments are manually calibrated, typically at least once every fuel cycle, in almost all nuclear power plants. The results of these calibrations over more than 20 years have shown that a majority of the instruments do not fall out-of-tolerance between refueling outages and therefore do not need a calibration. This has motivated the nuclear industry to try to extend the instrument calibration intervals through on-line drift monitoring. This work involves recording and analyzing the steady-state output of instruments during plant operation to identify drift and other abnormal problems in instrument outputs. For redundant instruments, this is accomplished by comparing the readings of the redundant instruments to distinguish between process drift and instrument drift. For non-redundant instruments, process empirical modeling using neural networks and pattern recognition principles, or other techniques such as physical modeling are used to estimate the process. This estimate is updated frequently and compared with the output of the corresponding instruments to detect any drift in the instrument output. Process modeling is also used with redundant instruments to account for common mode drift. The details of the modeling techniques are presented in [1].

Another important benefit of process modeling is in verifying the calibration of instruments over their entire range. More specifically, the models can provide the reference for verifying the calibration of instruments during plant startup and shutdown periods when the instruments are exposed to inputs that cover a wide range.

The on-line calibration approach was approved by the Nuclear Regulatory Commission (NRC) in July 2000 based on a topical report by the Electric Power Research Institute (EPRI). The NRC approval was provided in a Safety Evaluation Report (SER) that is included in EPRI report number TR-104965.

3 On-Line Detection of Venturi Fouling

In addition to on-line verification of calibration of process instrumentation channels, process empirical modeling, pattern recognition, and neural network techniques can provide an effective tool for on-line detection of performance problems in individual instruments or the plant. For example, Venturi flow elements can become fouled and result in erroneous flow indication. This has both safety and economical implications. Until recently, there has been no effective way to monitor for Venturi fouling. The modeling techniques mentioned above provide effective diagnostics of flow measurement problems using the existing plant signals and thereby helps in plant efficiency improvements and megawatt recovery.

Because of Venturi fouling problems, in some plants, new ultrasonic sensors are installed to measure the flow. Although the cost of the ultrasonic sensors for nuclear power plants can be as high as one million dollars, many plants have already installed these sensors because of the importance of accurate flow measurements in determining the reactor power. In fact, some ultrasonic flow sensors have been approved by the NRC for plant power uprates and megawatt recovery.

4 In Situ Response Time Testing of Pressure Transmitters

Accuracy and response time are two of the most important indicators of performance of pressure transmitters. As such, on-line methods have been developed to monitor the calibration and response time of pressure transmitters in nuclear power plants. The on-line calibration verification was mentioned already. For on-line measurement of response time of pressure transmitters, the noise analysis technique is used as described in [2]. This method is based on recording the random process fluctuations (noise) that exists naturally at the output of most process sensors while the plant is operating. The noise can be analyzed in the frequency domain and/or time domain to give the response time of the transmitter. As documented in [2], this method has been validated for response time testing of pressure, level, and flow transmitters in nuclear power plants. Also, a method referred to as the power interrupt (PI) test has been developed for in-situ response time testing of pressure transmitters. This method is applicable only to force balance pressure transmitters. The details are presented in [2, 3]

5 On-Line Detection of Clogging in Pressure Sensing Lines

Pressure sensing lines (also called impulse lines) are the small tubes that bring the pressure signal from the process to the sensor. Typically, the length of the sensing lines are 30 to 300 meters, depending on the service in the plant, and there are often isolation valves, root valves, snubbers, or other components on a typical sensing line. The malfunction in any valve or other component of the sensing line can cause partial or total blockage of the line. In addition, and more importantly, sensing lines can become clogged due to sludge and deposits that often exist in the reactor coolant system. The clogging of sensing lines can cause a delay in sensing a change in the process pressure, level, or flow. In some plants, sensing line clogging due to sludge or valve problems has caused the response time of pressure sensing systems to increase from normally about 0.2 seconds to as much as 5 seconds. This problem can be identified while the plant is on-line using the noise analysis technique as described in [2, 3]. Basically, if the response time of the pressure, level, or flow transmitter is measured with the noise analysis technique;

the results will include any delay due to the sensing line length, any blockages, voids, and other restrictions.

6 RTD and Thermocouple Cross Calibration

Redundant RTDs and thermocouples in nuclear power plants can be in-situ calibrated at isothermal conditions using the cross-calibration technique. This involves a multichannel data acquisition system to quickly record the temperature indications of the redundant RTDs and thermocouples. These temperatures are then averaged and the deviation of each RTD or thermocouple from the average of all RTDs (excluding any outliers) is calculated. Once the outlier RTDs are identified, they are excluded from the data and the data is corrected for plant temperature fluctuations and any temperature differences between the loops or between the hot legs and cold legs. After these corrections are implemented, a new average temperature is identified for the RTDs and the deviation of each RTD and thermocouple from this new average is calculated. The details of the cross-calibration technique are presented in [4].

The cross-calibration tests are often performed at several temperatures during plant startup or shutdown periods. With this approach, if any RTD is out-of-tolerance, a new calibration table can be developed for the RTD using the cross-calibration data taken at three or more temperatures. Also if large deviations for thermocouples are identified, they can be adjusted to bring the thermocouples in line with each other and with the RTDs.

The cross calibration data can be collected either at temperature plateaus or under ramp temperature conditions during startup and shutdown. Analysis and Measurements Services Corporation (AMS) has shown that the ramp data collection produces equivalent results to plateau data collection.

7 Response Time Testing of RTDs and Thermocouples

The response time of RTDs and thermocouples can change with ageing of the sensor. Many factors can contribute to this ageing degradation. For example, vibration can cause RTDs and thermocouples to move out of their thermowell and result in an increase in response time. Even a very small movement can cause a large change in response time. Exposure to high temperatures can also cause changes in response time. For example, inherent voids in sensor insulation materials can expand or contract and cause the response time to change. For these and other reasons, response time of RTDs and thermocouples are measured periodically in nuclear power plants. The measurement is made using the Loop Current Step Response (LCSR) method as described in [5]. Nearly all pressurized water reactor (PWR) plants in the USA perform the LCSR test on their primary coolant RTDs.

The LCSR test is performed remotely from the control room area while the plant is operating. It provides the in-service response time of RTDs and accounts for all installation and process condition effects on response time. If the RTD is used in a thermowell, the response time that is obtained from the LCSR test includes the dynamic response of the RTD and the thermowell combined. Therefore, any gap in the RTD/thermowell interface is also accounted for in the LCSR test.

The LCSR method can also be used to measure the response time of thermocouples as described in [6]. However, this requires higher heating currents and a different test procedure than RTDs. Therefore, in some plants, the response time of thermocouples are often tested using the noise analysis technique as was described earlier for pressure transmitters.

8 LCSR Test to Verify the Attachment/Bonding of Sensors to Solid Materials

Sensors such as thermocouples, thin-film RTDs, strain gages, and other resistive devices are used in nuclear power plants and other applications for measurement of surface conditions of pipes, vessels, and other components. In these applications, the sensors are bonded to a solid material. For example, strap-on or cemented RTDs are used to measure surface temperature on pipes such as pressure sensing lines in nuclear power plants.

As a result of long-term exposure to heat, humidity, vibration, and other process conditions, the bonding of sensors can deteriorate and cause the sensor to become detached from the solid surface and float in the air resulting in an erroneous indication. As such, new methods mostly based on LCSR testing have been developed to characterize the quality of bonding between sensors such as RTDs, thermocouples, and strain gages and a solid material. Reference [7] describes how the LCSR is used for these applications.

9 LCSR Testing for Sensor and Circuit Diagnostics

In addition to sensor response time testing and detection of sensor-to-solid bonding, the LCSR method can be used for diagnostics of wiring and circuit problems in instrumentation systems. For example, in the early 1980s, the LCSR method was being used for thermocouple response time testing on experimental nuclear fuel assemblies at the Argonne National Laboratory (ANL). In addition to providing dynamic response information, LCSR tests identified a number of reverse-connected thermocouples. The problem was manifested in unusual LCSR transients that were obtained during the response time measurements. In another instance, in a NASA project, rocket nozzle thermocouples were found by the LCSR test to be reverse connected; an event similar to that of ANL. In the same NASA project, the LCSR method identified thermocouples whose measuring junctions were not at the normal point at the tip of the thermocouple, or had developed secondary junctions. These problems would have resulted in erroneous temperature measurements in a very important application [7].

Thermocouple inhomogeneity can also be detected using the LCSR method as described in [8].

10 Testing of Cables and Connectors

The condition of nuclear power plant cables, especially I&C cables, is tested in some plants for a number of reasons such as troubleshooting to identify or describe problems, and baseline measurements for predictive maintenance and ageing management. There are electrical tests, mechanical tests, and chemical tests that can be used to monitor or determine the condition of cables. The electrical tests have the advantage of providing the capability to perform the tests in-situ, often with no disturbance to the plant operation.

The electrical tests involve impedance measurements and Time Domain Reflectometry (TDR) tests. The TDR test is popular in nuclear power plants for identifying the location of a problem along a cable. Particularly, it is often crucial to determine if a cable problem is in the containment or outside the containment. For example, RTD circuits that have shown erratic behavior have been successfully tested by the TDR method to give the maintenance crew proper directions as to the location of the problem. The TDR technique has also been helpful in troubleshooting motor and transformer windings, pressurizer heater coils, nuclear instrumentation cables, thermocouples, Motor Operated Valve (MOV) cables, etc.

To determine the condition of cable insulation or jacket material, in addition to TDR, electrical parameters such as insulation resistance, DC resistance, AC impedance, and series capacitance are measured. It should be pointed out that determining the condition of cable insulation material is a very challenging task. The lack of a suitable ground plane for making reliable electrical measurements hampers the success of the tests.

In mechanical testing of cables, the ductility of the cable insulation or jacket material is measured to determine if the material has become dry, brittle, or prone to crack. The test equipment is referred to as a Cable Indentor. Basically, the device is used to squeeze the cable and measure its relative hardness [9].

In chemical testing of cables, a small piece of the cable insulation material is peeled off for chemical analysis in a laboratory.

11 Management of Ageing of Reactor Internals

In the last ten years, predictive maintenance through vibration analysis has become one of the most prevalent practices in industrial processes. Using accelerometers and similar sensors, the vibration of operating machinery is measured passively while the process is on-line. The results of these tests are then trended to identify deviations from expected, normal, or historical behavior. This practice has proven to successfully identify the onset of many problems with industrial equipment, especially rotating machinery, and is believed to save billions of dollars every year by preventing equipment failures and plant downtime. In some cases, the vibration test equipment is installed in the plant permanently and data is collected and analyzed continuously; and in other cases, portable vibration equipment is used to test periodically.

In nuclear power plants, measurement of vibration of reactor internals can be performed very effectively using the existing neutron detectors. For example, the ex-core neutron detectors in PWRs can measure the vibration of the reactor vessel and the reactor vessel internals to better than one mil resolution (1 mil = 1/1000 of an inch or 0.025 millimeters). Furthermore, through cross-correlation of neutron signals and other existing sensors such as the core exit thermocouples or the reactor vessel level sensors, the flow through the reactor can be characterized to detect flow anomalies, flow shifting, flow blockages, and other problems. The details are presented in NUREG/CR-5501 report which AMS wrote for the NRC [10].

Accelerometers are also used in nuclear power plants for vibration measurements and loose parts monitoring. However, for the measurement of vibration of reactor vessel and its internals, neutron detectors have much better resolution and accuracy than accelerometers.

12 Conclusions

New instrumentation and maintenance technologies are available with great potential to benefit the safety and economy of nuclear power plants, facilitate plant life extension, and aid in the management of ageing of critical plant equipment. Some of these techniques were summarized in this paper. The details of these techniques are provided in the references identified in this paper.

References

1. Hashemian, H.M., "On-Line Testing of Calibration of Process Instrumentation Channels in Nuclear Power Plants," U.S. Nuclear Regulatory Commission, NUREG/CR-6343, (November 1995).

2. Hashemian, H.M., "Long Term Performance and Ageing Characteristics of Nuclear Plant Pressure Transmitters." US Nuclear Regulatory Commission, NUREG/CR-5851, March 1993.

3. Hashemian, H.M., et al., "Effects of Ageing on Response Time of Nuclear Plant Pressure Sensors." US Nuclear Regulatory Commission, NUREG/CR-5383, June 1989.

4. Hashemian, H.M., et al., "Ageing of Nuclear Plant Resistance Temperature Detectors." US Nuclear Regulatory Commission, Report Number NUREG/CR-5560, June 1990.

5. Hashemian, H.M., and K.M. Petersen, "Loop Current Step Response Method for In-Place Measurement of Response Time of Installed RTDs and Thermocouples," (American Institute of Physics) Seventh International Symposium on Temperature, 6:1151-1156, Toronto, Canada, May 1992.

6. Hashemian, H.M., "New Technology for Remote Testing of Response Time of Installed Thermocouples," United States Air Force, Arnold Engineering Development Center, Report Number AEDC-TR-91-26, Volume 1 - Background and General Details, January 1992.

7. Hashemian, H.M., C.S. Shell, and C.N. Jones, "New Instrumentation Technologies for Testing the Bonding of Sensors to Solid Materials," National Aeronautics and Space Administration, Marshall Space Flight Center, NASA/CR-4744, May 1996.

8. Hashemian, H.M., and K.M. Petersen, "Measurement of Performance of Installed Thermocouples," Proceedings of the Aerospace Industries and Test Measurement Divisions of the Instrument Society of America, 37th International Instrumentation Symposium, pp. 913-926, USA Paper #91-113, San Diego, California, May 1991.

9. DOE/EPRI Report, "Ageing Management Guideline for Commercial Nuclear Power Plants – Electrical Cables and Terminations," SAND96-0344, September 1996.

10. Hashemian, H.M., et al., "Advanced Instrumentation and Maintenance Technologies for Nuclear Power Plants," U.S. Nuclear Regulatory Commission, NUREG/CR-5501, (August 1998).

4 Overview of Recent KFKI AEKI Activities in the Field of Plant Surveillance and Diagnostics

Sándor Kiss, Sándor Lipcsei, and János Végh

KFKI Atomic Energy Research Institute, Applied Reactor Physics Department, H-1525 Budapest 114, P. O. Box 49, Hungary

{kisss, lipcsei, vej}@sunserv.kfki.hu

1 Introduction

Application of noise diagnostics techniques for the regular surveillance of the four VVER-440/213 type reactors of the Paks NPP (Hungary) has long time traditions at KFKI AEKI. In the eighties (shortly after the commissioning of the four units) the AEKI has designed, manufactured, installed and utilised a reactor noise diagnostics system (PDR, see [5]), which was capable of measuring a limited number of selected in-core or ex-core detectors simultaneously. Later the PDR system was modified to incorporate new services (e.g., acoustic leakage detection and vibration monitoring), this new version was installed at Units 3-4 of the Paks NPP in 1993 (CARD, see [6]). However, an important limitation still existed: the number of available in-core input signals for noise diagnostics was limited by the cumbersome hardware coupling. This problem was eliminated by the installation of a new signal coupling system connected to the VERONA core monitoring system [3]. This noise signal coupling system offers the possibility to connect any in-core signal group (arbitrary selected from 252 SPN detectors, 210 assembly outlet thermocouples, etc.) to the noise analysis system through well-conditioned data channels, passing AC and DC signal components alike. The actual number of signals depends on the hardware used for primary noise signal processing (e.g., for A/D conversion), routinely 18 or 32 signals are handled simultaneously by the present noise analysis system. This powerful in-core noise signal coupling system is in use still today, regular measurements are performed by two noise analysis configurations: one system is devoted to Units 1-2, while the other is used for Units 3-4. Actually the hardware components and software modules of these configurations can be considered as a mixture of those originating from the CARD system [6] and the JEDI system [7]. The JEDI was an ambitious project conducted by KFKI AEKI in the mid-nineties: the basic aim of the project was to build a system incorporating most of the proven noise analysis and signal validation methods in a unified shell. However, only the prototype of the system was developed, its application in the NPP was only partly realised.

The present paper is devoted to the description of those regular reactor noise diagnostics measurements that are routinely performed by KFKI AEKI at the Paks NPP in co-operation with experts from the power plant. Details of the in-core coolant velocity mapping and vibration monitoring of core internals (e.g., control rods) are presented, emphasising their importance in practical plant operation. Recent plans for noise analysis system upgrading and for introducing new on-line process analysis and signal validation services are given as well.

2 Core Coolant Velocity Mapping

From 1996 on, a gradual decline of noise analysis activities was observable at the Paks NPP, mainly due to the fact, that the plant has lost its interest in maintaining this work as a long-term operation support tool. This situation was accompanied by staffing problems and by that world-wide experience, that reactor noise analysis generally could not live up to those great expectations, which were mainly generated by the noise analysis experts themselves. Due to these problems regular noise measurements were interrupted for 1-1.5 years despite the fact, that a systematic data collection program was going on for all units between 1984 and 1996. Noise measurements were revived only in October 1997, when the NPP requested KFKI AEKI to perform a systematic study on Unit 2. Our task was to measure and evaluate the coolant velocity in those fuel assemblies, which were equipped with SPN detectors. The urgent reason for the study was a characteristic and significant asymmetry in the core outlet temperature distribution, which was detected during the start-up of cycle 14 by the VERONA core monitoring system. The detailed evaluation revealed a ± 5-6% asymmetry in the assembly temperature rise (ΔT) field between different parts of the core. As it turned out later, the asymmetry was caused by local crud depositions resulting coolant flow reduction in a number of fuel assemblies (see [1] for details). The phenomena was analysed in detail by various methods, but the plant (and the nuclear safety authority) required a regularly applicable and reliable method to monitor the magnitude and change tendency of the asymmetry in order to keep Unit 2 running until it was prematurely shut-down for refuelling in early 1998. In the following sections the applied noise measurement and evaluation methods are outlined, together with our results.

A considerable part of the fluctuations observable in the in-core neutron detector (SPND) signals is induced by the perturbations occurring in the primary coolant flowing through the fuel assembly. By inducing changes in the macroscopic cross sections, various coolant inhomogeneities (i.e., density or boron concentration fluctuations) cause perturbations in the neutron flux as well. Therefore a perturbation - running through a channel equipped with several detectors – appears in the signals of the detectors as a small transient: the time delay is proportional with the distance between the detectors.

During the operation of a PWR the coolant flow through the assemblies can be determined by the combination of measurements and calculations. The net flow through the reactor can be determined as a sum of individual loop flows, which can be calculated on-line by using main circulating pump Δp measurements and the pumps' characteristic Q = f(Δp) (i.e., delivered flow – pump head) curves. Core bypass flow can be determined by appropriate methods or a nominal bypass can be used. The net core flow is then obtained and individual assembly flows can be assigned with an accuracy of 1.1% (2.1% for control rods) in nominal, stationary reactor operation states. However, no special in-core measurements are available to determine individual fuel assembly flows, therefore one must use the standard fixed in-core instrumentation to evaluate local flow differences.

In order to estimate the coolant flow (or more precisely, the coolant velocity) in those assemblies, which are equipped with SPN detectors we have utilised a technique based on the correlation function between the signals of two detectors. The method is quite simple and it is based on the fact, that the correlation function shows a local maximum at the place corresponding to the transit time between the two detectors. Actually - for practical purposes - the so called impulse-response function between detectors A and B (IMP_{AB}) is used, this is defined by

$$IMP_{AB} = FFT^{-1}\left(\frac{CPSD_{AB}}{APSD_{A}}\right) \quad (1)$$

Here FFT^{-1} indicates inverse Fourier transformation, $APSD_A$ is the autospectrum of A and $CPSD_{AB}$ is the cross-correlation between A and B. The advantage of this method is that it is less influenced by the so-called global reactor noise component (see [2] for a detailed treatment of global-local perturbations). Figure 1 illustrates the impulse-response function of two rhodium SPN detectors: the transit time of the perturbation is c/a. 0.40sec.

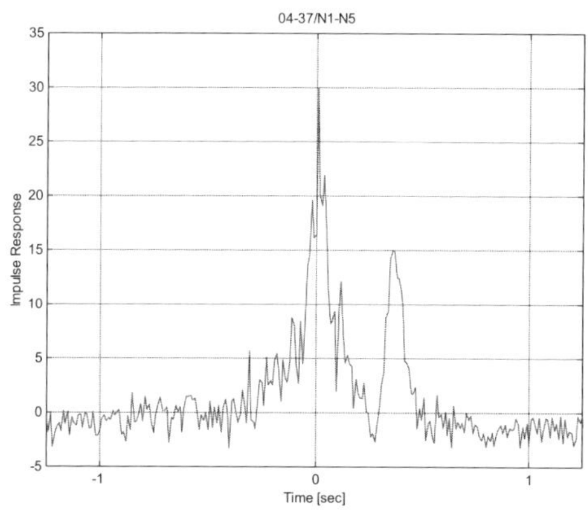

Fig. 1. Impulse response function of two detectors with 0.40sec transit time

Returning to the asymmetry at Unit 2, we have conducted regular core coolant velocity mapping (CCVM) for the whole cycle 14, which was some months less, than usually. Coolant velocity measurements were performed in those assemblies, which were furnished with SPNDs (altogether 36 such assemblies are present in a

VVER-440/213 core). Noise measurements have shown a significant difference between two parts of the core, this is illustrated on Figs. 2 and 3 below:

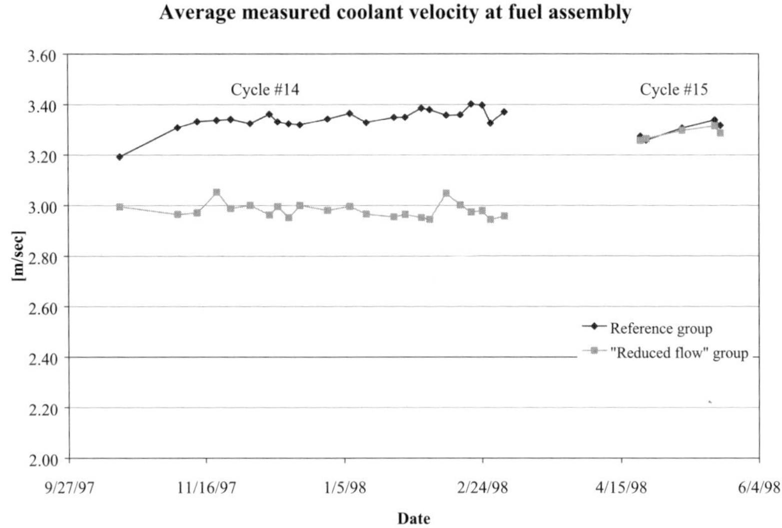

Fig. 2. Coolant velocity trends for the normal and reduced flow channels

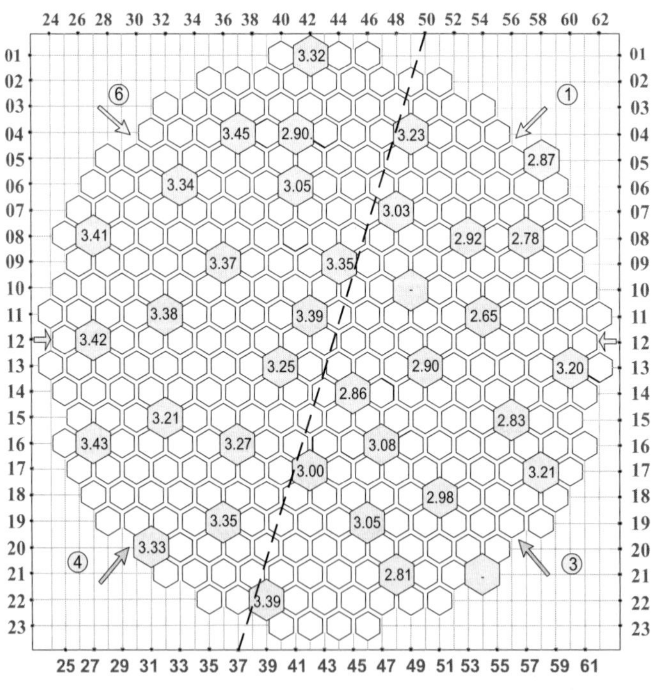

Fig. 3. Average coolant velocities [m/s] in the measured core positions

In March 1998 Unit 2 was shut down for refuelling and all the fuel assemblies were investigated in an off-core equipment to measure and analyse the nature and extent of crud deposition. It has to be noted, that our CCVM measurements were in a good agreement with the off-core hydraulic measurements. After Unit 2 was started up in cycle 15, AEKI has performed additional measurements to analyse the situation in the new cycle. As Fig. 2 shows clearly, the core was cleaned up perfectly and no asymmetry could be found in the coolant velocity distribution.
From 1998 on CCVM measurements are performed regularly for **all** units (once in each three weeks). The reason for these relatively frequent measurements is the gradual ageing of the Paks units: as they approach their originally predicted mid-lifetimes, more and more problems may arise with steam generator (SG) radiation levels. Keeping SG maintenance dose limits requires SG decontamination more frequently, but decontamination can later cause certain problems in water chemistry and may produce extra corrosion. These phenomena require continuous monitoring in order to detect possible problems in time.

3 Control Rod Vibration Monitoring

A vibrating control rod causes significant perturbations even in the signals of those SPN detectors, which are located 2-3 assembly rows away from the rod (see [4] for details). In our regular control rod vibration monitoring (CRVM) measurements we utilise the fact, that at the Paks NPP all 36 SPND chains can be used for diagnostic purposes. Actually a core map of SPND noise measurements can be constructed and this map can be easily used to localise a vibrating control rod. Fig. 4 shows such a situation at Unit 4 in 2001, where a 1.1Hz peak was observable due to a slight vibration of the control rod at core position 18-49. Figure 5 shows characteristic spectra of SPND 18-51 (located in the vicinity of the vibrating control rod), the peak at 1.1Hz is clearly detectable in the APSD, if frequency range between 0 and 5Hz is plotted.

CRVM measurements now form a standard part of our regular noise diagnostics program at the NPP: in VVER-440/213 type reactors control rods require a special attention due to their special construction (the bulk boron-steel rod is coupled to a fuel "follower" part, which is normally fully inserted into the core).

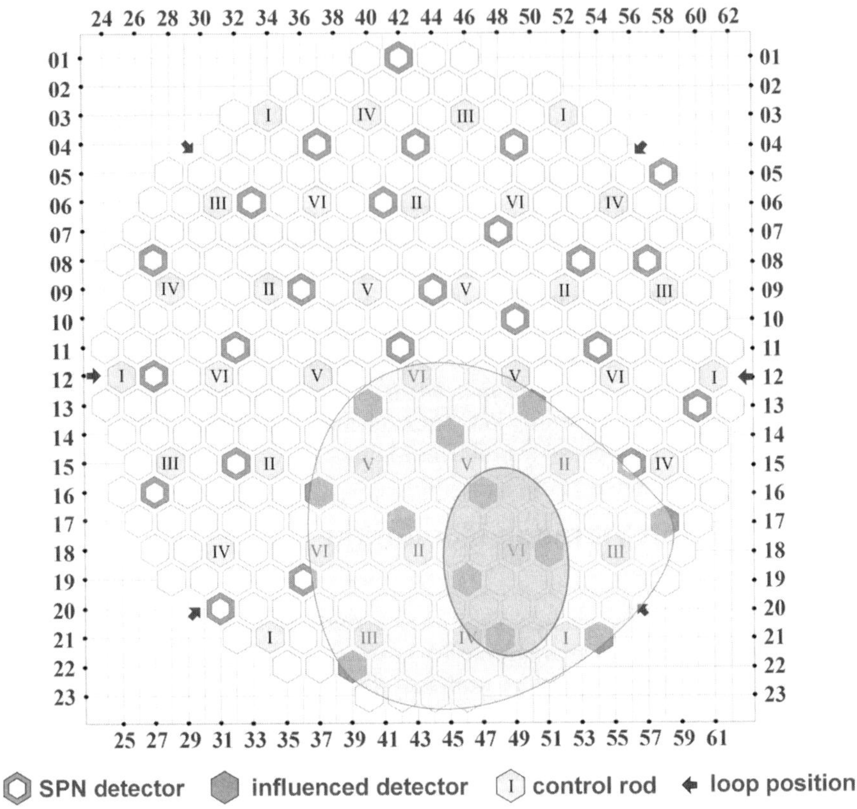

Fig. 4. Localisation of a vibration source at the core map

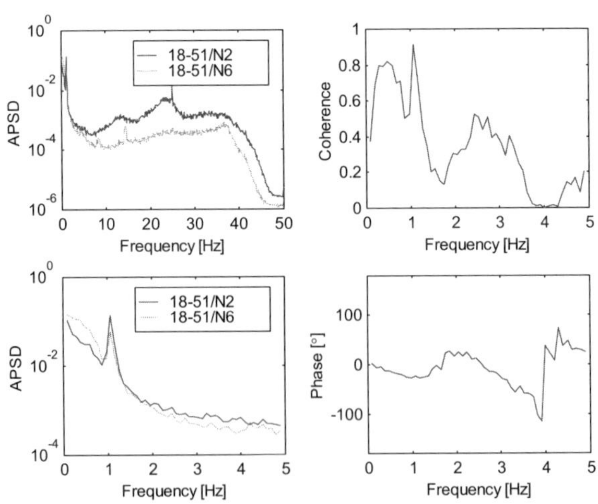

Fig. 5. Noise spectra showing a vibration source in the vicinity of an SPN detector

4 Further Developments

4.1 Noise Measuring System Upgrading

As it was mentioned above, a mid-aged NPP requires increased attention if on-line process and component monitoring is considered. In case of the Paks NPP there are further recent developments emphasising the need for improved surveillance methods and diagnostics systems in the future. These factors are as follows:

- An ambitious plant lifetime extension program is in preparation.
- A reactor power uprating program is planned in the near future.
- New fuel types are gradually introduced to optimise fuel economy.

The above listed three major projects (to be completed in the coming 4-5 years) will demand new or upgraded on-line monitoring systems with increased capacity, both in data acquisition and signal interpretation / modelling capabilities. In order to meet these new demands a gradual upgrading of the present in-core noise diagnostics system is planned, the architecture of the new system is shown in Fig. 6:

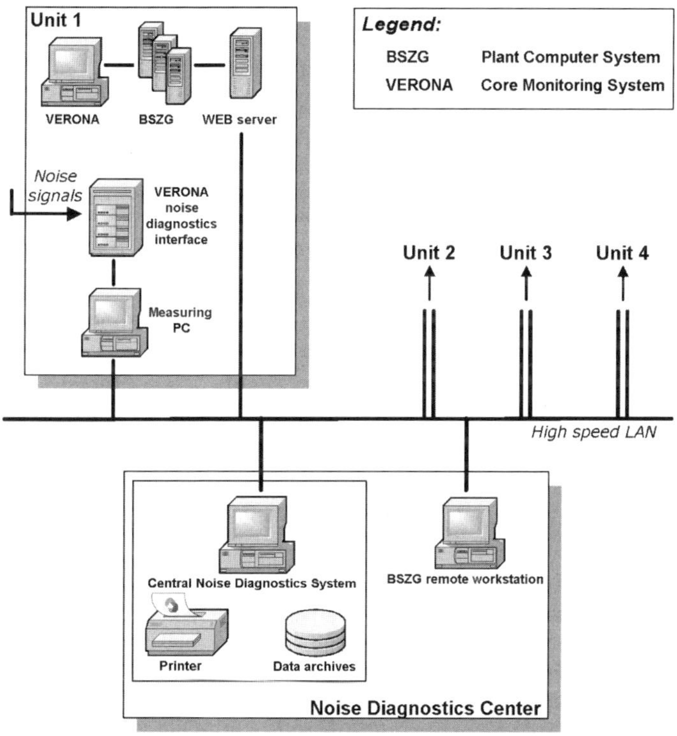

Fig. 6. Planned architecture of a distributed in-core noise diagnostics system

The new system will basically use the present in-core noise signal coupling equipment, but its signal conditioning and data acquisition parts will be replaced by standard high-precision industrial devices. The "Noise Diagnostics Center" part can be used for remote control of measurements performed at the units and for ensuring long term storage of filtered analysis results. On-line process information from the new Paks plant computer system will be available by using a remote workstation connected to the external server node of the new plant computer. The basic aim of the upgrading is to create a flexible and user-friendly system, which is able to handle the AC and DC components of at least 352 signals: 252 SNPDs, 36 SPND compensation cables, some in-core thermocouples plus some ex-core detectors.

4.2 COSMOS, a New Industrial Process Information System

COSMOS (Computerised Operation Support for Management, Optimisation and Surveillance) is a large-scale, three-year R&D project supported by the Hungarian Government through the National Research and Development Project (NRDP) under contract No. 2/016/2001. The project is to be launched in the autumn of 2001 and it is due to end in 2004. Main contributors to the R&D work are:

- Computer and Automation Research Institute (MTA SZTAKI),
- KFKI Atomic Energy Research Institute (KFKI AEKI),
- Dept. of Measurement & Information Systems (Univ. of Technology, BUTE),
- Dept. of Control Engineering & Information Technology (BUTE).

Two major industrial companies are contributing to the project and will utilise its results: Paks Nuclear Power Plant Company Ltd. (Paks NPP) and the Hungarian Oil and Gas Plc. (MOL Plc.).

The COSMOS project will result a knowledge-intense information technology applicable in large and complex industrial systems for process parameter monitoring, early detection and diagnosis of equipment and process faults, as well as for fast detection of malfunction root-causes. The schematic architecture of the planned general purpose software shell is shown in Figure 7.

Operation support is provided by advises related to the optimal operation of process equipment. On-line tasks are supported by off-line modules, performing the following tasks: establishment of a structured storage of the engineering-technical information associated with the components of the industrial system; preliminary analysis of the safety, operational and ergonomical consequences of planned reconstructions and technology changes, in order to select the optimal solution. The kernel of the modular system is represented by a novel central database, which ensures the storage of the various technical information and technology-related knowledge in a structured manner. The system provides user-friendly and efficient interactive tools for storing expert knowledge, for on-line modelling of process phenomena, for the visualization of diagnostics information and for on-line displaying of advisory documentation. In the following sections the **on-line** COSMOS modules are introduced briefly.

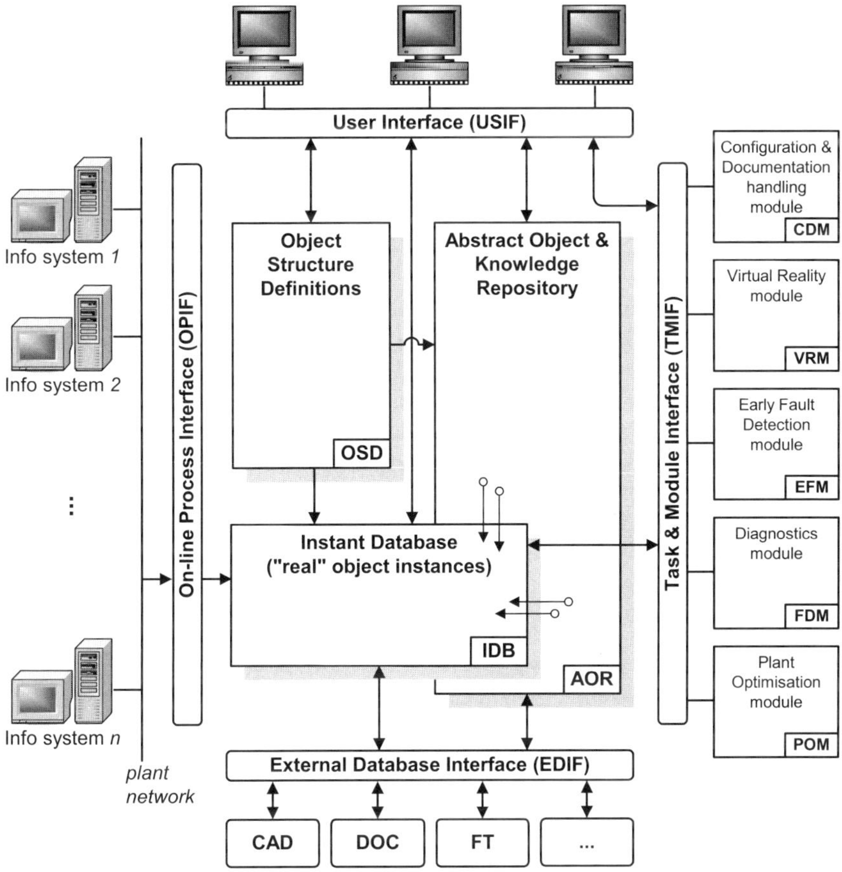

Fig. 7. COSMOS: the planned architecture of the general purpose software shell

Early Fault Detection Module

This module can be used to detect abnormal equipment behaviour in an early phase, the detection is based on process measurements, tests an on-line calculations. The module is able to monitor process equipment state (measuring and control devices, various active and passive components), and the state of associated supervision & control systems (these are primarily digital computers). In case of monitoring a technological process, a deviation from the normal state is identified according to the principle of analytic redundancy, i.e., by using process measurements and on-line calculations based on the components' model. In case of a digital system, the deviation is identified by using the components' fault-model and the results obtained from on-line tests. The results of the deviation analysis are displayed by the module on decision trees, "off-normal" exit points of the trees provide direct advises to the operators. For process components the early fault detection is "model-based," where the model is represented by formal and analytic calculations of sufficient details, describing the ideal operation state of a subsystem. This method can be used well for the identification of small leakage, slow measurement drifts, data transducer malfunctions and subsystem pressure anomalies (caused e.g., by a pipe break). For digital systems the formal model is represented by the system components' fault-model plus the on-line tests. Component faults can be detected by using the tests and the fault-model, the fault semantics (i.e., the behaviour of the faulty system) can be determined, as well.

On-line fault-detection models will be created by an interactive graphic tool, which can generate the executable code of the models, as well. Experts can use this tool to define the topology of the model graphically, i.e., to give the scheme of the subsystem to be modelled: the piping network and the associated equipment, the type of substance flowing in the pipes, and (for digital systems) the connection topology between different components. System specific expert knowledge is represented by decision trees related to specific equipment and components. Decision trees can be defined interactively by using a fault-tree editor (decision trees are considered as fault-trees of special type), their on-line evaluation is performed by the fault diagnostics module.

Fault Diagnostics Module

This on-line processing module can be used to identify the root-causes of equipment faults or system malfunctions in a fast and reliable manner. Possible equipment (system) fault modes can be defined by the experts in the form of fault-trees, this activity is supported by an interactive fault-tree editor. For subsystems of limited size the root-causes can be arranged in a table according to their weighted probabilities and afterwards this table can be used for on-line fault diagnosis. In case of large subsystems the determination of root-causes is carried out by an online fault-tree evaluation procedure. Decision conditions residing in the fault-tree nodes can be evaluated by using measured and calculated signals, characterising the actual state of the industrial process (for digital systems this information is available from the periodic on-line tests). The result of the analysis is displayed by the module on dynamically animated fault-trees, ensuring a fast and consistent information presentation. Based on the root-causes identified by the above described diagnostics procedure the system provides direct support to the operators: the advisory function is realised by displaying the corresponding emergency operating procedures or normal operating manuals from the document database.

Process Optimisation Module

In addition to ensuring conditions for the continuous safe operation of a large industrial system, another equally important task is to operate the complex equipment optimally: this is generally equivalent to operate the system at its maximum efficiency. The on-line process efficiency monitoring (optimisation) module determines the actual operation conditions of selected technological subsystems and the equipment efficiencies from characteristic on-line process measurements and from calculations performed in the background. The results of the analysis are displayed by the module on decision trees, where "off-normal" tree exit points provide direct operational advises to the operators. By using this module experts can have a more clear view about losses in the technology and they can assess the position of the actual working points with respect to the optimal states. With the help of this module e.g., net savings in the electricity production can be achieved, by a timely detection of deviations from optimal equipment operation. In case of the NPP application measured signals are taken from the new process computer. The realisation of this module is primarily based on the software tools listed at the fault-detection and fault diagnostics modules.

5 Summary

Methods and results of regular noise diagnostics measurements performed by AEKI at the Paks NPP were outlined in this document. Practical use of process surveillance for plant operation purposes was illustrated by core coolant velocity mapping and by control rod vibration monitoring measurements. Future plans for upgrading the present diagnostics system were treated shortly, together with the some design details of a planned large-scope industrial process monitoring system.

References

1. Adorján F, Czibók T, Kiss S, Krinizs K, and Végh J (2000) Core Asymmetry Evaluation using Static Measurements and Neutron Noise Analysis, Annals of Nuclear Energy 27: 649-658

2. Katona T, Meskó L, Pór G, and Valkó J (1982) Some Aspects of the Theory of Neutron Noise due to Propagating Perturbances, Progress in Nuclear Energy 9.

3. Lux I, Végh J, Adorján F, and Bürger L (1994) Experiences with the Upgraded VERONA-u VVER-440 Core Monitoring System, Proceedings of the IAEA Specialists Meeting on Advanced Information Methods and Artificial Intelligence in Nuclear Power Plant Control Rooms, Halden, Norway

4. Pázsit I and Glöckler O (1988) On the Neutron Noise Diagnostics of PWR Control Rod Vibration, III. Application at Power Plant, Nuclear Science and Engineering 99.

5. Pór G *et al.* (1986) Sophisticated Systems for Analysing Standard Signals of a PWR NPP for Diagnostic Purposes, Proceedings of the 6th Power Plant Dynamics, Control and Testing Symposium, Knoxville, USA

6. Pór G (1992) CARD, Computer Aided Reactor Diagnostics for Experts, Proceedings of the OECD/NEA-IAEA International Symposium on NPP Instrumentation and Control, Tokyo, Japan

7. Rácz A *et al.* (1995) A Family of Integrated Information and Expert Systems for Plant Noise Analysis, Proceedings of SMORN VII, Avignon, France

5 Adaptive Model-Based Control of Non-linear Plants Using Soft Computing Techniques

Patricia Melin and Oscar Castillo

Department of Computer Science, Tijuana Institute of Technology, P.O. Box 4207, Chula Vista CA, USA

{pmelin, ocastill}@tectijuana.mx

We describe in this paper adaptive model-based control of non-linear plants using soft computing techniques. First, the general concept of adaptive model-based control is described. Second, the use of fuzzy logic for adaptive control is described. Third, a neuro-fuzzy approach is proposed to learn the parameters of the fuzzy system for control. A specific non-linear plant is used to test the hybrid approach for adaptive control. A particular stepping motor was used as test bed in the experiments. The results of the neuro-fuzzy approach were good, both in accuracy and efficiency.

1 Introduction

Adaptive control is a method of designing a controller with some adjustable parameters and an embedded mechanism for adjusting these parameters [4]. Adaptive controllers have been used mainly to improve the controller's performance online. For each control cycle, the adaptive algorithm is normally implemented in three basics steps:

1. Observable data are collected to calculate the controller's performance,
2. The controller's performance is used as a guidance to calculate the adjustment to a set of controller parameters,
3. The controller's parameters are then adjusted to improve the performance of the controller in the next cycle.

Normally, an adaptive controller is designed based on one of the available techniques. Each technique is originally designed for a specific class of dynamic system. The controller is then adjusted as data are collected during run time to extend its effectiveness to control a large class of dynamic systems.

A typical application of adaptive control was to calibrate a system at startup. In this case, a controller is also designed for a specific class of dynamic systems. However the parameters that characterize the dynamic behavior of a particular system might not be known in advance. A controller is then designed and arbitrary values are assigned to initialize these parameters. After a few control cycles, parameters are adjusted to converge to the actual parameters of the system. This approach is often used for cases in which a system is designed to handle a variable payload. The payload is different each time, e.g., a crane is used to pick up a sizeable object. The payload will alter the basic dynamic behavior of a dynamic system. Adaptive control is normally used to calibrate these parameters that characterize such dynamic behavior.

Traditionally, there are four basic approaches for adaptive control: (i) gain scheduling, (ii) model reference adaptive system, (iii) self-tuning regulator, and (iv) dual control. Gain scheduling is a method of adjusting the control signal based on a known look-up table describing changes of a dynamic system. The model reference adaptive system is a method of comparing the performance of the actual system against an assumed mathematical model that describes the actual system, and designing control input to drive this comparison error to zero. Self-tuning regulator is a method of updating the parameters of a model that describes the plant based on observed data, and channeling the updated information into the controller that is designed based on these parameters. Dual control is a method of extending adaptive control to stochastic model dealing with uncertainties.

In this paper, an extension to the principle of adaptive control using soft computing techniques is proposed. First, the concept of adaptive fuzzy control is described. A neuro-fuzzy approach is used to learn the parameters of the fuzzy system using real data from the plant [5]. The proposed adaptive control was tested with a specific non-linear plant, to evaluate the performance of this approach. A stepping motor was used as the non-linear plant in this study. The simulation results clearly show the advantage of using soft computing techniques to improve the concept of adaptive control.

2 Fundamental Adaptive Fuzzy Control Concept

Adaptive fuzzy control is an extension of fuzzy control theory to allow the fuzzy controller, extending its applicability, either to a wider class of uncertain systems or to fine-tune the parameters of a system to accuracy [4]. In this scheme, a fuzzy controller is designed based on knowledge of a dynamic system. This fuzzy controller is characterized by a set of parameters. These parameters are either the controller constants or functions of a model's constants.

A controller is designed based on an assumed mathematical model representing a real system. It must be understood that the mathematical model does not completely match the real system to be controlled. Rather, the mathematical model is seen as an approximation of the real system. A controller designed based on this model is assumed to work effectively with the real system if the error between the actual system and its mathematical representation is relatively insignificant. However, there exists a threshold constant that sets a boundary for the effectiveness of a controller. An error above this threshold will render the controller ineffective toward the real system.

An adaptive controller is set up to take advantage of additional data collected at run time for better effectiveness. At run time, data are collected periodically at the beginning of each constant time interval, $t_n = t_{n-1} + \Delta t$, where Δt is a constant measurement of time, and $[t_n, t_{n-1})$ is a duration between data collection. Let D_n be a set of data collected at time $t = t_n$. It is assumed that at any particular time, $t = t_n$, a history of data $\{D_0, D_1, \ldots, D_n\}$ is always available. The more data available, more accurate the approximation of the system will become.

At run time, the control input is fed into both the real system and the mathematical model representing the system. The output of the real system and the output of that mathematical model are collected and an error representing the difference between these two outputs are calculated. Let x(t) be the output of the real system, and y(t) the output of the mathematical model. The error $\varepsilon(t)$ is defined as:

$$\varepsilon(t) = x(t) - y(t). \quad (1)$$

Figure 1 depicts this tracking of the difference between the mathematical model and the real dynamic system it represents.

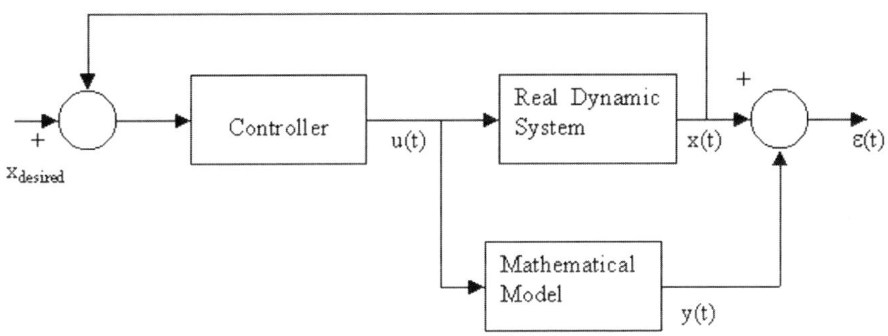

Fig. 1. Tracking the error function between of a real system and a mathematical model

An adaptive controller will be adjusted based on the error function $\varepsilon(t)$. This calculated data will be fed into either the mathematical model or the controller for adjustment. Since the error function $\varepsilon(t)$ is available only at run time, an adjusting mechanism must be designed to accept this error as it becomes available, i.e., it must evolve with the accumulation of data in time. At any time, $t = t_n$, the set of calculated data in the form of a time series $\{\varepsilon(t_0), \varepsilon(t_1), \ldots, \varepsilon(t_n)\}$ is available and must be used by the adjusting mechanism to update appropriate parameters.

In normal practice, instead of doing re-calculation based on a lengthy set of data, the adjusting algorithm is reformulated to be based on two entities: (i) sufficient information, and (ii) newly collected data. The sufficient information is a numerical variable representing the set of data $\{\varepsilon(t_0), \varepsilon(t_1), \ldots, \varepsilon(t_{n-1})\}$ collected from the initial time t_0 to the previous collecting cycle starting at time $t = t_{n-1}$. The new datum $\varepsilon(t_n)$ is collected in the current cycle starting at time $t = t_n$.

An adaptive controller will operate as follows. The controller is initially designed as a function of a parameter set and state variables of a mathematical model. The parameters can be updated any time during operation and the controller will adjust itself to the newly updated parameters. The time frame is usually divided into a series of equally spaced intervals $\{[t_n,t_{n+1})|\ n = 0,1,2,\ldots;\ t_{n+1} = t_n + \Delta t\}$. At the beginning of each time interval $[t_n,t_{n+1})$ observable data are collected and the error function $\varepsilon(t_n)$ is calculated. This error is used to calculate the adjustment in the parameters of the controller. New control input $u(t_n)$ for the time interval $[t_n,t_{n+1})$ is then calculated based on the newly calculated parameters and fed into both the real dynamic system under control and the mathematical model upon which the controller is designed. This completes one control cycle. The next control cycle will consist of the same steps repeated for the next time interval $[t_{n+1},t_{n+2})$, and so on.

A dynamic system can be described by a set of fuzzy if-then rules that correlate the input and the output. These rules include a set of parameters that is used to uniquely calculate the estimated output of a system given the inputs and current states of the system.

A dynamic system is mathematically modeled as a set of fuzzy if-then rules:

R^1: IF $x_1(t_n)$ is X_{11} AND …AND $x_i(t_n)$ is X_{1i}

$$\text{THEN } y_1(t_{n+1}) = a_{11}x_1(t_n) + a_{12}x_2(t_n) + \ldots + a_{1i}x_i(t_n) + b_{11}u_1(t_n) + b_{12}u_2(t_n) + \ldots + b_{1i}u_j(t_n)$$

R^2: IF $x_1(t_n)$ is X_{21} AND …AND $x_i(t_n)$ is X_{2i}

$$\text{THEN } y_2(t_{n+1}) = a_{21}x_1(t_n) + a_{22}x_2(t_n) + \ldots + a_{2i}x_i(t_n) + b_{21}u_1(t_n) + b_{22}u_2(t_n) + \ldots + b_{2i}u_j(t_n)$$

… … …

R^k: IF $x_1(t_n)$ is X_{k1} AND …AND $x_i(t_n)$ is X_{ki}

$$\text{THEN } y_k(t_{n+1}) = a_{k1}x_1(t_n) + a_{k2}x_2(t_n) + \ldots + a_{ki}x_i(t_n) + b_{k1}u_1(t_n) + b_{k2}u_2(t_n) + \ldots + b_{ki}u_j(t_n)$$

where $x_1, x_2, \ldots, x_i$ are the observable state variables of the real system, $y_1, y_2, \ldots, y_k$ are the calculated state variables of the mathematical model, and $u_1, u_2, \ldots, u_j$ the control inputs to both the real system and the mathematical model.

The above fuzzy mathematical model is characterized by a set of parameters $\{a_{ki}, b_{kj}\}$ These parameters will determine the behavior of the mathematical model. Theoretically, the model will approximate the real system. The more accurately the model approximates the real dynamic system, the better the controller designed based on this model behaves.

Similarly, a fuzzy controller, if designed based on the fuzzy mathematical model above, will have the following form of fuzzy rules:

R^1: IF $x_1(t_n)$ is X_{11} AND …AND $x_i(t_n)$ is X_{1i}
THEN $u_1(t_n) = k_{11}x_1(t_n) + k_{12}x_2(t_n) + \ldots + k_{1i}x_i(t_n)$

R^2: IF $x_1(t_n)$ is X_{21} AND …AND $x_i(t_n)$ is X_{2i}
THEN $u_2(t_n) = k_{21}x_1(t_n) + k_{22}x_2(t_n) + \ldots + k_{2i}x_i(t_n)$

R^j: IF $x_1(t_n)$ is X_{k1} AND …AND $x_i(t_n)$ is X_{ki}
THEN $u_j(t_n) = k_{j1}x_1(t_n) + k_{j2}x_2(t_n) + \ldots + k_{ji}x_i(t_n)$

In this case, the set of parameters $\{k_{ji}\}$ characterizes the controller. It can be understood intuitively that the parameter set $\{k_{ji}\}$ is designed as a function of the parameter set $\{a_{ki}, b_{kj}\}$, so that the state variables of the fuzzy model are driven to a target point with stability.

Therefore, it is natural for designing the controller to either adjust the system parameters $\{a_{ki}, b_{kj}\}$, which will lead to the adjustment of the control parameters $\{k_{ji}\}$, or directly adjust the controller parameters $\{k_{ji}\}$ as a scheme to implement adaptive strategies. The adjustment of the system parameters is sometimes called system identification. This system identification process is often applied to a system that can be described by a time-varying parameter set in the fuzzy model.

Adjusting the system parameters is normally carried out to improve the performance of the overall closed-loop system, or to improve accuracy of the assumed model upon which the controller is design. The process of calculating the adjustment to the parameters is the adjusting mechanism mentioned as part of the adaptive control problem.

3 Basic Concepts of Stepping Motors

Stepping motors are widely used in robotics and in the numerical control of machine tools where they have to perform high-precision positioning operations. However, the variations of the mechanical configuration of the drive, which are common to these two applications, can lead to a loss of synchronism for high stepping rates. Moreover, the classical open-loop speed control is weak and a closed-loop control becomes necessary. In this paper, fuzzy logic is applied to control the speed of a stepping motor drive with feedback. A neuro-fuzzy hybrid approach is used to design the fuzzy rule base of the intelligent system for control. In particular, we used the ANFIS methodology to build a Sugeno fuzzy model for controlling the stepping motor drive. An advanced test bed is used in order to evaluate the tracking properties and the robustness capacities of the fuzzy logic controller.

Stepping motors can be viewed as electric motors without commutators [8]. Typically, all windings in the motor are part of the stator, and the rotor is either a permanent magnet or, in the case of variable reluctance motors, a toothed block of some magnetically soft material. All of the commutation must be handled externally by the motor controller, and typically, the motors and controllers are designed so that the motor may be held in any fixed position as well as being rotated one way or the other. Most stepping motors can be stepped at audio frequencies, allowing them to spin quite quickly, and with an appropriate controller, they may be started and stopped "on a dime" at controlled orientations.

For some applications, there is a choice between using servomotors and stepping motors. Both types of motors offer similar opportunities for precise positioning, but they differ in a number of ways. Servomotors require analog feedback control systems of some type. Typically, this involves a potentiometer to provide feedback about the rotor position, and some mix of circuitry to drive a current through the motor inversely proportional to the difference between the desired position and the current position. In making a choice between stepping motors and servomotors, a number of issues must be considered; which of these will matter depends on the application. For example, the repeatability of positioning done with a stepping motor depends on the geometry of the motor rotor, while the repeatability of positioning done with a servomotor generally depends on the stability of the potentiometer and other analog components in the feedback circuit.

Stepping motors can be used in simple open-loop control systems; these are generally adequate for systems that operate at low accelerations with static loads, but closed loop control may be essential for high accelerations, particularly if they involve variable loads [1]. If a stepping motor in an open-loop control system is overtorqued, all knowledge of rotor position is lost and the system must be reinitialized; servomotors are not subject to this problem.

In this paper, the application of fuzzy logic is proposed to control the speed of a stepping motor drive. The closed-loop control scheme entails in incorporating engineering knowledge into the automatic control system by using the intuition and experience of the designer. This strategy was proposed by Zadeh [14], to describe complicated systems, which are hard to analyze using traditional mathematics. Indeed, Mamdani [10] was the first to report on the application of fuzzy logic to control a small laboratory steam engine. The success of this study led many scientists to attempt to control industrial processes such as chemical reactors, automatic trains, or nuclear reactors using fuzzy algorithms. The results of these experiments showed that, fuzzy controllers perform better, or at least as well as, classical controllers. Moreover, this technique offers the advantage of requiring only a simple mathematical model to formulate the algorithm, which can easily be implemented by a digital computer. These features are appreciated for nonlinear processes for which there is no reliable model and complex systems where the model is useless due to the large number of equations involved. Additionally, fuzzy logic is used more frequently for the control of electrical machines such as direct current or induction motors [9]. Nevertheless, the main problem with fuzzy logic is that there is no systematic procedure for the design of a fuzzy controller. For this reason, we propose in this paper the use of the ANFIS methodology [7] to adapt the parameters of the fuzzy system for control [2, 3, 11]. We use this neuro-fuzzy approach to train the controller with real data about the problem.

4 Fuzzy Logic Controller of the Stepping Motor

The fuzzy logic controller provides an algorithm, which converts the linguistic control, based on expert knowledge into an automatic control strategy. Therefore, the fuzzy logic algorithm is much closer in spirit to human thinking than traditional logical systems. The main problem with fuzzy logic controller generation is related to the choice of the regulator parameters. For this reason, we apply the ANFIS methodology to adapt the parameters of the fuzzy controller according to real data about the problem. The experiments were made on a system described in Figure 2. A program in the PC generates the step input to the system and stores the response. The fuzzy controller is also contained in the PC and acts on the system according to the corresponding responses. We now describe briefly the modules of the system.

1. Motor/Driver: Micro-step motor Vexta PK569BUA with five phases and 500 steps by turn. Power driver Vexta DFR1514A with multi-resolution (Minimum: 500 steps by turn; Maximum: 125000 steps by turn).
2. Encoder: Optical encoder Bourns of 40000 steps by turn. This encoder generates two square signals with 90 degree difference. With these signals the magnitude of motor movement is determined. We use the programmable logic device (PLD) of Altera (EP5032) to determine the movement of the motor.
3. Data Acquisition Card: PCL-818 of Advantech with 8 analog inputs and 2 analog outputs (12 bits), 16 digital inputs, 16 digital outputs. The sampling time used is 0.25 ms.
4. Computer/Software: Pentium III with 733 MHz. We design a small real time kernel in C language for control and data acquisition, and the fuzzy controller was programmed in MATLAB [12].

The linguistic control rules are established considering the dynamic behavior of the stepping motor drive and analyzing the error and its variation. These control rules are expressed as follows:

If Error is LP and Change_Error is LP then Speed = p_1*Error +q_1*Change_Error + r_1

If Error is LP and Change_Error is MP then Speed = p_2*Error +q_2*Change_Error + r_2 ...

This is a Sugeno fuzzy model [13] for controlling the stepping motor. We used the ANFIS methodology to estimate the parameters of the membership functions and the consequent functions. We used a fuzzy model of 9 rules and 3 membership functions for each linguistic variable. This was the fuzzy controller that gave the best results.

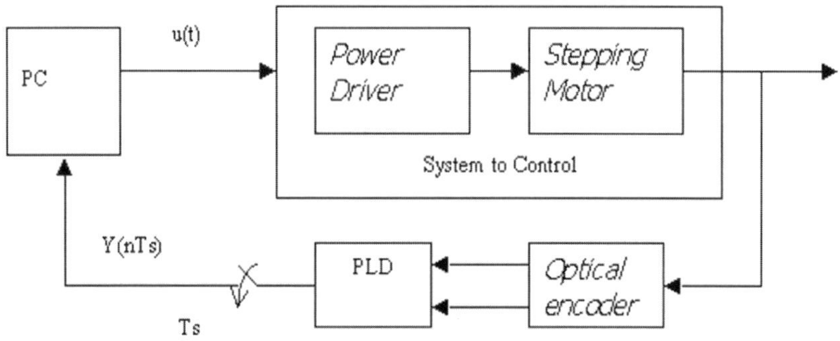

Fig. 2. Block diagram of the experimental system

5 Experimental Results

In this section, the tracking and adaptability features of the fuzzy control applied to the stepping motor are tested using simulation and experimentation. We first show in Figure 3 the response of the stepping motor to a sequence of step input signals (we use 400 samples). We show in Figure 4 the results of applying the ANFIS methodology with the training data and with the testing data. We used 20 epochs for training and the final error was of 0.000001, which is very good for this application. In Figure 5 we plot the predicted values by the fuzzy model and the real values for the system, and the curves are practically indistinguishable. Finally, we show in Figure 6 a plot of the difference between both the real and the estimated signal by the fuzzy model. Again, we can note that we have very good results for this application.

We also compared our results with a classical PID controller and with a fuzzy Mamdani controller, to measure how much the adaptive fuzzy approach could improve the performance. Of course, our fuzzy controller (designed with ANFIS) was better in tracking and adaptability than the other controllers. Still, we think that our results could be improved by optimizing the architecture of the controller using genetic algorithms [7].

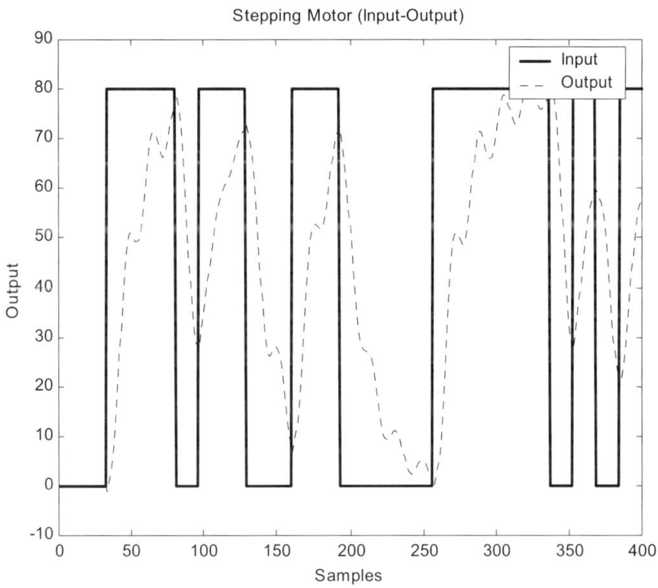

Fig. 3. Response of the stepping motor to a sequence of input signals

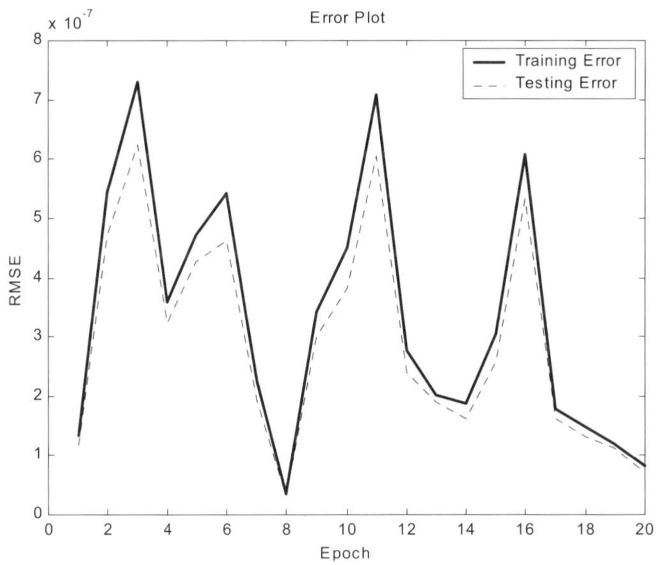

Fig. 4. Results of applying the ANFIS methodology to the training and testing data

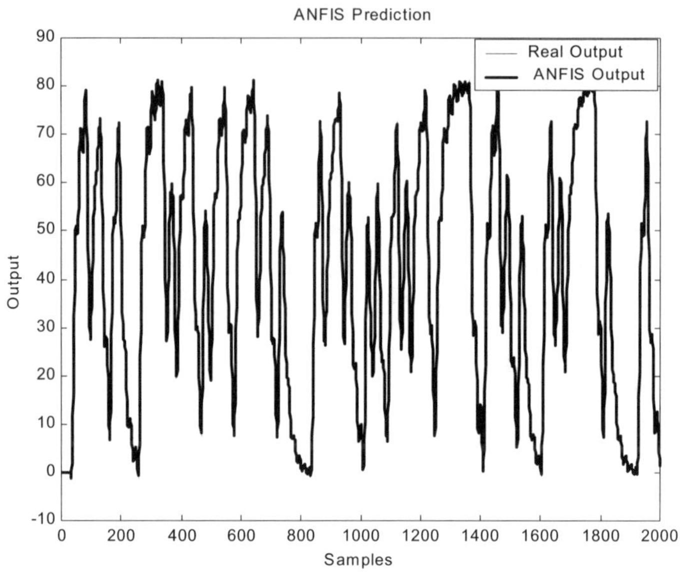

Fig. 5. Predicted values of the fuzzy model against the real values

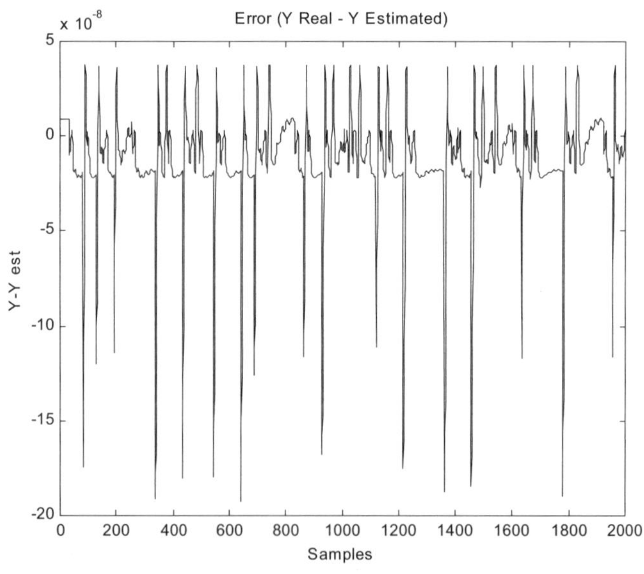

Fig. 6. Difference between the real and the estimated signal by the fuzzy model

6 Conclusions

In this paper, adaptive model-based control of non-linear plants was described. The implementation of an adaptive fuzzy controller for a specific plant was described. A neuro-fuzzy approach was applied to learn the parameters of the fuzzy system for adaptive control. The feasibility of fuzzy control for stepping motor drives has been proved and illustrated by simulation and experimentation. The best parameters for the fuzzy controller were determined by using the ANFIS methodology and also by using simulations of the stepping motor dynamics. An experimental system was used to validate experimentally the tracking ability and the insensibility to plant parameter changes.

The fuzzy controller presented very interesting tracking features and was able to respond to different dynamic conditions. Also, the fuzzy control computation is very inexpensive, and this regulator could be used for the control of machine tools and robotics manipulators [6] without significantly increasing the cost of the drive. The only extra cost is for the optical encoder. Another advantage of this method over classical quantitative controllers is that it does not require a fixed sampling time. Therefore, the proposed design confirms the fact that fuzzy control is relevant to the control fast of nonlinear processes such as stepping motor drives where quantitative methods are not always appropriate.

Acknowledgments

The authors would like to thank the Research Grant Committee of CONACYT (under grant 33780-A) for the financial support given to this research project. The authors would also like to thank the Department of Computer Science of Tijuana Institute of Technology for resources and time given to this project.

References

1. Betin, F., Pinchon, D., and Capolino, G. A. (2000). "Fuzzy logic applied to speed control of a motor drive," *IEEE Trans. Industrial Electronics*, Vol. 47, pp. 610-622.

2. Castillo, O. and Melin, P. (1998). "A new fuzzy-fractal-genetic method for automated mathematical modelling and simulation of robotic dynamic systems," Proceedings of FUZZ'98, Anchorage, Alaska, Vol. 2, pp. 1182-1187.

3. Castillo, O. and Melin, P. (1999). "A new fuzzy inference system for reasoning with multiple differential equations for modelling complex dynamical systems," IOS Press, Vienna Austria, pp. 224-229.

4. Castillo, O. and Melin, P. (2001). *Soft Computing for Control of Non-Linear Dynamical Systems*, Springer-Verlag, Heilderberg, Germany.

5. Chen, G. and Pham, T.T. (2001). *Introduction to Fuzzy Sets, Fuzzy Logic, and Fuzzy Control Systems*, CRC Press, Boca Raton, Florida, USA.

6. Fu, K.S., Gonzalez, R.C. and Lee, C.S.G. (1987). *Robotics: Control, Sensing, Vision and Intelligence*, Mc Graw-Hill.

7. Jang, J.S.R., Sun, C.T., and Mizutani E. (1997). *Neuro-Fuzzy and Soft Computing*, Prentice Hall.

8. Kenjo, T. and Sugawara, A. (1994). *Stepping motors,* Oxford Science Publications.

9. Li, Y.F. and Lau, C.C. (1989). "Developments of fuzzy algorithms for servo systems," *IEEE Control System Magazine*, Vol. 9, pp. 65-72.

10. Mamdani, E. H. (1974). "Application of fuzzy algorithms for control of a simple dynamic plant," *Proc. Of IEEE,* Vol. 121, pp. 1585-1588.

11. Melin, P. and Castillo, O. (1998). "An adaptive model-based neuro-fuzzy-fractal controller for biochemical reactors in the food industry," Proceedings of IJCNN'98, Anchorage Alaska, USA, Vol. 1, pp. 106-111.

12. Nakamura, S. (1997). *Numerical Analysis and Graphic Visualization with MATLAB*, Prentice-Hall.

13. Sugeno, M. and Kang, G.T. (1988). "Structure identification of fuzzy model," *Fuzzy Sets and Systems*, Vol. 28, pp.15-33.

14. Zadeh, L. A. (1975). "The concept of a linguistic variable and its application to approximate reasoning," *Information Sciences*, 8, pp. 43-80

6 Bayesian Networks in Decision Support

Tomas O. Carlsson, Ronald Wennersten

Royal Institute of Technology, Department of Chemical Engineering, Industrial Ecology and Process Safety, SE-100 44 Stockholm, SWEDEN

toc@foi.se

1 Introduction

Since the beginning of industrialism, the complexity of industrial processes has been continuously increasing. In modern plants during normal operation this complexity is handled by the computerised control system. However, some process conditions outside the design of the control system still have to be managed by the human operator. These conditions can be due to equipment malfunction, unknown inputs and disturbances. Due to the complexity and the limited direct process interaction that operators experience, they often have difficulties making the right decisions in these abnormal situations.

In order to resolve this task, various computerised decision support tools have been suggested. The aim of these decision support tools have often been to aid the operator in searching for the cause of the observed process symptoms. There have been several approaches to accomplish this.

In the quantitative model-based approach (Chen and Patton 1999, Gertler 1998), the process input signals are fed into a mathematical process behaviour model. The calculated process output of the model is compared to the actual behaviour of the process. The residual between observed and calculated process outputs is evaluated in order to detect and diagnose abnormal behaviour. The drawback of this method is that numeric mathematical process models rarely are available.

Qualitative model-based approaches (Iri et al. 1979, Mo et al. 1997, Vendam and Venkatasubramanian 1997, Catino et al. 1991, Shen et al. 1993) have been suggested in order to overcome the lack of quantitative models. In this case, the influence of process variables on each other is expressed as increasing and decreasing. A disadvantage of the model-based approach is that it gives ambiguous results. Another qualitative approach is to use expert systems (Kramer 1987), in this case operator experience and other knowledge sources are used to build a data base of rules. These rules are then used to classify process symptoms. The drawback is that the rule database is expensive to develop and maintain (Wennersten et al.

1996). One way to overcome this is to link the expert system to the routinely updating of the safety analysis (Wennersten et al. 1997).

An alternative source of information, besides process models and operator experience, are the recordings of process data. In neural networks (Fantoni 2000 and Frank 1996) and principal component-based approaches (McGregor et al. 1996 and Zhang et al. 1996), this data is utilised to build process models which can either classify fault patterns or predict process behaviour. The process predictions are used similarly to quantitative-model based approaches, to detect differences between predicted and observed behaviour. Drawbacks of the neural network approach are that it is a non-transparent method and that it does not always perform well outside the training set. In the case of principal component analysis, one shortcoming is the difficulty to model non-linear processes.

A problem common to all fault diagnosis approaches is that they have to deal with uncertainty. Bayesian networks provides probability theory to deal with this uncertainty. In addition, it is also possible to train Bayesian networks on process data. The aim of this chapter is to discuss the potential of Bayesian networks to resolve some of the current problems of the decision support tools as outlined above. It is organised as follows.

First, Bayesian networks will be given a quick introduction. This is followed by a problem statement, in order to clarify the fault diagnosis task. Then, various approaches to fault diagnosis utilising the benefits of Bayesian methods will be discussed. The discussion will include both early applications, the current approach of the authors of this paper, as well as suggestions on potential applications. The chapter is not a complete review on fault diagnosis with Bayesian networks, but should rather be considered as an introduction to the benefits of Bayesian methods.

2 Bayesian Networks

This section briefly describes the principles and different variants of Bayesian networks. To some extent, the description is organised according to the documents of the Bayes Net Toolbox for Matlab (Murphy 2001a, 2001b). An example of an inference algorithm is given in the section on fault diagnosis applications.

2.1 Discrete Models

In Bayesian networks, variables are represented by nodes. The nodes are linked by directed arcs. Only nodes that directly influence one another are connected. Indirect influence is represented by the path from one node to another. The arcs are directed from the *parent* node to the *child* node.

The relation between two variables is expressed by a conditional probability. The state of a child node is conditional on the state of it's parent nodes. This is expressed in a conditional probability table, as illustrated in Fig. 1. In the conditional probability table of the Rain node in Fig. 1, the probability that it is raining is conditional on whether or not it is cloudy.

P(C=F)	P(C=T)
0.5	0.5

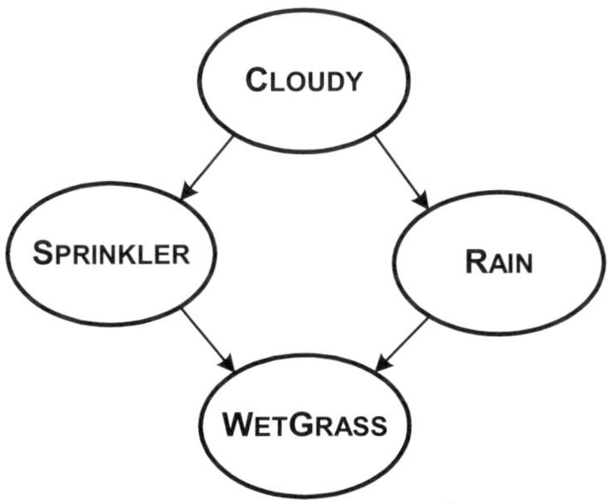

C	P(S=F)	P(S=T)
F	0.5	0.5
T	0.9	0.1

C	P(R=F)	P(R=T)
F	0.8	0.2
T	0.2	0.8

S	R	P(W=F)	P(W=T)
F	F	1.0	0.0
T	F	0.1	0.9
F	T	0.1	0.9
T	T	0.01	0.99

Fig. 1. Lawn sprinkler example (Russell and Norvig 1995)

The probability that the grass is wet is conditional on both parents, Sprinkler and Rain. In Fig. 1 the probability that the grass is wet, if the sprinkler is on and it is not raining is

$$P(W = T | S = T, R = F) = 0.9 \quad (1)$$

where P is probability, W is WetGrass, S is Sprinkler, R is Rain, T is true and F is false. Now, assume that the grass is observed to be wet, but we do not no if it is raining. The *a posteriori* probability that it is raining is inferred by applying Bayes' rule:

$$P(R=T|W=T)=\frac{P(W=T|R=T)\times P(R=T)}{P(W=T)} \tag{2}$$

thus whether it is raining or not will depend on the conditional probability that it is wet if it is raining, the *a priori* probability that it is raining and the overall probability that the grass is wet if nothing else is observed. Hence, given a set of observations, the probability that a hypothesis is true is directly proportional to the likelihood of the observations, given the hypothesis. Thus, inference in Bayesian networks is the search for a hypothesis whose estimate maximises the likelihood.

If new evidence is added to the network, e.g., that the sprinkler is on. Then, the denominator would change from $P(W=T)$ to $P(W=T|S=T)$, which is much larger, hence the probability that it is raining would decrease. This can be interpreted as the sprinkler "explains away" the fact that the grass is wet (Pearl 1988). This feature is especially useful in fault diagnosis reasoning.

2.2 Mixed Discrete, Continuous Models

In the previous example the variables were binary discrete, with the two states, false or true. However the network can be equally applied to variables with several states, e.g., low, normal and high. In addition, a mixture of both continuous-valued and discrete models can be used.

The relation between continuous-valued variables is expressed by a linear model with gaussian noise:

$$y=\mu+wx\pm\sigma \tag{3}$$

If a continuous node has discrete parents, then for each possible combination of the parents state Q_i; the mean μ the loading factor W and the variance σ of the linear gaussian model changes. The probability of a certain value of Y, is expressed by the probability density function *N*.

$$P(Y=y|X=x,Q=i)=N(y;\mu_i+W_ix,\sigma_i) \tag{4}$$

2.3 Dynamic Models

It is also possible to make dynamic models, allowing arcs to nodes in adjacent time slices. In Fig. 2 a *prior network* describing the relation between variables at the initiation of the sequence, and a *transition network*, describing the relation between nodes in adjacent time slices is defined. The dynamic models can be used for smoothing, filtering and prediction.

2.4 Learning

If the structure of the network is known, the conditional probability parameters can be learned by statistics on observed cases. If the nodes are continuous, Bayesian regression analysis can be used to obtain the parameters of the linear

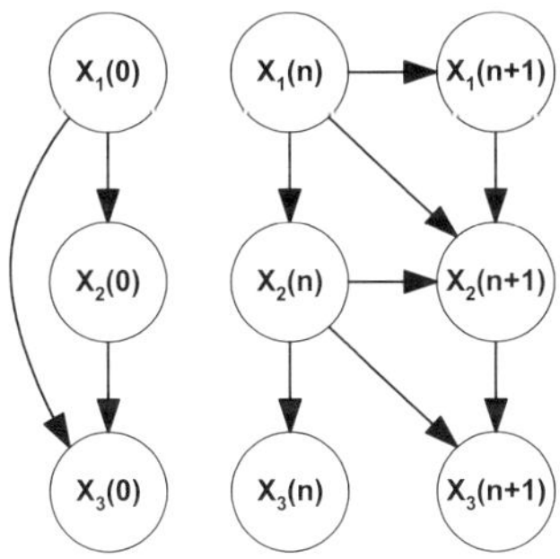

Fig. 2. Prior and transition network of a dynamic Bayesian network

expressions. Though more difficult, methods to derive the structure of the network also exists (Friedman et al. 1998).

3 Fault Diagnosis Problem Statement

In this section the fault diagnosis task will be clarified by using an explanatory example: the Tennessee Eastman simulator (Downs and Vogel 1993). The process is illustrated in Fig. 3 with the control scheme suggested by McAvoy and Ye (1994). Frank (1996), excluding faults in control units, computer hardware and software, has made the following list of typical faults:

- Defect constructions, ruptures, fractures, lost and loose parts.
- Actuator faults, such as damage in bearings, deficiencies in force or momentum, defects in gears and ageing affects.
- Sensor faults, such as scaling errors, hysteresis, drift, dead zone, short cuts and contact failures.
- Abnormal parameter variations in the system.
- External obstacles, such as collisions and clogging of outflows.

All these types of faults apply to the Tennessee Eastman process. In addition, operator errors, mounting errors and measurement of the wrong entity could also be added to the list.

In the Tennessee Eastman process, a cascade control system is applied with several layers of feedback loops, both control loops and process stream refluxes. In this environment, a failure in one part of the process will affect a large set of process variables. To illustrate this a failure in the cooling water control valve in the condenser during a process state transition has been simulated. The difference between normal behaviour and the behaviour due to the faulty cooling valve is shown in Fig. 4. It clearly illustrates the difficulties that face process operators during an abnormal situation.

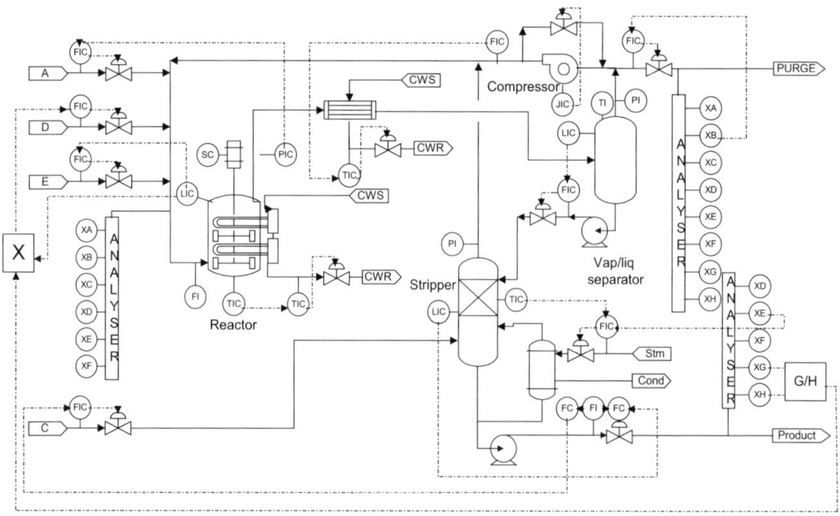

Fig. 3. The Tennessee Eastman process

4 Fault Diagnosis with Bayesian Networks

Bayesian networks can be used for fault diagnosis purposes in several ways. In this section we explore some of the suggested approaches. Benefits and drawbacks regarding capabilities, implementation and upholding of the systems are discussed.

4.1 Qualitative Model-Based Bayesian Fault Diagnosis

As mentioned previously, a feature of automated reasoning in Bayesian networks is the ability to "explain away" observations by new evidence. This feature can be utilised by the inclusion of fault nodes in a Bayesian network. In this way one fault may explain away a set of deviations, hence avoiding a large set of potential diagnoses, which is the case for qualitative models. This concept has been applied to both discrete and mixed discrete/continuous-valued Bayesian networks.

In the discrete case, first the structure of the network is defined from a qualitative plant model represented by a directed digraph between process parameters, e.g., *A* to *B* in Fig. 5. For each fault scenario node, an invalidation node *F* is added to the network and linked to the variables the fault directly affects. During normal operation an increase in variable A lead to increase in variable B. In the conditional probability table of node B, the effect of A on B will be conditional on the state of node F. During inference both A and B are observed. If there is a deviating behaviour between node A and B, F will explain away this difference hence indicating a fault.

The benefit is that it allows a formal way of handling the uncertainty in qualitative models. The drawbacks are that a qualitative model still has to be developed and that fault symptom knowledge is necessary. The method has been applied to both space mission control (Horowitz and Barry 1995) and chemical processes (Leung and Romagnoli 2000, Tan and Kramer 1993, Carlsson and Wennersten 2001a).

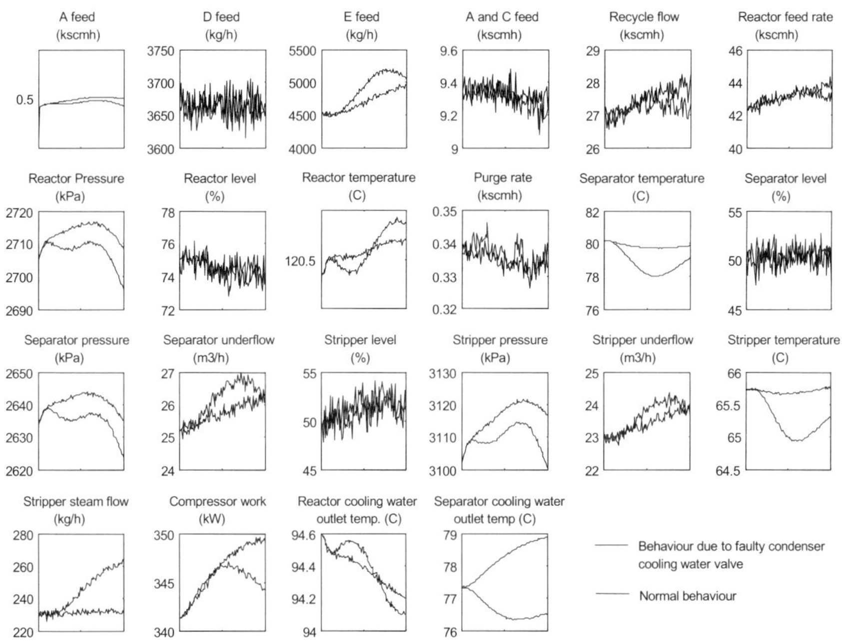

Fig. 4. Fault symptoms due to a faulty condenser cooling water valve during reactor setpoint change

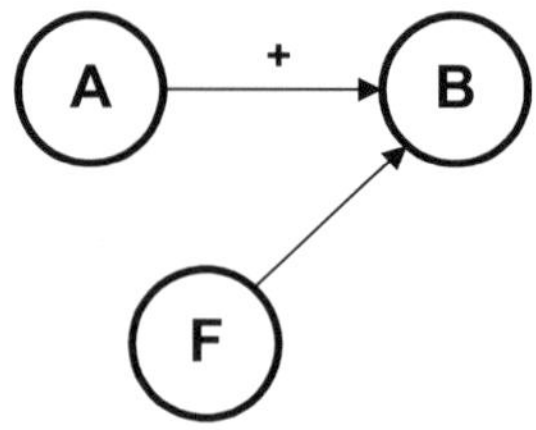

Fig. 5. Network structure of a qualitative model representation. A increases B, conditionally on fault node

4.2 The Generalised-Likelihood Ratio (GLR) Approach

Stochastic linear dynamic models can be considered a special case of dynamic Bayesian networks, consisting of continuous-valued nodes only. These models, often referred to as Kalman filters, are expressed on the following form.

$$\mathbf{x}_{t+1} = \mathbf{A}\mathbf{x}_t + \mathbf{G}\mathbf{u}_t + \mathbf{w}_t \quad (5)$$

$$\mathbf{y}_t = \mathbf{C}\mathbf{x}_t + \mathbf{b}_t + \mathbf{v}_t \quad (6)$$

where the k-vector $\mathbf{x}$ are the internal process variables, $\mathbf{A}$ is the $k \times k$ state transition matrix, the r-vector $\mathbf{u}$ is the process input variables, $\mathbf{G}$ is the $k \times r$ gain matrix, $\mathbf{y}$ are the p process measurements and $\mathbf{C}$ is the $p \times k$ observation matrix. The k-vector $\mathbf{w}$ and p-vector $\mathbf{v}$, are random variables representing the state evolution and observation noises, respectively. Specially for fault diagnosis is $\mathbf{b}_t$, a fault term adding a deviation to the sensor reading $\mathbf{y}$.

The linear dynamic model without the fault term $\mathbf{b}_t$ represents normal operation. The value of $\mathbf{b}_t$ represents different fault hypothesises, e.g. sensor jump or sensor step depending on the application. In the diagnosis algorithm the maximum likelihood estimate is searched for, by applying Bayes rule (Tsafestas and Watanabe 1990, Willsky 1976). For normal operation, the hypothesis that b is zero will have the highest likelihood score, whereas for a faulty situation, including b in the filter will gain a higher score.

The method has been applied to sensor validation in robotics (Nicholson and Brady 1994) and mission control systems (Bickford et al. 1997). The dynamic linear model can be derived either from mathematical models or by system identification techniques utilising stored process data. A benefit of the method is that it does not necessary require any modelling. A drawback is that the method only applies to sensor validation.

4.3 ARX-Model-Based Actuator and Sensor Fault Diagnosis

Recently a crossover of the two previous methods was suggested (Carlsson and Wennersten 2001b).

In this approach a dynamic Bayesian network consisting of a mixture of continuous-valued, control and sensor signals, and discrete nodes, faults, is used. The linear dynamic models of the continuous-valued nodes are conditional on the discrete fault nodes. The linear dynamic models are represented in the form of multivariate autoregressive models with an exogenuous varible. It differs from the Kalman filter by only containing the process output variables **y**, and the process input variables **u**. It has the following form:

$$\mathbf{y}_{t+1} = \mathbf{w}_x \mathbf{x}_t + \mathbf{w}_u \mathbf{u}_t + \boldsymbol{\sigma} \tag{7}$$

where **w** is the linear load factor and $\boldsymbol{\sigma}$, gaussian noise. All variables are adjusted to zero mean and unit variance.

Fig. 6 shows the univariate case of a switching ARX-model. In this example the exogenous variable u is the actuator signal, and y is a sensor signal. There are two binary fault nodes, one representing actuator failure, the other sensor failure.

During normal operation the fault nodes are in their false state and the relation between the process variables is represented by the ARX-model. If the actuator failure node F_u is true, the value of the control signal is replaced by zero in the linear model of the sensor signal and the variance of the sensor signal is increased by the actuator signal's load factor (w_u). If the sensor failure node is true, then it's value in the child nodes is replaced by increased variance analogous to the actuator failure. In the sensor node itself, mean is set to zero and variance is set to unit, in this case one.

In this way the probability density functions of the continuous-valued variables are flattened out when a failure occurs. During normal operation the non-manipulated ARX-model will have the highest likelihood score; where as, during deviating behaviour, the flatter probability density function will have a higher likelihood-score.

The ARX-model can be derived by regression analysis on stored process data at normal operation. The method has been tested on the Tennessee Eastman process (Carlsson and Wennersten 2001b). In these tests it seemed especially efficient at diagnosing actuator failures. This is beneficial, as other data mining-based approaches, e.g., neural networks, often only handle sensor failures.

To illustrate the algorithms behind the reasoning in Bayesian networks, a more in depth description on the above mentioned method is given below.

As shown in Figure 7, during diagnosis, evidence in the form of a sequence of process data $\mathbf{x}_{1:T}$, consisting of process input and output signals during the interval 1:T, is fed to the inference engine. The output of the inference engine is a list of probabilities of all possible faults during the sequence P(F_1(1:T)),…, P(F_n(1:T)).

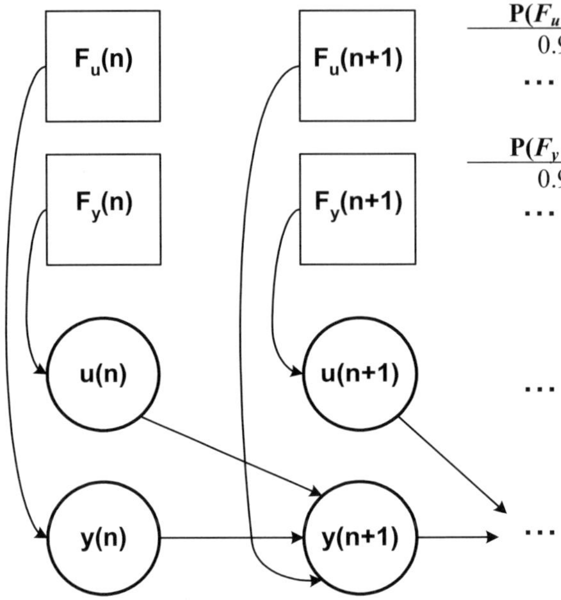

$P(F_u = F)$	$P(F_u = T)$
0.99	0.01
...	

$P(F_y = F)$	$P(F_y = T)$
0.99	0.01
...	

F_u	F_y	$y(n+1)$
F	F	$N(y(n+1); w_y y(n) + w_u u(n), \sigma_y)$
T	F	$N(y(n+1); w_y y(n), \sigma_y + w_u)$
F	T	$N(y(n+1);0,1)$
F	F	$N(y(n+1);0,1)$

Fig. 6. Structure and conditional ARX-models of dynamic Bayesian network for fault diagnosis

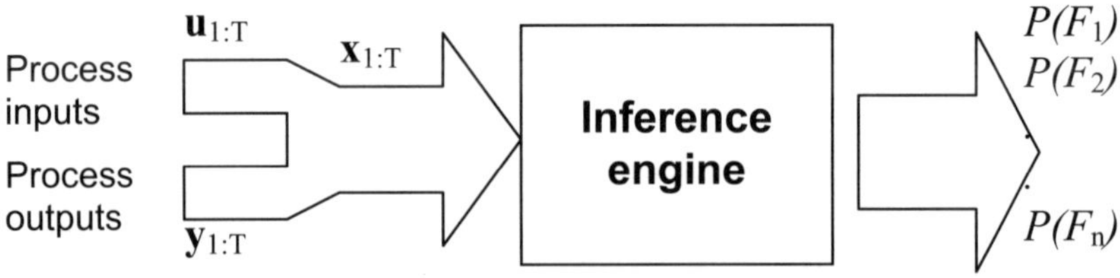

Fig. 7. Principle of the inference process

The probability of a sequence of states $S_{1:T}$, of the fault nodes, $S_t=F_1(t),F_2(t),F_3(t),\dots,F_n(t)$ can be calculated with the recursive method below (Murphy 1998). It is derived by applying Bayes rule both to the forward and backward chain and works as follows.

In the forward chain:

$$\begin{aligned}
&P(S_t = j|\mathbf{x}_t, \mathbf{x}_{1:t-1}) \\
&= \frac{1}{c} P(\mathbf{x}_t|S_t = j, \mathbf{x}_{1:t-1}) P(S_t = j|\mathbf{x}_{1:t-1}) \\
&= \frac{1}{c} P(\mathbf{x}_t|S_t = j, \mathbf{x}_{t-1}) \sum_i P(S_t = j|S_{t-1} = i, \mathbf{x}_{1:t-1}) \\
&\times P(S_{t-1} = i|\mathbf{x}_{1:t-1}),
\end{aligned} \tag{8}$$

where

$$\begin{aligned}
c = P(\mathbf{x}_t|\mathbf{x}_{1:t-1}) &= \sum_j P(\mathbf{x}_t|S_t = j, \mathbf{x}_{t-1}) \\
&\times \sum_i P(S_t = j|S_{t-1} = i, \mathbf{x}_{1:t-1}) M_{t-1|t-1},
\end{aligned} \tag{9}$$

and $M_{t-1|t-1}$ is the probability of each state in the previous time slice.

In the backward chain:

$$\begin{aligned}
&P(S_t = j|\mathbf{x}_{1:T}) \\
&= \sum_k P(S_t = j|S_{t+1} = k, \mathbf{x}_{1:T}) P(S_{t+1} = k|\mathbf{x}_{1:T}) \\
&= \sum_k P(S_t = j|S_{t+1} = k, \mathbf{x}_{1:t}) P(S_{t+1} = k|\mathbf{x}_{1:T}) \\
&= \sum_k \frac{P(S_t = j|\mathbf{x}_{1:t}) P(S_{t+1} = k|S_t = j, \mathbf{x}_{1:t})}{P(S_{t+1} = k|\mathbf{x}_{1:t})} \\
&\times P(S_{t+1} = k|\mathbf{x}_{1:T}).
\end{aligned} \tag{10}$$

4.4 Multiple-Model Approach

Similar to the previously described GLR-approach, the multiple-model approach also uses the Kalman filter representation to model system behaviour:

$$\mathbf{x}_{t+1} = \mathbf{A}\mathbf{x}_t + \mathbf{G}\mathbf{u}_t + \mathbf{w}_t \tag{11}$$

$$\mathbf{y}_t = \mathbf{C}\mathbf{x}_t + \mathbf{v}_t \tag{12}$$

In the multiple-model approach a bank of models is fed with process signals **u** and **y**. Each model represents either the fault-free process or an actuator or sensor failure, and has it's own set of parameters **A**, **G**, **C**, **w**, **v**. In the diagnosis algorithm, Bayes rule is applied in order to identify the model that best predicts the observed process behaviour (Berec 1998, Eide and Maybeck 1996, Tzafestas and Watanabe 1990, Willsky 1976).

The method has been applied to navigational systems, flight control systems and electrocardiograms (Tzafestas and Watanabe 1990). The Kalman filters for the fault models can either be derived from quantitative models or by system identification on process data representing various failure scenarios (Ljung 1999).

In the latter case, the multiple-model approach could be used similarly to neural and PCA-based classifiers for frequently occurring disturbances. This way an operator could quickly distinguish a new faulty scenario from a similarly scenario.

4.5 Probabilistic Expert Systems

Expert systems have long been used for fault diagnosis applications (Kramer 1987). By linking process symptoms to equipment failures through a set of *if – then* rules, it is possible to deduce the cause of a process upset.

Bayesian networks can also be used as probabilistic expert systems (Chong and Walley 1996, Kirsch and Kroschel 1994). In this case symptoms are linked to causes similarly to rule-based expert systems. But, instead of using *if – then* propositions, a symptom is linked to it's causes by making it conditional on the causes in a Bayesian network. The benefits in comparison to traditional expert systems are that it is possible to express the uncertainty of a proposition. Also as the Bayesian network, in fact, is a joint probability distribution, any new evidence will effect the diagnosis and bi-directional reasoning is possible.

In comparison to fuzzy expert system, the logic and mathematical basis of Bayesian networks is more straightforward. However, the vagueness that is conveniently handled in fuzzy logic has to be dealt with differently in a Bayesian network. Either and entity is high or low, dual membership is not allowed. However, it is possible to express duality by e.g. claiming that the probability that an entity is high is

60% and that it is low is 40%. This is called soft evidence (Murphy 2001). The pros and cons regarding Bayesian versus fuzzy approaches is in many cases a matter of opinion.

The general drawbacks of expert systems equally apply to probabilistic expert systems, i.e., they are expensive to develop and difficult to maintain.

4.6 Alarm Analysis and Filtering

As illustrated previously for the Tennessee Eastman process, when a malfunction occurs in one part of a process, several process upsets occur all-round the plant often causing a so-called alarm flood. Identifying the direct alarm among all the "phantom alarms" can often be cognitively demanding.

Previous work to resolve this task has been done by using expert systems or alarm sequence pattern matching, to detect errors (Choi et al. 1995, Davison and Hamilton 1997). This approach equally as the expert systems for fault diagnosis, suffers from the difficulties of developing and maintaining the rule data base.

However, as the alarms often are correlated, clearly there is a potential for data mining. In telecommunication networks, which is another application that often faces floods of alarms, data mining techniques, e.g., case-based reasoning and neural networks, have been suggested to identify alarm patterns with similar causes (Gardner and Harle 1998). Other applications are in power distribution systems (Teo et al. 1992) and vehicles (Guo et al. 2000).

As mentioned in the section on Bayesian networks, it's possible to learn both structure and the conditional probability parameters of a network, using training sets of data. This topic is currently of great interest to the bioinformatics community and drug developers. However, the methods (Friedman et al. 1998) could equally be applied to the alarm floods of the process industry. This way it would be possible to filter out phantom alarms, and group alarm together according to similar causes. New alarms would be able to distinguish from similar. The applicability of the approach would of course depend on how similar different alarm situations are, and how well they can be distinguished.

5 Concluding Remarks

In this chapter, different approaches to fault diagnosis with Bayesian networks have been discussed. The key benefits of using Bayesian networks in fault diagnosis are as follows.

- Through the use of probability theory it is possible to reason with and quantify the uncertainty of a diagnosis.

- Due to the a priori probability factor, prior information, e.g., failure rates or expert judgement, can be included in the diagnosis. Linking it to safety analysis routines, such as Hazard and Operability studies (HAZOP) would provide a platform for the merging of process knowledge and process data.
- The method allows user interaction, e.g., evidence that a component is not faulty can be taken into account in the diagnosis.
- It is possible to train a Bayesian network on recorded process data, hence process knowledge or models are not necessary.

Several of the methods described in this chapter still remains to be implemented in real processes. How implementation should be done, for which processes and what purposes remains an interesting topic for further study.

References

Berec L (1998) A multi-model method to fault detection and diagnosis: Bayesian solution. An introductory treatise. International Journal of Adaptive Control and Signal Processing 12:81-92

Bickford RL, Bickmore, TW, Caluori VA (1997) Real-time sensor validation for autonomous flight control. AIAA/ASME/SAE/ASEE Joint Propulsion Conference & Exhibit, 33rd, pp 1-10

Carlsson TO, Wennersten R (2001a) Graphical process models for risk assessment and operator decision support. In: 10th International symposium on Loss prevention and safety promotion in the process industries. Elsevier, Amsterdam, pp 699-711

Carlsson TO, Wennersten R (2001b) Sensor and actuator fault diagnosis with a Bayesian network based on dynamic linear models. Submitted to: Computers and Chemical Engineering

Catino CA, Grantham SD, Ungar LH (1991) Automatic generation of qualitative models of chemical process unit. Computers and Chemical Engineering 15:583-599

Chen J, Patton RJ (1999) Robust model-based fault diagnosis for dynamic systems. Kluwer Academic, London

Choi SS, Kang KS, Kim HG, Chang SH (1995) Development of an on-line fuzzy expert system for integrated alarm processing in nuclear power plants. IEEE Transactions on nuclear science 42:1406-1418

Chong HG and Walley WJ (1996) Rule-based versus probabilistic approaches to the diagnosis of faults in wastewater treatment processes. Artificial intelligence in engineering 1:265-273

Davison S, Hamilton J (1997) Intelligent alarm handling and analysis. In: IEE Colloquium on Stemming the Alarm Flood (Digest No: 1997/136), pp 9/1-9/4

Downs JJ, Vogel EF (1993) A plant-wide industrial process control problem. Computers and Chemical Engineering 17:245-255

Eide P, Maybeck P (1996) An MMAE failure detection system for the F-16. IEEE Transactions on Aerospace and electronic systems 32:1125-1136

Fantoni PF (2000) Neuro-fuzzy model applied to full range signal validation of PWR nuclear power plant data. International Journal of General Systems 29:305-320

Frank, P.M. (1996) Analytical and qualitative model-based fault diagnosis – A survey and some new results. European J. of Control 2:6-28

Friedman N, Murphy K, Russell S (1998) Learning the structure of dynamic probabilistic networks. In: Proceedings of the 14th Conference on Uncertainty in Artificial Intelligence, pp 139-147

Gardner RD, Harle DA (1998) Pattern discovery and specification techniques for alarm correlation. In: IEEE Network Operations and Management Symposium. Vol. 3, pp 713-722

Gertler JJ. (1998) Fault detection and diagnosis in engineering systems. Marcel Dekker, New York

Guo H, Crossman JA, Murphey YL, Coleman M (2000) Automotive signal diagnostics using wavelets and machine learning. IEEE Transactions on Vehicular Technology 49: 1650-1662

Horovitz E, Barry M (1995) Display of information for time-critical decision making. Proceedings of Eleventh Conference on Uncertainty in Artificial Intelligence, pp 296-305

Iri M, Aoki K, O'Shima E, Matsuyama H (1979) An algorithm for diagnosis of system failures in the chemical processes. Computers and Chemical Engineering 3:489-493

Kirsch H, Kroschel K (1994) Applying Bayesian networks to fault diagnosis. In: Proceedings of the Third IEEE Conference on Control Applications 2, pp 895-900

Kramer M.A. (1987) Expert systems for process fault diagnosis: a general framework. In: Foundations of Computer Aided Process Operations. Proceedings of the First International Conference, pp 557-587

Leung D, Romagnoli J (2000) Dynamic probabilistic model-based expert system for fault diagnosis. Computers and Chemical Engineering 24:2473-2492

Ljung L (1999) System identification: theory for the user. Prentice Hall, N.J.

MacGregor JF, Kourti T, Nomikos P (1996) Analysis, monitoring and fault diagnosis of industrial processes using multivariate statistical projection methods. In: IFAC 13th Triennal World Congress M, pp 145-150

McAvoy TJ, Ye N (1994) Base control for the Tennessee Eastman problem. Computers and Chemical Engineering 18:383-413

Mo JK, Lee G, Nam DS, Yoon YH, Yoon ES (1997) Robust fault diagnosis based on clustered symptom trees. Control Eng. Practice 5:199-208

Murphy K (1998) Learning Switching Kalman Filter Models. Compaq Cambridge Research Lab Tech Report 98-10

Murphy K (2001a) The Bayes Net Toolbox for Matlab. In: Computing Science and Statistics: Proceedings of Interface, 33: *in press*

Murphy K (2001b) Bayes Net Toolbox for Matlab, University of California, Berkeley, http://www.cs.berkeley.edu/~murphyk/Bayes/bnt.html

Nicholson AE, Brady JM (1994) Dynamic Belief networks for discrete monitoring. IEEE transactions on systems, man and cybernetics 24:1593-1610

Pearl J (1988) Probabilistic reasoning in intelligent systems : networks of plausible inference. Morgan Kaufmann, CA

Russell S, Norvig P (1995) Artificial Intelligence: A Modern Approach. Prentice-Hall, Englewoods NJ

Shen Q, Leitch RR, Coghill GM (1993) Fuzzy qualitative modelling. In: IEE Colloquim on Two decades of fuzzy control-Part 2 Digest No. 1993/118, pp 3/1-4

Tan JS, Kramer MA (1993) On-line hazard aversion: A Bayes approach. IEEE Transactions on Reliability 42:317-326

Teo CY, Kwok WK, Lee CC (1992) Application of AI techniques for fault diagnosis in power distribution system. In: First International Conference on Intelligent Systems Engineering, pp 53-58

Tzafestas S, Watanabe K (1990) Modern approaches to system/sensor fault detection and diagnosis. Journal A 31:42-57

Vedam H, Venkatasubramanian V (1997) Signed digraph based multiple fault diagnosis. Computers and Chemical Engineering 21:S655-S660

Wennersten R, Narfeldt R, Gränfors A, Sjökvist S (1995) In: Proc. of the 11th symposium on Loss prevention and safety promotion in the process industries, pp 661-670.

Wennersten R, Narfeldt R, Gränfors A, Sjökvist S (1996) Process modelling in fault diagnosis. Computers and Chemical Engineering 20:S665-S6670

Willsky AS (1976) A survey of design methods for failure detection in dynamic systems. Automatica 12:601-611

Zhang J, Martin EB, Morris AJ (1996) Fault detection and diagnosis using multivariate statistical techniques. IChemE Transactions of the Institute of Chemical Engineers, Part A. Chemical Engineering Research and Design 74: 89-96

7 Hidden Markov Model Based Transient Identification in NPPs

Kee-Choon Kwon

Korea Atomic Energy Research Institute, P.O. Box 105, Yuseong, Daejeon 305-600, Korea,

kckwon@kaeri.re.kr

In this chapter, a transient identification based on a hidden Markov model (HMM) has been suggested and evaluated experimentally for the classification of transients in nuclear power plants. The transient can be identified by its unique time dependent patterns related to the principal variables. The HMM, a double stochastic process, can be applied to transient identification which is a spatial and temporal classification problem under a statistical pattern recognition framework. The HMM is created for each transient from a set of training data by the maximum-likelihood estimation method. The transient identification is determined by calculating which model has the highest probability for the given test data. Several experimental tests have been performed with normalization methods, clustering algorithms, and a number of states in HMM. Several experimental tests have been performed including superimposing random noise, adding systematic error, and untrained transients. The proposed real-time transient identification system has many advantages, however, there are still a lot of problems that should be solved to apply to a real dynamic process. Further efforts are being made to improve the system performance and robustness to demonstrate reliability and accuracy to the required level.

1 Introduction

Transient identification in nuclear power plants (NPPs) means classifying the type of transients by interpreting the major plant variables and operating status of the equipment. The transient is defined as when a process proceeds to an abnormal state from a normal state. The term identification, as applied to an engineering system or process, means the classification of the cause which brought an undesirable state or failure of the system. The identification can be done at several different levels, e.g., component, subsystem, function or event [1]. For the proposed transient identification system, identification is made at the event level to determine which transient has occurred in NPPs. It is necessary to identify the type of transient through continuous monitoring of NPPs during its early stage to provide sufficient information to the human operators in order to assist for proper operator action selection to prevent a more severe situation or to mitigate the accident consequence [2].

Typical transients in NPPs are associated with unique, time-dependent patterns of major variables and equipment status. This time dependent pattern may be used to identify the transient; hence, identification can be treated as a pattern classification problem [3]. It is difficult to identify the transient by a human operator when the preceding patterns of some transients are very similar and the patterns change further with time. Recently, attempts have been made to solve transient identification problems using a computer-based system.

In this transient identification problem, the classification may involve spatial and temporal patterns. Temporal patterns usually involve ordered sequences of data appearing with time. Spatial patterns mean the unique pattern of each transient, variations of the same transient, which may occur under different operating modes or at different break sizes.

The transient identification systems for NPPs have been developed using techniques such as an Artificial Neural Network (ANN) [4, 5], fuzzy logic [6], nearest neighbors modeling optimized by using a genetic algorithm [7], adaptive template matching [8], and observer-based residual generation [9]. All of these systems are still considered as a prototype or are under evaluation and have not yet been applied to real operating NPPs.

The ANN and fuzzy logic approaches can absorb spatial variations, but cannot provide proper solutions for temporal variations. So it is reasonable to adopt a double stochastic approach for the classification of the patterns. The Hidden Markov Model (HMM), a double stochastic process, enables modeling of not only spatial phenomena but also time scale distances. The HMM can be used to solve classification problems associated with time series input data such as speech signals or plant process signals, and can provide appropriate solutions by its modeling and learning capabilities, even though it does not have the exact knowledge to solve the problems. Most of the HMM applications for pattern classification in dynamic processes have a typical architecture to solve a spatial-temporal problems, but the target systems are different as in dynamic obstacle avoidance of mobile robot navigation, radar target, human action, American sign language, heart signals, sonar signals, two-handed actions, conditions of an electrical machine, deep space network antenna, moving light displays, environmental noise, and human genes in DNA. But the HMM has never been applied for transient identifications in NPPs.

The goal of this paper is to provide a transient identification system for use in NPPs that demonstrates the spatial and temporal modeling and learning capabilities with hidden Markov models. To demonstrate these capabilities, simulated NPP parameters are collected and clustered to build hidden Markov models for transient identification purposes.

2 Hidden Markov Model for Transient Identification

The problem of transient identification is defined as the recognition of transient types, ω_j given the sequential input patterns X_t at time *t*. The input pattern X_t is mathematically defined as an object described by a sequence of features at time *t* [10].

$$X_t = (x_1, x_2, \cdots, x_i, \cdots, x_d) . \tag{1}$$

The space of input pattern X_t consists of the set of all possible patterns:

$X_t \subset \Re^d$; $\Re^d$ is a *d*-dimensional real vector space.

The *k* observed input data up to time *t* is defined as Φ_{t-k} :

$$\Phi_{t-k} = \{X_{t-k+1}, \cdots, X_{t-1}, X_t\} . \tag{2}$$

The set of possible transient classes ω_j at time *t* forms the space of classes Ω :

$$\Omega = \{\omega_1, \omega_2, \cdots, \omega_c\} \text{ ; c is the number of classes} \tag{3}$$

The classes Ω are assumed to be mutually exclusive and exhaustive. The recognition task can be considered to be the finding of function f, which maps the space of input patterns Φ_{t-k} to the space of classes Ω :

$$f : \Phi_{t-k} \rightarrow \Omega \tag{4}$$

A dynamic process often exhibits sequentially changing behavior. If the one short-time period is defined as *frame*, the probability of the *frame* transition is different from each transient in NPPs. Therefore, the probability of *frame* existence and the transition between *frames* can be statistically modeled. The probability of a transient occurring in an NPP is already given and is called the *a priori* probability. When a transient has occurred in NPPs, the type of transient can be determined only by selecting the type of transient ω_j with the highest *a priori* probability $P(\omega_j)$. This decision is obviously unreasonable. It is more reasonable to determine the type of transient after observing the trend of time-series major variables, namely, to get the conditional probability $P(\omega_j \mid \Phi_{t-k})$. This conditional probability is called the *a posteriori* probability. Decision-making based on the *a posteriori* probability is more reliable, because it employs both *a priori* knowledge together with observed time-series data [11].

Classification of the unknown pattern X_t corresponds to finding the optimal model $\hat{\omega}$ that maximizes the conditional probability, $P(\omega_j \mid \Phi_{t-k})$, the probability that the system is in class ω_j at time t given that Φ_{t-k} was observed at time t, $P(\omega_j \mid \Phi_{t-k})$ over the type of transient ω_j. We can apply Bayes rule to calculate *a posteriori* probability,

$$P(\hat{\omega} \mid \Phi_{t-k}) = \max_{\omega} \frac{P(\Phi_{t-k} \mid \omega_j) P(\omega_j)}{P(\Phi_{t-k})}, \tag{5}$$

where

$$P(\Phi_{t-k}) = \sum_{j=1}^{c} P(\Phi_{t-k} \mid \omega_j) P(\omega_j). \tag{6}$$

The conditional probability, observing Φ_{t-k} given that the system is in class ω_j at time t, $P(\Phi_{t-k} \mid \omega_j)$ comes from comparing the shapes of the transient models with input observations, while the *a priori* probability $P(\omega_j)$ comes from the transient probability which represents how often the transient appears in the NPP. Since $P(\Phi_{t-k})$ is independent of $\hat{\omega}$, we get

$$\begin{aligned} P(\hat{\omega} \mid \Phi_{t-k}) &\propto P(\Phi_{t-k} \mid \hat{\omega}) P(\hat{\omega}) \\ &= \max_{\omega} \left[P(\Phi_{t-k} \mid \omega_j) P(\omega_j) \right]. \end{aligned} \tag{7}$$

In fact, *a priori* probability $P(\omega_j)$ can be calculated in NPPs, and should satisfy the following equation,

$$\sum_{j=1}^{c} P(\omega_j) = 1. \tag{8}$$

But the transient identification system does not cover all of the transients occurring in NPPs, and consequently cannot satisfy Eq. (8). Therefore, the present observed data controls the decision. $P(\Phi_{t-k} \mid \omega_j)$ is called the likelihood of $\hat{\omega}$ with respect to the set of samples. In real implementation, *a priori* probability $P(\omega_j)$ can be assumed that the occurring probabilities of all transients are equal [12]. The maximum likelihood estimate of $\hat{\omega}$ is, by definition, that value of $\hat{\omega}$ that maximizes $P(\Phi_{t-k} \mid \omega_j)$. In this identification problem, HMM is used to estimate the conditional probability $P(\Phi_{t-k} \mid \omega_j)$. By using HMM, the pattern variability in parameter space and time can be modeled effectively [11, 13].

Using HMMs in a classification problem with the Bayes rule and maximum likelihood training requires two things: the evaluation of $P(\Phi_{t-k} \mid \omega_j)$ for the implementation of the Bayes rule and the maximization of the likelihood for the training of the classifiers. Fortunately, there exist computationally efficient procedures for these two tasks. HMM parameters are estimated from the Baum-Welch algorithm and guarantee a finite improvement on each iteration in the sense of maximization of likelihood. An HMM is trained for each transient from a set of training data, and an iterative maximum likelihood estimation of model parameters from observed time-series data. Incoming observations are classified by calculating which model has the highest probability for producing that observation. The detailed definition of HMM is described in reference [14].

3 Transient Identification System

3.1 Preprocessing

Training and test data from the test simulator should be converted to a proper codebook which is input data for HMM identifier. Preprocessing means to make a codebook, and this method is divided into two techniques, normalization and vector quantization.

The input symptom vector to be used for vector quantization have same value as displayed in a real plant instrument because it is directly received from the test simulator or the plant computer in NPPs. The range of input data values are significantly different. Therefore, it should be normalized before using the input data of the clustering algorithms. Normalization is one of several transformation techniques. It has the effect of reducing the parameters to a common range. This provides a measure that allows the relative importance of any factor or interaction to be identified more clearly. It improves the numerical accuracy of the regression and the computation of significance.

If the distribution of the variable's values are of normal distribution, it is reasonable to be normalized by maximum value. But, most of the values of the specified variable are distributed in upper region in case of normalized by maximum value. It is undesirable to use the input data of the clustering algorithms. It is more reasonable to normalize between its minimum and maximum values. In similar way, it can be also considered that mean centered normalization. In the mean centered normalization method, data between mean and minimum or maximum range is linearly distributed. In other way, we can consider nonlinear data distribution of between mean and minimum or maximum range considering the standard deviation.

Features may be represented by continuous, discrete, or discrete-binary variables. It is expected that the feature vector contains most of the classification information available from the object. Feature extraction is an important task for classification or recognition. In feature extraction, data can be transformed from high-dimensional pattern space to low-dimensional feature space. Vector quantization (VQ), the process of approximating a block of continuous amplitude signals by a discrete signal is one method of feature extraction. The idea is to quantize each continuous vector to one of a relatively small number of template vectors, which together comprise what is called a codebook. The sequence of codebook indices obtained in this way forms the desired sequence of discrete symbols. In this paper, two VQ methods, k-means algorithm and Self-Organizing Map (SOM) [15] are introduced to compare the identification capability.

3.2 Real-Time Test Environment

It is more realistic to receive training and test data directly from the operating NPPs. But severe transient or accident seldom occur in real NPPs and it is almost impossible to make transient conditions only for experimental purpose. Therefore it is necessary to use the simulator for operator training or simulation code ready for safety analysis to implement and test the transient identification system. In this implementation, the test simulator was modified from the compact nuclear simulator which is installed at the Korea Atomic Energy Research Institute Nuclear Training Center for training non-operator personnel to fit testing the transient identification system. The test simulator is divided into two major parts: a mathematical modeling program, which executes the plant dynamic modeling program in real-time; and a supervisory program that manages user instructions.

The mathematical modeling programs consist of static and dynamic parts. The initial state, a 100% full power condition is set up in the static calculation, which is performed once before the start of the dynamic calculation. The dynamic calculation is performed every 0.2 second to represent a real-time mathematical modeling simulation. The test simulator provides the function to activate 79 predefined malfunctions. This function realizes the transient or accident condition to get training data and to test the transient identification system [16].

3.3 Data Collection

The nine typical transients are selected among different postulated transients that may occur in NPPs in consideration of the simulation capability of the test simulator because the purpose of this paper is to demonstrate the capability of hidden Markov models that apply to the transient identification problem. The description of the target transients are as follows:

ATWS (Anticipated Transient Without Scram)

FWLB (FeedWater Line Break inside containment)

LOCA (small Loss Of Coolant Accident)

LSLC (Loss of Steam generator Level Controller signal)

MSIV (Main Steam Isolation Valve closure)

MSLI (Main Steam Line break Inside containment)

MSLO (Main Steam Line break Outside containment)

PORV (Power Operated Relief Valve stuck open)

SGTR (Steam Generator Tube Rupture)

The input symptom vector is a collection of the principal variables and the status of major equipment from the transient simulation in the test simulator. The major variables and equipment status used to identify the nine different types of transients and one normal state are summarized in Table 1.

Table 1. List of input vector variables

No.	Variable description	Unit
1	Pressurizer pressure	kg/cm^2
2	Pressurizer level	Normalized
3	Reactor coolant average temperature	Deg C
4	Steam generator pressure	kg/cm^2
5	Steam generator level	Normalized
6	Reactor power	%
7	Reactivity	%dk/k
8	Average fuel temperature	Deg C
9	Feedwater line flow	m^3/hr
10	Main steam line flow	m^3/hr
11	Steam flow from steam generator	m^3/hr
12	Steam pressure from steam generator	kg/cm^2
13	Secondary radiation monitoring	MicroC/cc
14	Containment pressure	kg/cm^2
15	Containment temperature	Deg C
16	Containment humidity	%
17	Pressurizer relief tank pressure	kg/cm^2
18	Pressurizer relief tank temperature	Deg C
19	Net electrical power	MWe
20	Position of main steam isolation valve	Normalized
21	Reactor trip signal	Digital

There are two types of data, i.e., training data and test data. The training data are needed to train the clustering algorithm and HMM identifier. To test the classification capability of the HMM identifier, test data are also needed. The training and test data are collected from the test simulator. It is needed to get more widely spread training data per transient to design a reasonable classifier. In this experiment, we considered different operating modes and different break sizes in each transient. The HMM may absorb the variations from the different operating modes and different break sizes in a transient.

The training data are provided off-line from the test simulator. Major variables and the equipment status are combined for the input symptom vector in each transient when the test simulator emulates a transient situation. The transients are simulated in the test simulator by activating the malfunctions during normal operation, then get major variables, such as temperature, pressure, flow, pump on/off status, or valve open/close status. The training data are collected from different operating modes, such as 50%, 55%, 60%, 65%, 70%, 75%, 80%, 85%, 90%, 95% of reactor power and full power and different break size in each transient and normal state. Each training data consist of around 60 time interval input vectors. Each time interval is 1 second, and this means around 1 minute of data are collected.

The test data are collected off-line and on-line. The off-line test data are collected by the same method as training data. But they are collected by different operating modes and different break sizes to generate diverse test cases such as trained operating modes and non-trained break sizes, non-trained operating modes and trained break sizes, non-trained operating modes and non-trained break sizes, and trained operating modes and trained break sizes. The on-line test data for any operating modes or any break sizes are gathered directly from the real-time test simulator through data communication between test simulator and the transient identification process. The test simulator is executed every 0.2 second and the calculated simulation variables are stored in the shared memory. The transient identification process receives on-line test data from the shared memory every one second

3.4 Real-Time Transient Identification System

The major components of transient identification system are vector quantizer and HMM identifier. Fig. 1 shows the block diagram of the implemented transient identification system. First, the collected training data are normalized, then the training data are used to train the vector quantizer with off-line. The trained vector quantizer will be used to vector quantize the test data. A vector quantized codebook of training data are the training input set of the HMM identifier.

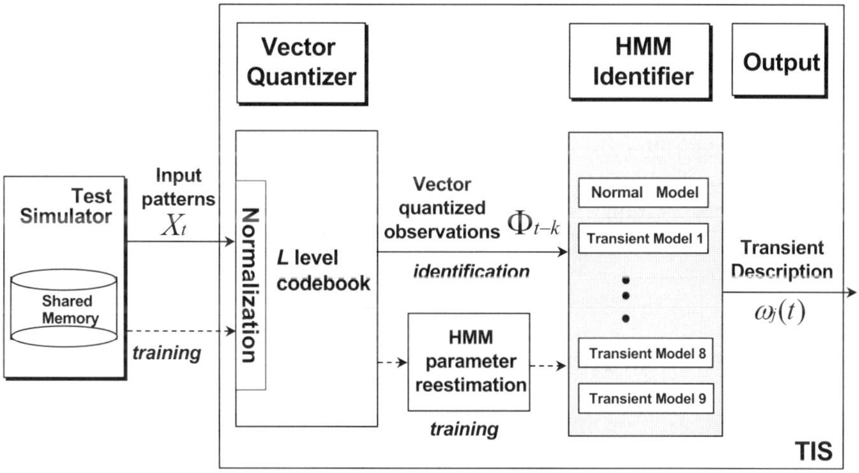

Fig. 1. A block diagram of transient identification system

The test symptoms are vector quantized to give input codebook of the HMM identifier. In this implementation, the *k-means* algorithm or SOM are used to cluster the input vector into L disjoint sets. In this implementation, the 300 clusters were chosen for an optimal solution after several attempts, meaning that every input vector is assigned to one of 300 clusters. The codebook size is 60, which means the system receives 60 time interval input vectors every one minute to classify the types of transients. In the initial one minute, the test results are incorrect because the codebook size is less than 60. The system should wait until it receives one minute amount of input vectors. This initial stage is called the "Ready mode." During the next time step, the system receives another 60 time interval input vectors as the sliding window method. To get input symptom vector every second, the system should be implemented with a real-time.

A left-to-right HMM has been considered appropriate for processing those signals whose properties change over time. The underlying state sequence associated with the model has the property that as time increases, the state index increases or stays in the same state. That is, the state always proceeds from left to right. The basic model consists of six states which have less than two direct transitions to the right state. Few initial conditions are given to this model, and these initial conditions are equivalent to all transient models. The re-estimation algorithm of the HMM may give a local minimum of the likelihood function. It is important to choose initial estimates of the HMM parameters so that the local minimum is the global minimum. Experience has shown that the uniform initial estimates work well. We also choose uniform initial estimates in this implementation by assuming that the observation symbol probability is equivalent to each state.

The training is performed by a forward algorithm and a backward algorithm, and making a re-estimate from the Baum-Welch algorithm in each model, given the multiple input observations [11]. The re-estimation is done until the convergence condition, $P(O|\bar{\lambda}) \geq P(O|\lambda)$, i.e., the new model estimates are more likely to produce the given observation sequence O, is satisfied in each model. The probability, $P(O|\lambda)$, is calculated by the optimal path which is obtained by the Viterbi algorithm for the given input observations in each model. The transients are classified by examining which model has the highest probability for the given input observations. The prototype of transient identification system was implemented in an HP747*i* industrial workstation and the programming was done using "C" language.

4 Experimental Results

In this section, experimental test results for a base model are described. Then, the selected model is suggested and the improved model is proposed to improve identification accuracy. The base model consists of Max-min normalization method, SOM clustered by training and test data, and 6 states HMM identifier. The experimental tests have been carried out after training of the vector quantizer and HMM identifier. Table 2 shows the results of the base model off-line test. In this experiment, there are the following four test cases to compare the trained or non-trained data for operating modes and break sizes.

- Case I: Trained operating mode, Non-trained break size.
- Case II: Non-trained operating mode, Trained break size.
- Case III: Non-trained operating mode, Non-trained break size.
- Case IV: Trained operating mode, Trained break size.

Table 2. Off-line test results for base model

Case / Transient	Case I	Case II	Case III	Case IV	Average Except Case IV
ATWS	-	100	-	100	100
FWLB	100	100	100	100	100
LOCA	27.3	100	30.0	100	51.6
LSLC	45.5	100	50.0	100	64.5
MSIV	-	100	-	100	100
MSLI	0.0	100	0.0	100	32.3
MSLO	72.7	80.0	70.0	100	74.2
PORV	100	100	100	100	100
SGTR	100	100	100	100	100
Average	63.6	97.8	64.3	100	76.8

As presented in this table, most of transients are correctly identified when given trained break size and non-trained operating mode. But in the case of non-trained break size, the recognition rate is remarkably lower than the trained break size. There is almost no difference when comparing the case of the trained and the non-trained operating mode. But the identification rate for the non-trained break size is lower than the non-trained operating mode. We estimated that the HMM can absorb these break size differences. But in real implementation, the HMM cannot completely absorb the break size differences. In the case of IV, with both trained operating mode and trained break size, the identification rate is 100%. This means the trained data can be completely identified.

The on-line test for base model was performed using on-line test data which covers various operating modes and break sizes for each transient. The on-line test results are depicted in Fig. 2. The total identification rate of the on-line test reaches 95.8% within 79 seconds, but the immediate detecting rate is only 44.3%. In the case of LOCA, MSLI, or PORV, the transients are correctly identified from incorrect transients. In the initial stage of these transients, there are no distinctive features between preceding patterns of identified and misidentified transients. The distinctive features appeared after a few seconds, then, the HMM identifier was able to identify the correct transient types. It takes several tens of seconds to detect a SGTR transient because its distinctive features appear after several tens of seconds in a real situation.

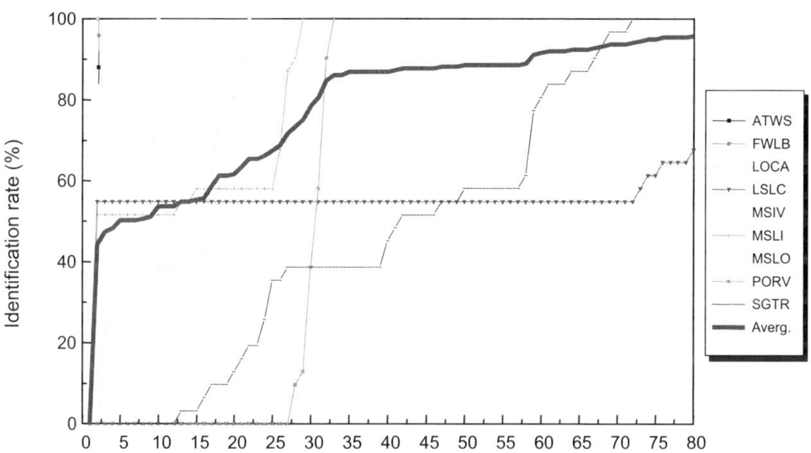

Fig. 2. On-line identification rate for base model

The selected model will be suggested according to the results of the previous experimental tests. The selected model means adopting a proper normalization method, clustering algorithm, and the number of HMM states. According to the results of previous experimental tests, the selected model has Max-min normalization method, k-means clustering algorithm, and six states of left-to-right HMM. The normalization methods are compared based on the assumption that the input normalization might influence the result of classification. The Max-min method has advantages in the present case. Consequently, the Max-min method is chosen for the selected model. To choose the proper clustering algorithm, two types of clustering methods are compared. Considering the results, the *k*-means algorithm has better performance than SOM. Finally, the *k*-means algorithm method with training data is chosen to implement the selected model. There are no effect on the identification rate based on the number of states in the HMM. The selected model adopts 6 states as the number of state in the HMM.

The total identification rate of the selected model is 100% within 17 seconds, and its immediate detecting rate is 54.0%. The total identification and immediate detecting rate of the selected model is higher than the base model. In the selected model, the detecting rate of correct transients from incorrect transients is significantly reduced from the base model. According to the results of the above experimental tests, it can be concluded that the selected model has good performance. But it cannot be said that the selected model is robust. Therefore, three more experimental tests such as superimposing random noise, adding systematic error, and testing by untrained transients were performed to verify a robustness of the selected model.

The random noise is superimposed on to the original test data except two variables such as position of the main steam isolation valve and reactor trip signal. The terminology "2% of random noise" means maximum ±2% of random noise is superimposed on to test data in the normalizing progress. The test results of superimposing random noise are depicted in Fig. 3. In the case of 2% to 8% of random noise, the identification rates are same as no random noise case. The identification rate decreases slightly at 10% of random noise. But, the identification rate is abruptly decreased more than 12% of random noise. It can be said that the proposed transient identification system is robust within 10% of random noise. In neural network applications, the recognition rate is linearly decreased by increasing the noise level [2], or the recognition rate is maintained to 16% of noise level, then the rate abruptly decreases [17].

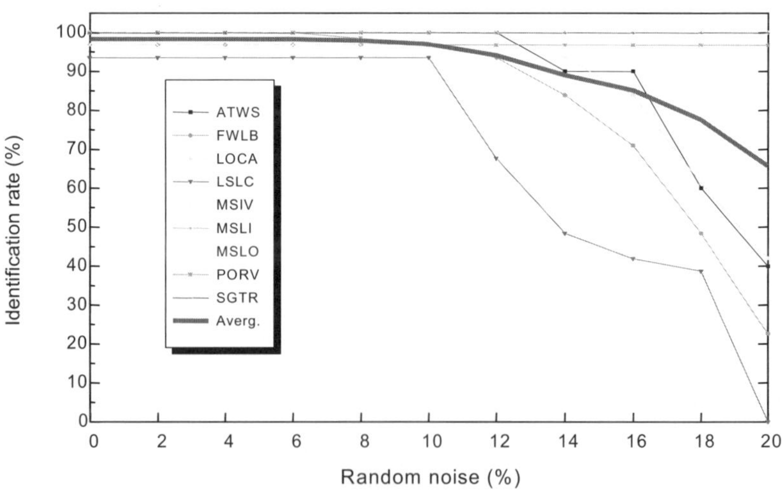

Fig. 3. Test results of superimposing random noise

Additional tests are performed where the systematic error signal is added differently from the random noise. The first error signal is a "loss of pressurizer level signal" which has large variations when the transients occur. In this case, the identification rate is significantly lower than with no error signal. From this test result, to ensure the performance of the transient identification system, valid signals through a signal validation process should be provided. In the application of a neural network case, the system can identify the transient properly even though some sensor signals are missed [5]. But the second error signal is a "loss of main steam flow signal" that has small variations when the transients occur. In this case, the identification rate is the same as if there was no error signal. This means the error signal which has small variations have almost no effect in identification rate.

It is desirable that the never-trained transient should be classified as an unknown transient. In particular, a severe accident can happen because of an inadequate operation due to incorrect identification. The implemented system can classify the unknown transient like the other neural network applications [3, 9, 12]. It is classified as unknown transient that the output path probability of the HMM identifier is less than the threshold which is the least output path probability from all test data. The test results of untrained transients are summarized in Table 3. As shown in Table 3, three untrained transients are classified as unknown transients. But four untrained transients are identified as normal states. In the case of the incorrect classification, the HMM identifier may not be exactly identified because the clustering results are very similar to the misclassified transient.

Table 3. Test results of untrained transient

Untrained transient description	Reactor power (%)	Test result
Drop of all control rods in control bank group A	95	Unknown
Reactor coolant pump trip	88	Unknown
Charging line rupture	78	Normal state
Turbine control valve failure (close)	90	Unknown
Steam line rupture (100cm^2)	100	Normal state
Failure in feedwater controller	100	Normal state
Failure in pressurizer controller: maximum failure	95	Normal state

In the selected model, there still exist temporary misclassified transients. Finally, two heuristic training approaches are attempted to improve the classification accuracy. One of these approaches is corrective training [18]. The heuristic corrective training method re-estimate model parameters using validation data set which are not included in training or test data set. After re-estimation of HMM parameters using training data set, perform transient identification on the validation data. If any transient is misclassified, adjust the estimated model parameters to reduce the probability of misclassified transients.

The other heuristic approach is principal component method. The principal component is applied to increase discriminating power between two transients which are expressed by similar patterns. This method adds weighting factor to important variables that have large impact on identification results when vector quantization is progressed. Consequently, the classification rate is slightly increased when the principal component method is applied. In conclusion, Fig. 4 shows the classification rate of three on-line tests depend on time. The classification rate of the selected model is much better than the base model, and the classification rate of improved model is a little better than the selected model.

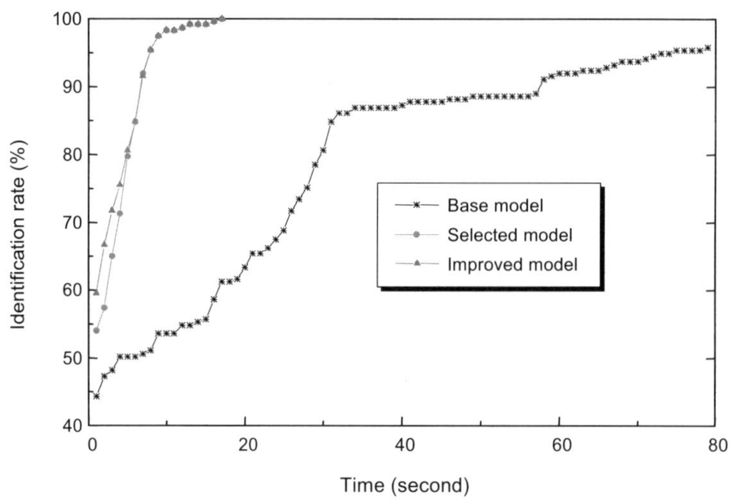

Fig. 4. On-line identification rate for three models

5 Conclusions

The proposed transient identification system has lots of advantages as described above. However, there are still a lot of problems that should be solved before it can be actually applied to an operating NPPs. Further efforts are being made to improve the system performance and robustness to demonstrate reliability and accuracy to the required level. It is a severe drawback that the training data are extremely rare compared with other applications. The HMM-based identifier is more difficult to train properly since they tend to require more training data. It is more desirable to get additional training data, however, it is a difficult job in this area. Therefore, research for the modified HMM structure or new algorithm for the estimation of HMM parameters is suggested. As shown in experimental tests, the identification performance heavily depends on clustering methods. It is important to find more efficient clustering method which is suitable for the proposed system. The HMM-based identifier cannot completely resolve the "don't know" issue which is the hot topic in this research field. It should be considered the "don't know" issue in the future. It also needs to seek a fitness measure to confirm the degree of belief. Besides improvements described in the improved model, it needs to be integrated with other features such as knowledge processing [19] or neural network [20] to improve its accuracy.

Acknowledgement

This work has been carried out under the nuclear research and development program by Ministry of Science and Technology.

References

[1] I. S. Kim, "Computerized Systems for On-line Management of Failures: A State-Of-The-Art Discussion of Alarm and Diagnostic Systems Applied in the Nuclear Industry," *Reliability Engineering and System Safety*, Vol.44, pp.279-295, 1994.

[2] A. Ikonomopoulos, R. E. Uhrig, and L. Tsoukalas, "A Hybrid Neural Network-Fuzzy Logic Approach to Nuclear Power Plant Transient Identification," *AI91 Frontiers in Innovative Computing for the Nuclear Industry*, pp.217-226, Jackson, Wyoming, Sep. 15-18, 1991.

[3] Y. Bartal, J. Lin, and R. Uhrig, "Transients Identification in Nuclear Power Plants Using Probabilistic Neural Networks and the Problem of Knowledge Extrapolation," *9th Power Plant Dynamics, Control & Testing Symposium*, pp.49.01-49.08, Knoxville, TN, USA, May 24-26, 1995.

[4] Z. Guo, R. E. Uhrig, "Accident Scenario Diagnostics with Neural Networks," *8th Power Plant Dynamics, Control & Testing Symposium*, pp.53.01-53.11, Knoxville, TN, USA, May 1992.

[5] S. W. Cheon, "Application of Neural Networks to a Connectionist Expert System for Transient Identification in Nuclear Power Plants," *Nuclear Technology*, Vol.102, pp.177-191, May 1993.

[6] T. Iijima, *Application Study of Fuzzy Logic method for Plant-State Identification*, HWR-432, OECD Halden Reactor Project, Dec. 1995.

[7] J. Lin, Y. Bartal, and R. Uhrig, "Using Similarity Based Formulas and Genetic Algorithms to Predict the Severity of Nuclear Power Plant Transients," *9th Power Plant Dynamics, Control & Testing Symposium*, pp.53.01~53.09, Knoxville, TN, USA, May 24-26, 1995.

[8] E. Jeong, K. Furuta, and S. Kondo, "Identification of Transient in Nuclear Power Plant Using Adaptive Template Matching with Neural Network," *Nuclear Plant Instrumentation, Control, and Human-Machine Interface Technologies*, The Penn State Univ. USA, pp.243-250, May 6-9, 1996.

[9] J. W. Hines, D. W. Miller, and B. K. Hajek, "A Hybrid Approach for Detecting and Isolating Faults in Nuclear Power Plant Interacting Systems," *Nuclear Technology*, Vol.115, pp.342-358, Sep. 1996.

[10] P. Smyth, "Tutorial Material; Machine Learning: Theory and Application," *The 3rd World Congress on Expert Systems*, Seoul, Korea, Feb. 5-9, 1996.

[11] X. D. Huang, Y. Ariki, and M. A. Jack, *Hidden Markov Models for Speech Recognition*, Edinburgh University Press, Edinburgh, 1990.

[12] Y. Bartal, J. Lin, and R. Uhrig, "Nuclear Power Plant Transient Diagnostics Using Artificial Neural Networks that Allow "Don't-Know" Classifications," *Nuclear Technology*, Vol.110, pp.436-449, Jun. 1995.

[13] C. Couvreur, *Environmental Sound Recognition: A Statistical Approach*, Ph. D Thesis, Faculté Polytechnique de Mons, Belgium, June 1997.

[14] L. R. Rabiner, "A Tutorial on Hidden Markov Models and Selected Application in Speech Recognition," *Proceedings of the IEEE*, Vol.77, No.2, pp.257-285, Feb. 1989.

[15] T. Kohonen, "The Self-Organizing Map," *Proceedings of the IEEE*, Vol.78, No.9, pp.1464-1480, Sep. 1990.

[16] K. C. Kwon, S. J. Song, W. M. Park, and S. P. Lyu, "The Real-time Functional Test Facility for Advanced Instrumentation and Control in Nuclear Power Plants," *IEEE Transactions on Nuclear Science*, Vol.46, No.2, pp.92-99, April 1999.

[17] E. Jeong, K. Furuta, and S. Kondo, "Identification of Transient in Nuclear Power Plant Using Neural Network with Implicit Time Measure," *Proceedings of the Topical Meeting on Computer-Based Human Support Systems*, pp.467-474, Philadelphia, PA, USA, June 25-29, 1995.

[18] T. Applebaum and B. Hanson, "Enhancing the Discrimination of Speaker Independent Hidden Markov Models with Corrective Training," *Proc. ICASSP-89*, pp.302-305, Glasgow, Scotland, May 1989.

[19] Y. Ohga and H. Seki, "Abnormal Event Identification in Nuclear Power Plants Using a Neural Network and Knowledge Processing," *Nuclear Technology*, Vol.101, pp.159-167, Feb. 1993.

[20] A. Kundu and G. Chen, "An Integrated Hybrid Neural Network and Hidden Markov Model Classifier for Sonar Signal Classification," *ICASSP*, pp.3587-3590, 1995.

8 Expert System-Based Implementation of Failure Detection

Jung-Taek Kim, Kee-Choon Kwon, In-Koo Hwang, Dong-Young Lee, Jung-Woon Lee, Sang-Jeong Lee*

Korea Atomic Energy Research Institute,
P.O. Box 105, Yusong, Taejon 305-600, Republic of Korea
*Chungnam National University,
220 Kung-Dong, Yusong, Taejon 305-764, Republic of Korea

jtkim@kaeri.re.kr

In light of the need to improve MMIS of NPPs, the advanced I&C research team of KAERI has embarked on developing an Alarm and Diagnosis-Integrated Operator Support System, called ADIOS. ADIOS was composed of two modules such as alarm process module and fault diagnostic module. The alarm process module can offer key alarms to operator using dynamic alarm filtering methods and the fault diagnostic module was able to diagnose the component failure caused by a sensor fault and a hardware fault. ADIOS has been built in an object-oriented AI environment of G2 expert system software tool. In the G2 environment, every alarm and diagnostic component have been treated as an object of class. The attributes of each alarm object include activation status, alarm message, process value, activated time, priority, acknowledgment state, interlocked equipment and caused alarm. If an alarm is activated, its icon color on an overview process mimic diagram changes corresponding to its priority, which can be set initially or determined dynamically by reasoning rules and procedures. The process conditions, such as plant or equipment status and correlated alarms' states have determined the priority of the activated alarm and the fault diagnostic condition. The knowledge base and the inference engines of ADIOS have been constructed by process analysis of the plant and discussion with operators and nuclear plant experts.

1 Introduction

When transitions are occurred in large and complex systems such as nuclear power plants or industrial process plants, it is often difficult to diagnose them.

Various computer-based operator-aiding systems have been developed in order to help operators diagnose the cause of transitions of the plants, such as CASH (Moum et al. 1996) and fault diagnosis systems such as PRODIAG (Reifman et al. 1999). However, most of the operator support systems carried out only a function of alarm processing and the other only a function of a fault diagnosis (Lupton et al. 1992, Kim 1994, O'Hara et al. 1994). ADIOS (Alarm and Diagnosis-Integrated Operator Support System) has been developed to perform functions from alarm process to fault diagnosis (Hwang et al. 1996; Kim et al. 1996, 1997, 2001). ADIOS has been composed of two modules: an alarm process module and a fault diagnosis module. The alarm process module can offer key alarms to operator using dynamic alarm filtering methods and the fault diagnostic module was able to diagnose the component failure caused by a sensor fault and a hardware fault. ADIOS has been built in an object-oriented AI environment of G2 expert system software tool (Ham et al. 1995).

The bases of implementing any expert system in G2 are production-system (IF-THEN-ELSE types of rules) and an object-oriented knowledge representation scheme. Various equipment of a nuclear power plant, such as condensate pumps or the pressurizer, and various alarms, like process alarms or temperature alarms, are hierarchically defined in G2 to take the advantage of inheritance of object properties. The attributes of each class or object are then determined to facilitate knowledge-based processing of alarm signals, sometimes augmented by raw process parameters.

This paper discusses the architecture of the knowledge base of ADIOS, focusing on alarm and diagnostic processing, along with the inference scheme. Also discussed herein are the advantages of, and some issues in, developing an improved alarm and diagnostic system in an object-oriented environment, and also using a real-time AI tool, such as G2.

2 ADIOS Configuration

ADIOS has been composed of two main modules such as an alarm process module and a fault diagnostic module and a few of display-units as shown in Figure 1. The two main modules have been implemented with the each expert development tool, G2 tool (Gensym, 1995); the alarm process module is implemented with G2-1 and the fault diagnosis module with G2-2. Each of ADIOS module receives the simulator data from I&C Functional Test Facility (FTF) through GSIs (G2 Standard Interface) (Kwon et al. 1999). I&C FTF simulates the process of Kori unit 4 nuclear power plant and sends the simulation data to each processing modules. The alarm process module checks each signal of the plant whether or not the signal exceeds the setpoint of it. The module sends a "diagnosis starting signal" to a fault diagnosis module through a G2-To-G2-Interface when alarms are fired. After performing the alarm process and the fault diagnosis, two modules send results to those display units.

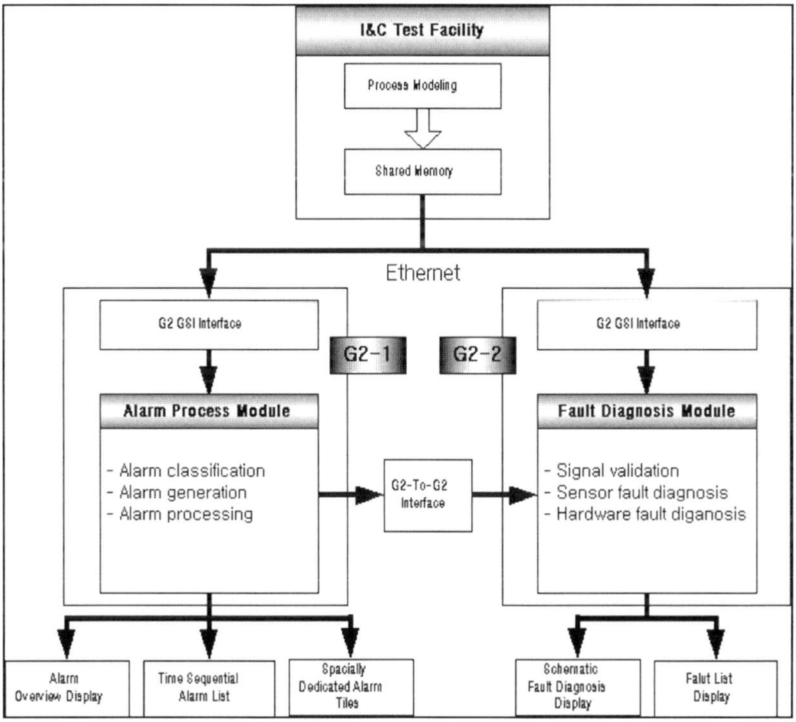

Fig. 1. ADIOS Configuration

3 ADIOS Alarm Process Module

3.1 Alarm Classification

Alarm Classification

The alarm management in ADIOS is based on an alarm classification scheme which clearly distinguishes the types of alarms based on their inherent characteristics. The alarms are classified into three major categories: 1) process alarms, 2) status alarms, and 3) plant alarms.

The process alarms are either main or auxiliary process alarms, depending on whether they belong to the main process (e.g., reactor coolant system, steam generators, condensate and feedwater systems, and main steam system) or to auxiliary systems (e.g., component cooling water system, nuclear service cooling water system, circulating water system, or turbine plant cross cooling water system). The status alarms are divided into equipment-related alarms, equipment-status alarms, equipment-status information, system-status information, and system-trouble alarms. An equipment-related alarm indicates an abnormality in the support function of equipment, for example, *pump bearing temperature high.* And, an equipment-status alarm is an indication that the equipment is at a state that it is not supposed to be in. *Pressure relief valve not opening* is an example of equipment-status alarm. For equipment-status alarms, a clear distinction was made between status information and status alarms. The equipment/system-status information is operating information of equipment or system such as a *Valve Open*, a *System Actuation*, etc. The status information and status alarms are presented to the operator separately, as opposed to the conventional alarm systems where they are intermingled. System trouble alarms include most of *Trouble/Disable* alarms which are being used in current conventional power plants. Important alarms at plant level, such as containment pressure, are classified to the plant alarms.

Definition of Alarm Objects

Every alarm is defined as an object of a subclasses of ***Alarm*** class, the attributes of which include message text, process value, set-point, activation status, priority, acknowledgment or reset status, causal alarm, level precursor, and so on, according to its class. Subclasses of alarms are defined for different use in the processing scheme of ADIOS. For example, process alarms, e.g., a pressure alarm in the main process line, and equipment alarms, e.g., a vibration high alarm.

Each alarm object with those attributes contains most of the information necessary for alarm processing and display control. Some attributes of the alarm object change their values dynamically during the run time of the alarm system. The process value of an alarm gets its value from the corresponding process variable of the plant or simulator. The attribute value of the acknowledgment or reset status is used to control the flashing display depending on the acknowledgment status of the alarm when it is activated or deactivated. Table 1 illustrates an attribute table of an alarm object. The attributes, causal alarm and level precursor, are used in prioritizing the alarms based on the relationship among alarms. The processing of alarms is discussed below in more detail.

For implementing the state-dependency, ***relation*** provided in G2 has been used. Any alarm object, which can be active as a result of any equipment state, for example, pump ON or pump OFF, is defined to have a relation to its corresponding equipment.

Table 1. Attribute table of an alarm object

PZR-PRESS-HI, a pressure-alarm	
Notes	OK
Item configuration	none
Names	PZR-PRESS-HI
System name	prmy-rcs-pzr
Alarm source	valid
Tile message	"Pressurizer Press High"
P value	83.3
Status	off
Acknowledge or reset	initialized
Priority	1
Default priority	1
Mode	all
Setpoint	165.2
Kind	hi-alarm
Interlocked equipment	none
Causal alarm1	none
Causal alarm2	none
Causal alarm3	none
Level precursor	pzr-press-dev-hi
Time on	"13:14:27"
Time off	"13:15:26"
Range	0-210

3.2 Alarm Processing

Alarm Prioritization

ADIOS uses various alarm processing techniques such as equipment-state dependency, plant-mode dependency, alarm generation, cause-consequence relationship, and multi-setpoint relationship. In addition, other methods including categorization of alarms into the process alarms (e.g., temperature or pressure alarms of the main process) and equipment-related alarms (e.g., vibration or lubrication alarms of a pump), presentation of status alarms (e.g., PORV not closed) on the process mimic, and representation of group alarms assimilating information from several related alarms are key features of ADIOS. Figure 2 illustrates how the alarms are processed and presented in ADIOS.

In conventional alarm systems, alarms are generated by setpoint checking. They are activated when the associated process values exceed the alarm setpoints, and deactivated when they return to their normal values.

The activated alarms then get into the prioritization phase to conclude their priority depending on several conditions related to them. Those conditions would be plant operation mode, equipment status, related alarm status, and so on. In the present version of ADIOS, all alarms are initially given their own default priorities, and those priorities can be decreased or increased dynamically by any processing algorithm during the run time of the alarm system.

The plant-mode dependency is used to de-emphasize those alarms that are activated as a consequence of the plant mode change. The equipment-state dependency is used to reduce the priority of those alarms that occur when equipment changes its status; e.g., the priority of the discharge pressure low alarm is suppressed or lowered if it occurs after a pump stops. The multiple setpoint relationship uses the relationship between several alarms on the same process parameter. For instance, when both the low and low-low level alarms of a steam generator are on, the priority of the low alarm can be lowered. The causality between alarms also allows us to prioritize alarms between causal and consequential alarms; the causal alarms require more attention than the consequential alarms.

Alarm Display

The presentation of alarms should conform to the operators' mental model of the process as much as possible. The alarm presentation in ADIOS aims at the operator's fast recognition of the overall plant or alarm situation, and also an easy access to any detailed alarm information. For this purpose, the prioritized alarms are displayed on the process overview mimic, Figure 3, and also the chronological list of alarms is presented on another dedicated video display unit (VDU), with those alarms categorized by systems shown on the third VDU as a spatially dedicated soft alarm panel. Priority 1, 2, or 3 alarms are shown differently in red, yellow, or white, respectively. The same color coding is applied to the alarm texts in the alarm list, and also to the window tiles on the soft alarm panel.

Intimately correlated process alarms are represented on the process overview mimic after being coalesced into groups. The group alarms, such as *"T_{avg}," "ΔT," "RX," and "L,"* represent several related alarms. For instance, the *"L"* group alarm indicates that it represents a deviation in the steam generator (SG) level, high-high, high, low, and low-low SG level alarms. The group alarms take the highest priority among the associated subsidiary alarms that have been activated. For instance, if both *"high-high"* and *"high SG Level"* alarms are active and their priorities are 1 and 2, respectively, then the priority of the *"L"* alarm is set to 1.

Contrary to the above case, some alarm information is combined into one window unit in a conventional alarm system. When *SG 1 Water Level Deviation High/Low* is activated, an operator should check the SG level indicator if he wants to know whether it is a high alarm or low alarm. If the alarm, *System AL Non TRN TROU/DISA,* is active, it is not easy to find out which component or actuator is *TROUBLE* or *DISABLE*. ADIOS resolves those kinds of the combined alarms and presents more detailed messages.

Status alarms (e.g., PORV not closed) are not directly shown on the overview mimic if no such alarms are active. However, for example, when the power operated relief valve (PORV) is open while it should be closed, the status alarm of PORV is indicated by changing the body color of the valve to red; the red color will blink until the operator acknowledges the alarm by clicking on the valve. The PORV status alarm is activated when the valve should have been closed because the pressurizer pressure decreased below its setpoint, but has not been closed, or vice versa.

When an equipment-related alarm, such as a lubrication alarm, is activated, the alarm is shown on the process overview mimic by changing the boundary color of the equipment to red, yellow or white. When the operator wishes to look at the specific alarms activated, he or she can click on the equipment icon after first acknowledging the alarmed condition. Then, the specific alarms are shown on a subsidiary window.

Alarm lists are provided on a dedicated screen in time sequence. The operator can choose only a certain priority of alarms that are of interest to him in a given situation, either single priority or multiple priorities. A soft alarm panel, namely, a computer-based panel of alarm tiles arranged in the similar way as in conventional alarm systems, is also provided to take advantage of the spatial dedication of alarm information. These alarms are arranged in systems, after some grouping and sorting of the conventional alarms.

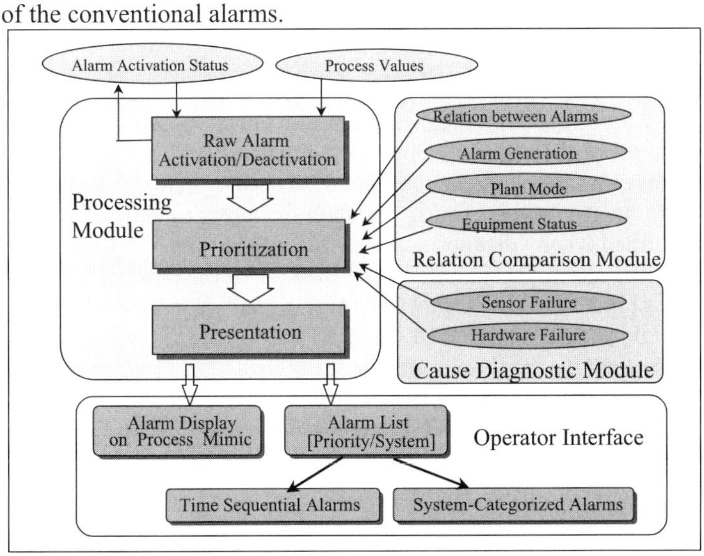

Fig. 2. Processing flow for ADIOS

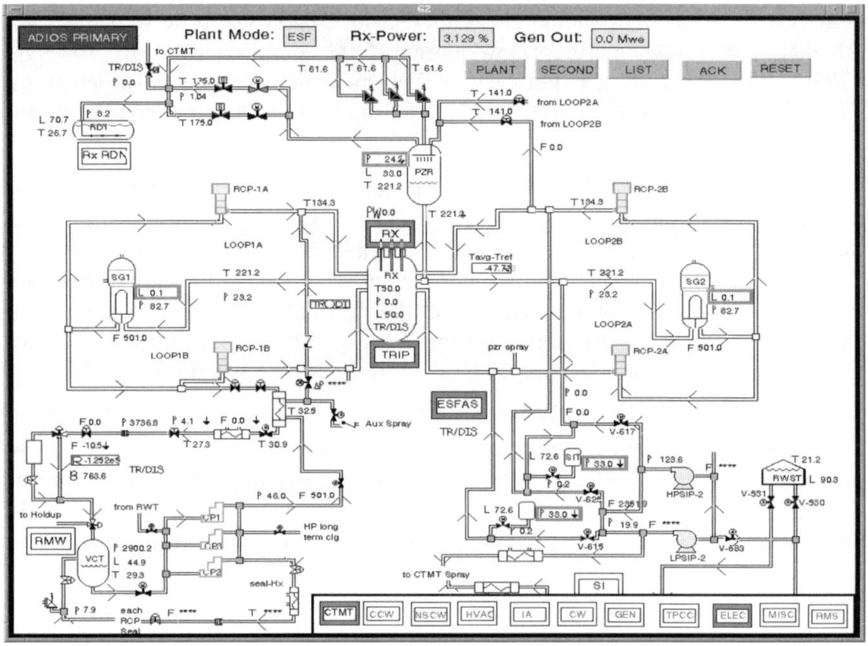

Fig. 3. Primary overview process schematic display

4 ADIOS Fault Diagnostic Module

The fault diagnosis module has two sub-modules such as a sensor fault diagnosis module and a hardware fault diagnosis module. For monitoring transient states of entire plants, all of the sensors belong to the plant should be inspected and diagnosed. However it is impossible substantially because there are too many sensors to monitor in nuclear power plants. The fault diagnosis module of ADIOS, therefore is performed when "diagnosis starting signal" attributed by the fired alarm is passed from the alarm process module. When an alarm is fired in alarm process module, the object named "diagnosis starting signal" is generated in the fault diagnosis module and the sensor fault diagnosis sub-module is started automatically. The hardware fault diagnosis sub-module is performed if there are not faulted sensors. We applied the fault diagnosis module to the feedwater system from Kori Unit 4 Plant in Korea, a 3-loop Westinghouse PWR as shown in Figure 4.

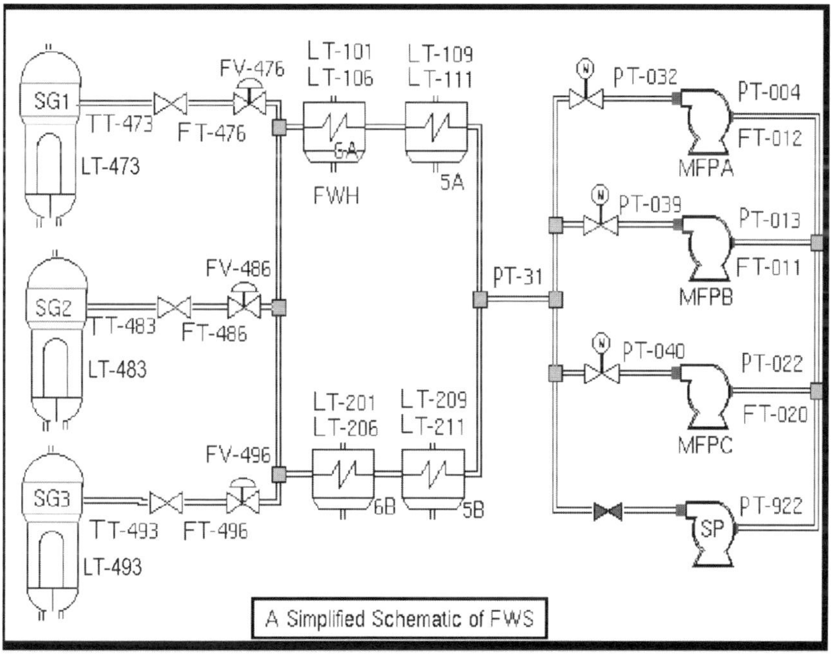

Fig. 4. A simplified schematic of FWS

4.1 Sensor Fault Diagnosis Module

The detection of sensor faults is based on the effective use of coherent relationships among the process variables and parameters, i.e. sensor validation criteria (SVC). SVCs are formulated by using the algorithms such as the thermal-hydraulic equations and redundancy among sensors. A defected sensor is determined automatically using the valid SVC and the invalid SVC. Fourteen SVCs are provided by the coherent relationships among the process and parameters of the FWS as shown in Figure 4. Table 2 represents the simplified expressions of SVCs, the bases, and the conditions necessary to apply the criteria.

Figure 5 shows one of the flow paths to detect sensor faults for pressure transmitter PT4. If SVC 5 is satisfied, the PT4 sensor is considered as "valid." If it is not met, then a diagnosis of sensor faults will be initiated. First of all, an operating condition will be checked. If MFPA (main feedwater pump A) and MFPB are operating, and SVC 2 and SVC 9 are "yes," then the candidates of false sensors, CFS = {PT32, PT4, FT2, STA} and the candidates of valid sensors, CVS = {PT31, PT32, PT39, FT476, FT486, FT496, FT2, FT11}. Finally, the false sensors, FS, are composed of {PT4, STA}. Therefore the suction pressure of MFPA (PT4) and the turbine speed of MFPA (STA) are considered as faulted sensors. Then, one of two sensors will finally be decided as sensor failure through a characteristic curve of MFPA.

Table 2. Sensor validation criteria

SVC	Expression	Basis	Condi-tion
1	PT31 PT32 PT39 PT40	Pressure Balance	A,B,C
2	PT31 PT32 PT39	Pressure Balance	A,B
3	PT31 PT32 PT40	Pressure Balance	A,C
4	PT31 PT39 PT40	Pressure Balance	B,C
5	PT32 PT4 FT2 STA	MFPA Curve	
6	PT39 PT13 FT11 STB	MFPB Curve	
7	PT40 PT42 FT22 STC	MFPC Curve	
8	FT476 486 496 FT12 11 20	Mass Balance	A,B,C
9	FT476 486 496 FT12 11	Mass Balance	A,B
10	FT476 486 496 FT12 20	Mass Balance	A,C
11	FT476 486 496 FT11 20	Mass Balance	B,C
12	LT473 LT474 LT 475 LT 475	Level Redundant	
13	LT483 LT484 LT 485 LT 485	Level Redundant	
14	LT493 LT494 LT 495 LT 495	Level Redundant	

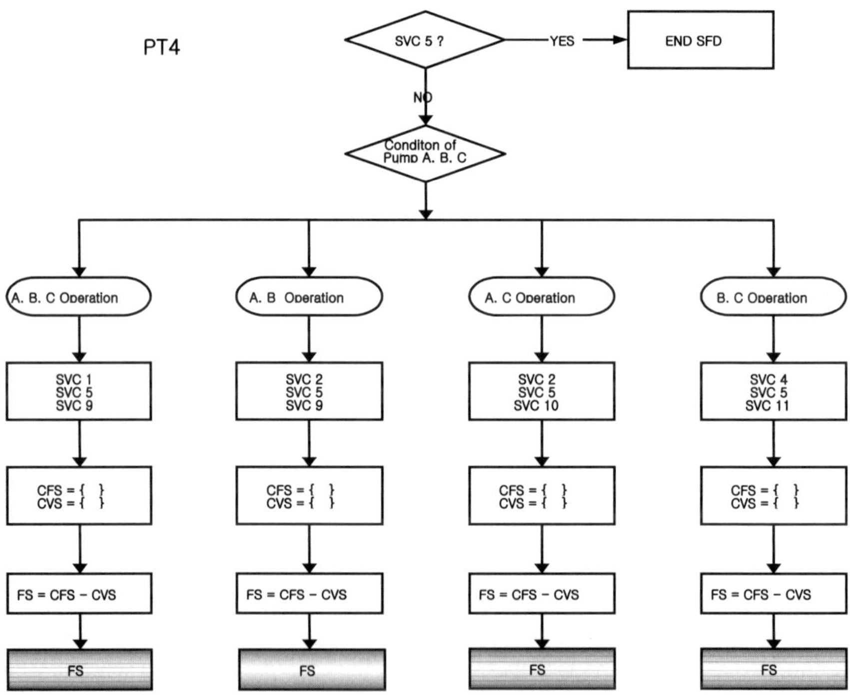

Fig. 5. Flow path to detect sensor fault, PT4

4.2 Hardware Fault Diagnosis Module

When the transient signal is inputted in this fault diagnosis module, first, the sensor fault is detected by using the sensor fault diagnosis module. If there are not defected sensors, the hardware fault diagnosis module is preformed in order to find the causal of hardware fault such as 'valve open' or 'tube break.'

The hardware fault is detected with the signed digraph (SDG) and the disturbance patterns of a sensor group. The SDG represents the fault propagation structure in terms of the essential causalities among the process variables and parameters. The SDG of FWS is shown in figure 6. The disturbance patterns of the sensor group represent the changes of a sensor group such as a pressure or a water level. Sensors, PT32, PT39, and PT40 can be gathered into a group in order to represent the situation of MFPs. The changes of each sensor in a group are presented to "High," "Normal," and "Low" according to the value of sensor. For example, the disturbance pattern, PT32-L PT39-N PT40-H, represents that PT-32 is low, PT-39 normal, and PT-40 high.

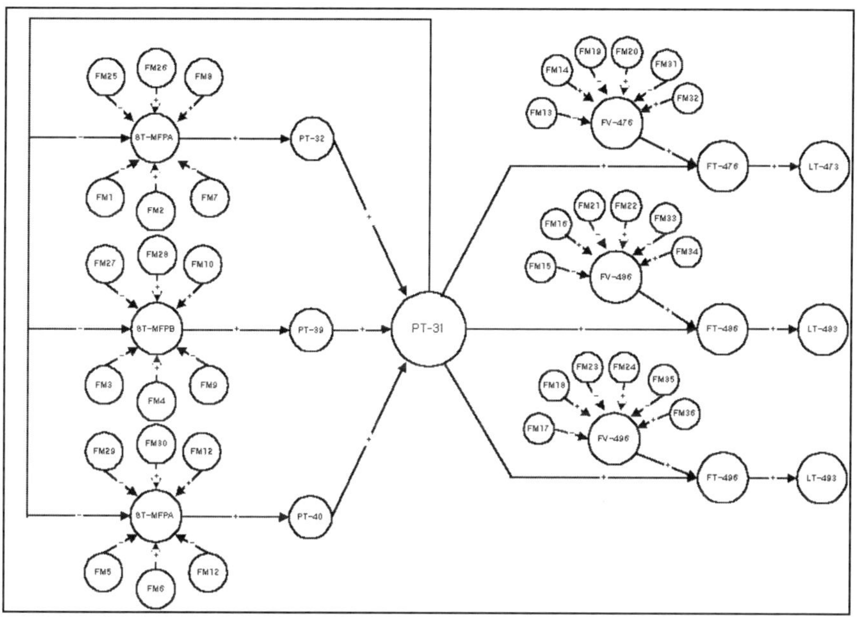

Fig. 6. SDG model of FWS

For diagnosis of hardware fault, we draw the SDG of a target system and define the failure modes of the system, and then connect the failure mode to components of SDG according to their relationships. If the failure mode effects negatively on the connected component, their relationship is "-," and if positively, "+." Then, we check the disturbance of sensor through fault propagation paths on SDG starting from a failure mode. And failure modes are classified into disturbance patterns of a group sensor. For example, Table 3 is shown the failure mode of pressure disturbance patterns PT32-L, PT39-L, PT40-L. The signal of sensors is changed to "H," "N," or "L" according to its disturbance when fault diagnosis sun-module is started. Failure modes matched to the disturbance pattern of a sensor group are searched in the sub-module by using the pattern matching method.

Table 3. Pressure Disturbance Pattern

PT32-L, PT39-L, PT40-L

FM	On-line verification	HFD message
1	SA auto; SB, SC manual; SA output low	Output of Speed Controller of MFPA is Low
3	SB auto; SA, SC manual; SB output low	Output of Speed Controller of MFPB is Low
5	SC auto; SB, SA manual; SC output low	Output of Speed Controller of MFPC is Low
7	SA auto; SB, SC manual; SA setpoint low	SP Speed Controller of MFPA is Low
9	SB auto; SA, SC manual; SB setpoint low	SP Speed Controller of MFPB is Low
11	SC auto; SB, SA manual; SC setpoint low	SP Speed Controller of MFPC is Low
25	SA auto; SB, SC manual; MFPA speed low	Speed of MFPA is Low
27	SB auto; SA, SC manual; MFPB speed low	Speed of MFPB is Low
29	SC auto; SB, SA manual; MFPC speed low	Speed of MFPC is Low

5 Benefits of G2 Object-Oriented Expert System Tool

G2 is composed of a knowledge-base, real-time inference engine, procedure language, development environment, operator interface, and interface to external data servers. In the expert system, the knowledge of the system is incorporated explicitly in a separated part of the program and it is readable and easy to modify because it is built incrementally, where as in every conventional programming language the knowledge is expressed in the ordinary program code (Ola Mattsson 1990).

For implementing ADIOS, G2 has been an effective tool in developing and programming an alarm and diagnostic processing concept. It was very helpful in constructing alarm objects and their attributes as they are treated as an object instance of *Alarm* class at design stage. The other typical advantageous features of using G2 were:

(1) Introductive Guidance of Commands: If a programmer tries to create a rule or procedure for the knowledge-base, G2 shows available texts of program commands or items which can follow next to the current statement. It makes programming much easier than conventional language coding.

(2) Developmental Environment in Graphics: G2 provides easy way to create objects, define classes, edit icons, and inspect the knowledge-base graphically. Therefore it gives the benefit of rapid prototyping in the development stage.

(3) User Interface: By use of buttons and workspaces it is easy to design and construct user interfaces to communicate with the system.

(4) Off-line Simulation: The built-in simulator or procedure can provide effective simulation or off-line testing without any difficulty.

(5) Interface with external systems: G2 exchanges data in an easy way with external systems by GSI (G2 Standard Interface) module.

(6) Distributed Environment: G2 can run an application knowledge base with Tele-windows for multi-users. It makes several users develop an application knowledge base and use it simultaneously.

6 Conclusions

The MMIS team of KAERI has embarked on developing an Alarm and Diagnosis-Integrated Operator Support System, called ADIOS, to filter or suppress unnecessary or nuisance alarms and to diagnose the sensor fault and the hardware fault. In the alarm process module of ADIOS, alarms are processed by several representative methods, including state dependency, mode dependency, and multi-setpoint relationship, which is upgraded in several ways. The processing of alarm signals is clearly seen on the process schematic constructed using the graphic facility of G2. The fault diagnosis module of ADIOS is developed and integrated in ADIOS in order to diagnose a sensor fault and a hardware fault. The sensor fault is detected by sensor validation criteria method and the hardware fault is diagnosed by the signed digraph method.

In developing a G-2 based alarm and diagnostic processing system, several issues need to be clarified. First, the construction of a reliable knowledge base is required for the practical application of the alarm and diagnostic system. The knowledge base should produce a reasonable processing result even for unforeseen plant transients or alarm situation. Secondly, the alarm and diagnostic system should be presented in accordance with the operator's mental model and cognitive processing, rather than a simple reduction in the number of alarms. It also appears important to provide a quick access to any information upon the operator's request, without suppressing or filtering out any alarms. Thirdly, the verification and validation of the expert system software tool is a concern and a barrier to the practical installation of the alarm and diagnostic system. Lastly, the G2 shell does not provide any flexible function to produce various tones of sound.

References

Moum BR, Decurnex C, Fordestrommen NT, Karlsen TV (1996) Olsen TV, CASH: The news alarm system in HAMMLAB, HALDEN Report, HWR-480

Reifman J, Wei RYC (1999) PRODIAG: A process-independent transient diagnositic system-I: Theoretical concepts, *Nuclear Science and Engineering* **131** : 329-347

Gensym Corporation, G2 Reference Manual, Version 4.0, Sep. 1995

Lupton LR, Lapointe PA, Guo KO (1992) Survey of international developments in alarm processing and presentation techniques, NEA/IAEA International Symposium on Nuclear Power Plant Instrumentation and Control, Tokyo, Japan

Ham CS et al. (1995) The development of basic technology for instrumentation and control. KAERI/RR-1503. Korea

Kim IS (1994) Computerized Systems for On-line Management of Failures: A State-of-Art Discussion of Alarm Systems and Diagnostic Systems Applied in the Nuclear Industry. *Reliability Engineering and System Safety* **44**: 279-295

O'Hara JM, Brown WS, Higgins JC, Stubler WF (1994) *Human Factors Engineering Guidance for the Review of Advanced Alarm Systems*, NUREG/CR-6105, U.S. Nuclear Regulatory Commission

Hwang IK et al. (1996) An Object-Oriented implementation to improve Annunciation IAEA Specialists' Meeting (IWG-NPPCI) on Experience and Improvements in Advanced Alarm Annunciation Systems in Nuclear Power Plants, Chalk River, Ontario, Canada

Kim JT et al. (1996) An Evaluation Approach for Alarm Processing Improvement. IAEA Specialists' Meeting (IWG-NPPCI) on Experience and Improvements in Advanced Alarm Annunciation Systems in Nuclear Power PlantsChalk River, Ontario, Canada

Kwon KC et al. (1999) The Real-Time Functional Test Facility for Advanced Instrumentation and Control in Nuclear Power Plants. *IEEE Transaction on Nuclear Science*, **46**, No. 2

Kim JT et al. (1997) An Analysis of the Causal Alarm in Alarm and Diagnosis-Integrated Operator Support System (ADIOS), MARCON97, Knoxville, Tennessee, USA

Mattsson O (1990) An Expert System for OSI, *Doc. No. CODEN:LUTFD2/(TFRT-5428)/1-89*, Department of Automatic Control Lund Institute of Technology, Sweden

9 Detection of Incipient Signal or Process Faults in a Co-Generation Plant Using the Plant ECM System

William Nieman

SmartSignal Corporation,
4200 Commerce Court, Suite 102, Lisle, IL 60532, USA
bnieman@smartsignal.com

Ralph Singer

Singer & Associates,
1602 Brestal Court, Naperville, IL 60565

A 161 megawatt co-generation plant operated by the US Steel Corporation is highly instrumented, providing real-time monitoring of turbine, extraction, generator, condenser and bearing functions, including pressure, valve position, temperature, amperage, voltage and displacement. An equipment condition monitoring (SmartSignal eCM™) system based on an empirical profile of normal operational conditions and utilizing pattern recognition techniques was used to monitor 60 signals on this plant at one-minute intervals. A study of plant data from a two month period shows that the Plant eCM™ fault detection and identification technology generates a high fidelity estimated value for each of the signals and enables early detection of significant, but subtle, variations in pressures, temperatures, flows, and energy level. Alerts annunciated by the system focus attention on groups of signals that can be early indications of equipment failure or process imbalance.

1 Introduction

A 161 megawatt co-generation plant operated by the US Steel Corporation is highly instrumented, providing real-time monitoring of turbine, extraction, generator, condenser and bearing functions, including pressure, valve position, temperature, amperage, voltage and displacement. An equipment condition monitoring (SmartSignal eCM™) system based on an empirical profile of normal operational conditions and utilizing pattern recognition techniques was used to monitor 60 signals on this plant at one-minute intervals. A study of plant data from a two month period shows that the Plant eCM™ fault detection and identification technology generates a high fidelity estimated value for each of the signals and enables early detection of significant, but subtle, variations in pressures, temperatures, flows, and energy level. Alerts annunciated by the system focus attention on groups of signals that can be early indications of equipment failure or process imbalance.

A decision was made to enhance normal equipment monitoring of the co-generation facility following a costly shutdown in the first quarter of 2000. The shutdown was made to implement manufacturer recommended modifications to the generator and to investigate abnormally high wheel chamber pressures in the steam turbine. Prior to the shutdown, plant personnel noticed a variety of abnormal symptoms, and attempted to diagnosis the problem. However, loads and operating conditions on this plant vary greatly as various plant processes change. These variations hid trends of increasing wheel chamber pressures until output capacity of the generator became restricted. Root cause suspects were faulty sensors, malfunctioning steam control valves, and sodium contamination of the turbine blades. Salt contamination was found to be the cause of the generator capacity reduction and was removed during a 4-week outage, which cost $2,000,000 in parts, man-hours, and electric power purchases.

This event caused the company to initiate a search for a technology that would provide early detection and identification of deteriorating conditions so that a problem of this magnitude could be avoided in the future. The equipment monitoring system selected and now in use is the SmartSignal Corporation Plant eCM™. This system monitors turbine, extraction, generator, condenser and bearing functions, including pressures, flows, valve positions, temperatures, amperages, voltages and vibrations (rms displacement). Most of the signals have absolute alarm limits, defined and annunciated by the control system. Signal inputs are sampled at one-minute intervals, acquired from the plant control system and stored on the plant-wide historian.

The operator of the co-generation facility chose SmartSignal plant *e*CM™ as a process monitoring technology for several reasons:

- It provides warning of deteriorating machine or process conditions well in advance of absolute alarm limits;
- The software is signal independent, - it works equally well on any kind of sensor input;
- The software interfaces easily to hardware and software systems already in place;
- The software has the capacity to analyze partially correlated signal sets of any size on a real-time basis;
- The technology was developed and first applied in the nuclear power industry, which has needs and signal environment similar to a co-generation plant

2 Purpose

The purpose of the present study is to demonstrate the uses and effectiveness of equipment condition monitoring using an empirical process model based upon an advanced and highly robust method of pattern recognition (Wegerich et al. 2001). This technology uses historical data to create a reference profile, representing the full range of normal process variability. In the current example, data used to generate a reference profile are limited to a short period of time immediately following the overhaul of the turbine and generator during the maintenance outage. Reference data are processed to determine a matrix of "learned states," representing the full range of variability for all 60 sensors (Gross et al. 1997). During monitoring, all sixty sensors are sampled at every time step and the pattern of relationships existing between the sensor values in this sampled vector is compared to patterns present in the learned state matrix. This comparison is then used to compute a vector of estimated signal levels.

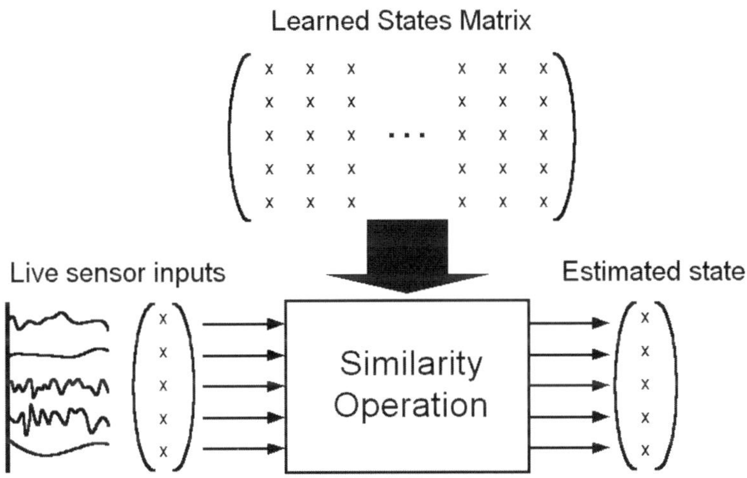

Fig. 1. Schematic description of the Plant eCM technology. Signal values at each time step are compared using a non-linear similarity operator to a reference profile matrix that represents the natural variation in the system. The comparison produces an estimate for each signal. An alert is annunciated if the difference between the signal and estimate are statistically different

The estimated state vector is then compared to the input signal vector and the difference (residual) is evaluated to determine whether the statistical variation of each signal residual is within expected limits or indicates an incipient disturbance. This decision is converted into a status indicator that is immediately available to an operator, enabling corrective actions to be taken that can prevent or mitigate problems. Abnormal states can be detected even when signals are well within absolute alarm limits and normal process ranges.

3 Methods

Because of the impracticality of performing controlled tests on the co-generation system, two periods of process upset from normal data over two months, November and December, 2000, have been selected. (While the data plotted are shown as hourly, the system is monitored at one-minute intervals.) The movement of key process variables and the warning activity of the equipment monitoring software are evaluated to illustrate equipment condition monitoring.

In the data discussed, the observation period begins at hour 0, November 11, 2000, 0:00 (12 AM). The data for November ends at hour 720, November 30, 2000, 23:00 (11 PM). The data for December begins at hour 721 (December 1,2000, 0:00) and continues to hour 1464, December 31, 2000, 23:00. Data for January 1, 2000 (hour 1465) to January 9, 2001, 23:00, hour 1680 finishes the data set. The events discussed use this relative scale from 0 to 1680 hours (70 days).

Event one is a 12-hour long "machine trip" event, from 2:00 to 14:00 on November 4 (hour 74 – 86). The event is characterized by an early "decay" phase, where the generator output reduced from normal output by approximately 25 percent, and a main "degenerate" phase, where generator output is zero for five hours; a two-hour "restart" phase occurs before the system returns to normal operational condition (Fig. 2).

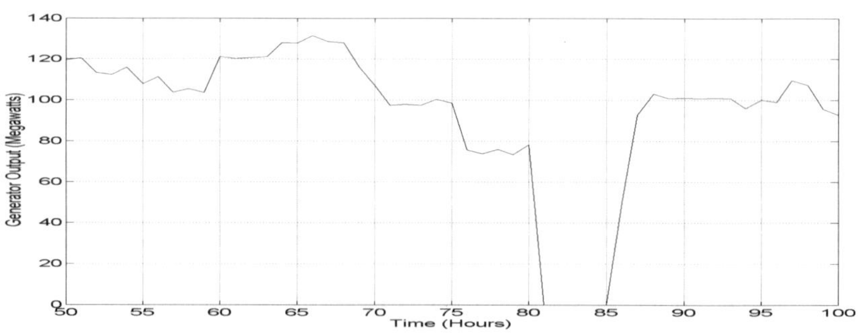

Fig. 2. Generator output (MW) data that includes the event one machine trip, hours 74 – 86

Event two is more complex and extends over a period of anomalous voltage level from December 1, 2000 at 13:00, to December 22, 2000 at 20:00. It includes a period of condenser maintenance, December 11, 14:00 to December 14, 20:00 (Fig. 3). The period defined by these two effects displays anomalous behavior in a number of other signals in the generator system, including extraction steam flow, HP wheel chamber pressure, last stage blade temperature, generator winding temperatures, and high pressure inlet valves (C1, C2, F, H) position. This period was selected to discussion these deviations from the profile norm.

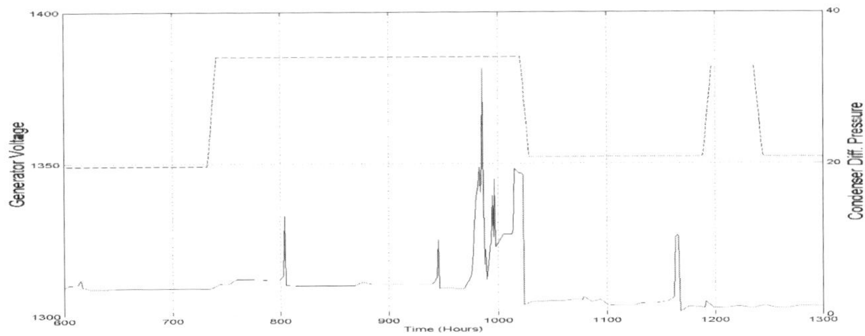

Fig. 3. Data for the generator over-voltage and condenser maintenance periods that define event two, hours' 700 – 1250

4 Description of Results

4.1 Event One

The data presented for this event illustrate the major features of the equipment condition monitoring provided by the Plant *e*CM modeling system in the US Steel co-generation application. These features include:

- High fidelity, real-time estimation of the data signal during normal operation;
- An estimate of the normal signal level when the signal deviates from the reference data normal level;
- Alerts to abnormal behavior on very small deviations from the model condition (Figs. 4, 5 and 7);
- Ability to focus attention on particular signals in the monitored system that may be showing early indications of equipment failure (Figs. 6, 7).

Generator data for event one are presented in Fig. 4. The event occurs between hours 74 and 86. In addition to generator output, high-pressure inlet valve B (Fig. 5), high pressure inlet valve G (Fig. 5), reactive (excitation) energy (VARS, Fig. 6), bearing 1 temperature (Fig. 7), and bearing 2 temperature (Fig. 7), plots are selected from the sixty signals sampled because they show representative, or particularly interesting features. Note that the time interval shown in these plots is expanded compared to that in Fig. 2 in order to show the extent of "normal" behavior that preceded the event of interest.

In each of the plots shown, the actual data signal is plotted as a dark trace; the estimated signal from the empirical model is shown as a light trace. Abnormal behavior is indicated by asterisks at the top of the plot, indicating that the data signal is statistically abnormal - lower or higher than the estimate.

In Fig. 4, the ability of the empirical profile to match the data signal is demonstrated over the interval of zero to 80 hours, after which the generator output level goes to zero, below the "normal" range represented in the profile. Indications of abnormal behavior, or alerts, begin at hour 81 when the data and model signals diverge. Alerts continue through hour 88, the point at which the slope of the data signal trend flattens to its normal output level. The point at hour 88 shows a small residual that still causes an alert at the end of the event.

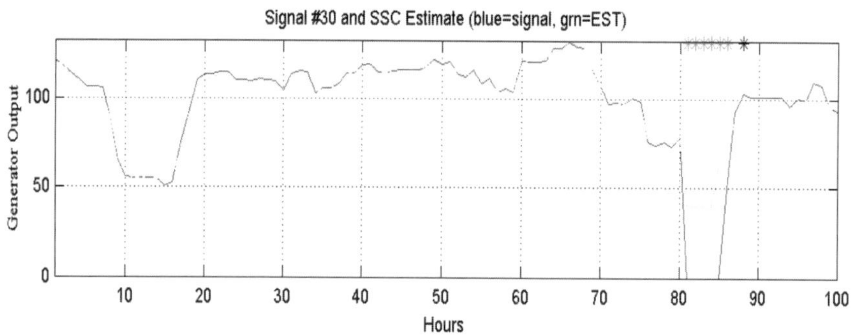

Fig. 4. Generator output (megawatts). Shows normal fidelity of the Plant eCM model estimate and the small residual magnitude that causes an alert to be annunciated between hours 80 and 90

The trends in Fig. 5 show the levels of high-pressure inlet valves B and G during the event, hours 74-86. The level of valve B decreases in nearly the same way that the generator output does, but valve G stays at its pre-shutdown level until the output goes to zero. Yet, in both cases alerts are annunciated ahead of the decrease of generator output to zero, because the inlet pressures are expected by the profile to drop with the generator output.

Compare the patterns of the signal data and estimates for the generator output and high pressure inlet valves B and G during the dip in generator output that occurred between hours 8 and 17, and the machine trip event (hrs 80-90). In the earlier case, no alerts annunciated for the generator output and inlet valve B; alerts are annunciated four hours ahead of the machine trip for valve G, and two alerts occur at the end of the trip before normal operation is restored.

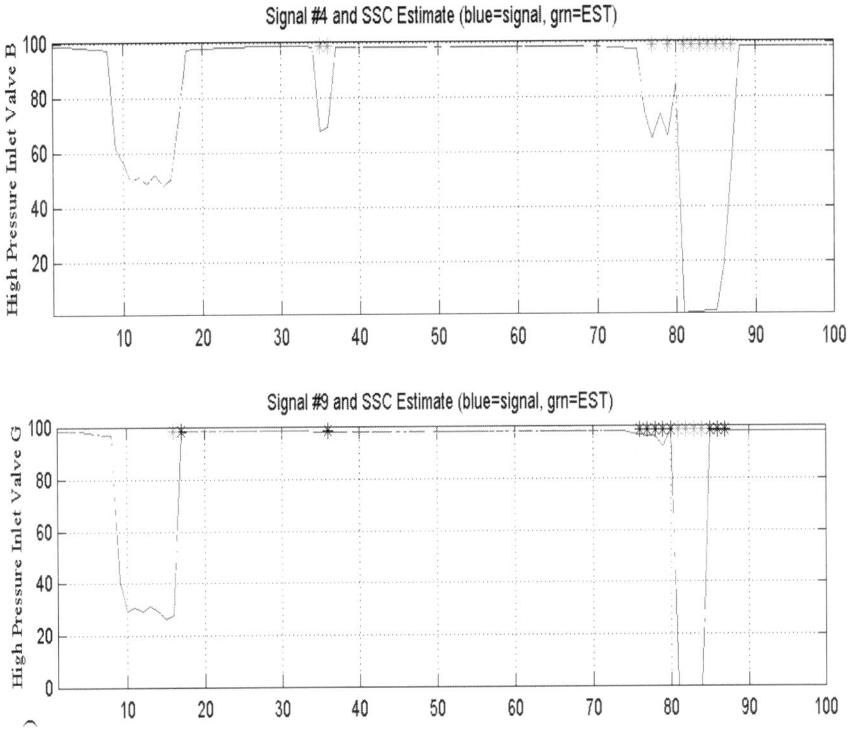

Fig. 5. High-pressure inlet valve positions (percent open), B and G. Shows annunciation of alerts prior to the decrease of generator output to zero. Note correspondence of signal and estimate between hours 75-90

The different response of the empirical model to the two periods demonstrates its ability to recognize normal and abnormal behavior at the selected level of probability – that is, that the magnitude of the residual is normal or abnormal. By estimating the level of a given signal from the level of all other signals in the model (1,2), the model identifies the early pattern as normal and the machine trip pattern as abnormal. Note that valve B apparently alerts because the magnitude of the dip in level is greater than the model predicted to be normal. Valve G did not dip at all until the output level reached zero, but the model expected a drop concomitant with the other two signals.

The variable VARS, reactive (excitation) energy, shows interesting behavior beginning more than ten hours prior to the machine trip. This variable oscillated between a level of 40-60 for the first sixty hours of the study period. It then displayed a mean level increase that was higher than the level recognized by the model as normal. Alerts occurred at 62 hours and continued intermittently up to the machine trip event. The importance of this plot is this: the VARS signal level was within the normal operating range up to hour eighty. Yet the level of the signal, with respect to the level of the other signals in the generator, was perceived to be too high and therefore alerts were annunciated by the modeling software.

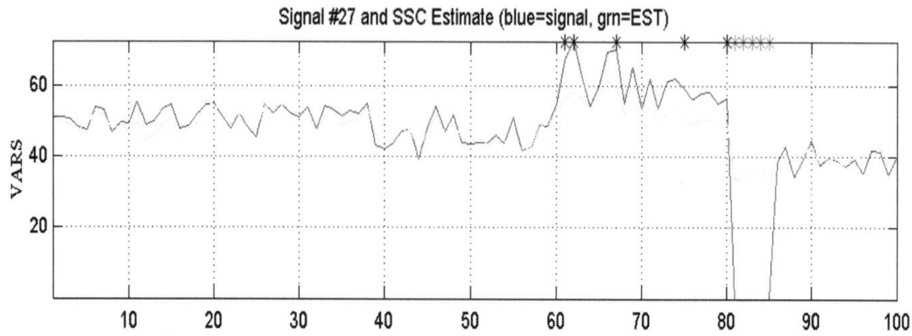

Fig. 6. Reactive energy (VARS). Shows annunciation of sporadic alerts, more than ten hours before the machine trip event, when mean signal level increases

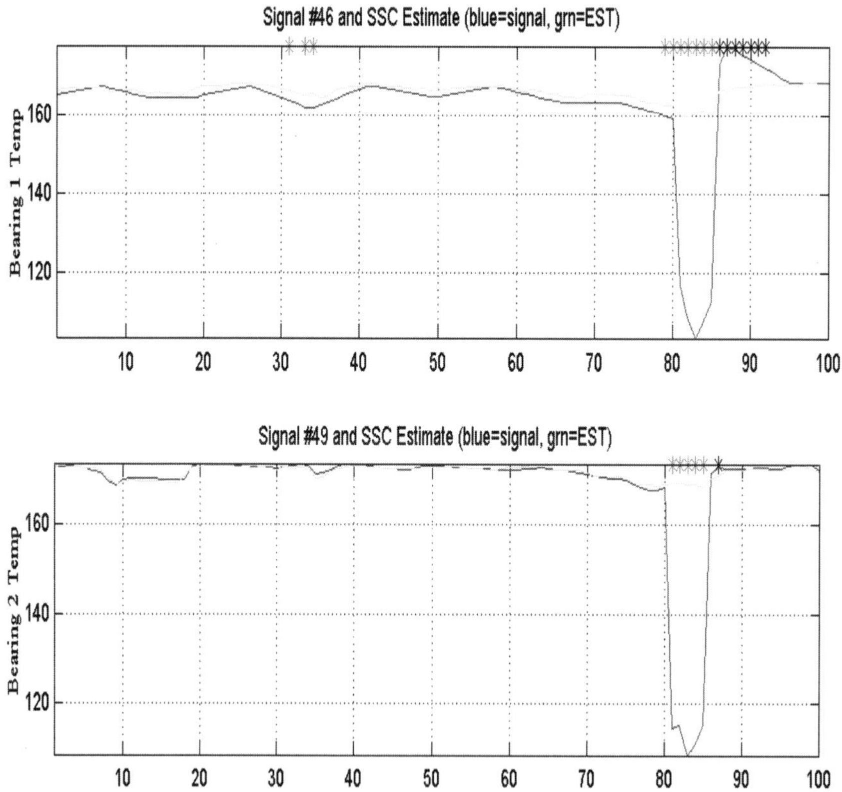

Fig. 7. Comparison of bearing temperatures (bearing 1 and bearing 2, deg. C) in the period of event one. Shows fidelity of the bearing profile estimation and annunciation of alerts during the transient increase in temperature for bearing one following the event, hours 80-90

One final example is given for event one. The trend of temperature for bearing one is shown in Fig. 7. In this plot the bearing temperature is seen to dip well below normal, starting at hour 79. What is notable in this plot is that the modeling software identifies as abnormal the brief (seven hours) excursion of the temperature to higher than normal values immediately after the generator output returns to normal. Bearing 2 temperature does not show such an extreme excursion. The modeling software in this case brings to the attention of the operator the anomalous temperature rise of this bearing one signal.

4.2 Event Two

The second event to be described includes a period of (1) an anomalously high voltage level and (2) a condenser maintenance period, which caused the condenser differential pressure to show anomalous behavior. These two events were controlled processes, rather than sub-system faults. But the analysis of the generator system and the empirical model during this "synthetic fault" period shows how groups of related signals behave during system perturbations. Such events can be recognized over time more easily when an empirical model of desired system conditions is available to help identify the sub-system location and signal magnitude of anomalies that develop when particular (combinations of) sensors are perturbed.

The data for event two are presented in Fig. 8, showing the voltage and condenser differential pressure (CDP) levels as dark lines; the estimates from the empirical model as light-colored lines, and alerts as asterisks, at the top of the plot. The event occurs between hour 950, before the condenser maintenance event, and hour 1250, following a second over-voltage event. In addition to these signals, extraction steam flow, exciter voltage, and HP wheel chamber pressure are chosen for examination (Fig. 9). These plots demonstrate interactions among the generator system variables and informative characteristics of the model estimates and alerts. Note that the time interval shown in these plots is expanded compared to that in Fig. 3 in order to show the extent of "normal" behavior that preceded the event of interest.

The voltage and CDP data presented in Fig. 8 illustrate that the residuals from the model were large enough to annunciate alerts on a number of signals during two periods of over-voltage and during most of the condenser maintenance period. But both voltage and CDP signals were normal following the CDP maintenance (hours 1030 - 1180). Voltage residuals became large again at 1188 hours through 1243 hours. Two other brief spikes in CDP occurred in the period shown that caused alerts to be annunciated.

Extraction flow and HP wheel pressure are shown together in Fig. 9 because they show similar patterns of alerts during event two. Both signals alert through most of the period from the beginning of the CDP maintenance, through the period of normal voltage and CDP, up to the second over-voltage event because the actual value is higher than indicated by the profile conditions. Note that the model estimate closely resembles the HP wheel chamber pressure signal, but is offset to a lower value between 975 and 1175 hours. Exciter voltage is shown to compare its dynamic pattern to HP wheel chamber pressure. In detail, the HP wheel chamber pressure shows peaks that correspond to peaks in exciter voltage and IP wheel chamber pressure (not shown), which do not alert through this period.

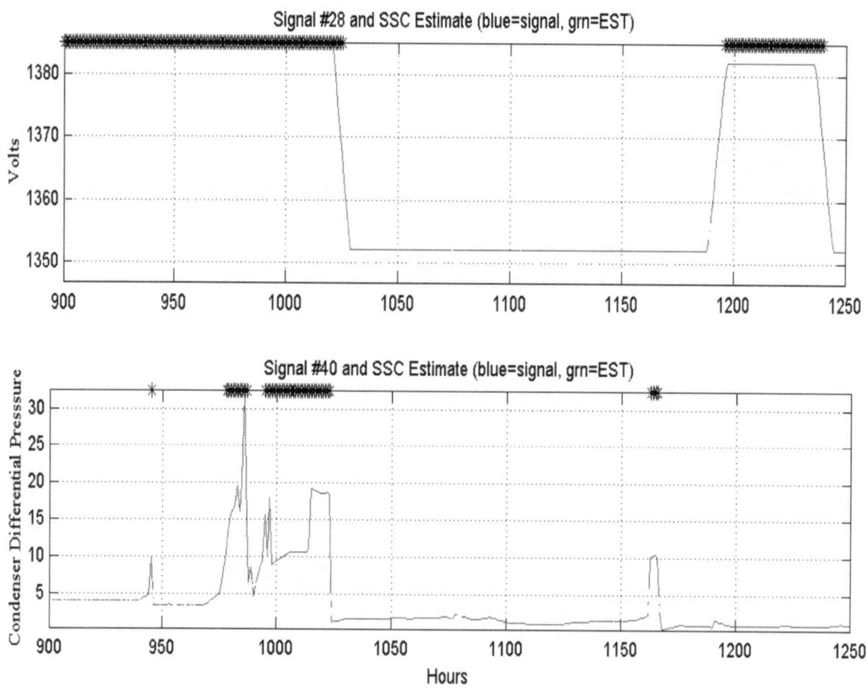

Fig. 8. The data for generator voltage (top plot) and condenser differential pressure show the actual signal levels compared to the model estimate levels

Extraction steam flow shows a broad maximum through the period of interest, compared to its level at the beginning of the condenser maintenance and the beginning of the second over-voltage at the end of the period. Flow level throughout this period is 40-75 units higher than the estimated level, which is flat around 350, so alerts are annunciated.

5 Discussion

The examples of two events at the USS GaryWorks co-generation plant illustrate the applicability of an empirically derived profile based upon advanced methods of pattern recognition for monitoring valuable equipment systems. Several points should be noted. High fidelity real-time estimation of the data signal by the empirical profile during normal operation provides a reliable basis for judging whether processes are at the level defined as normal in the profile of the system. The profile provides an estimate of the normal level of a signal when the signal itself deviates from the normal operating level. Alerts to abnormal behavior are provided on very small, but statistically meaningful deviations from the model (estimated) condition. When alerts appear on a group of signals simultaneously, this association stands out, adding information for process engineering or equipment troubleshooting in systems that can have complex interactions among variables.

The high fidelity of the model is accomplished by applying a non-linear similarity operator to all modeled signals simultaneously, giving an estimate of the normal signal levels (1,2), as defined by the profile. The normal patterns of partially correlated groups of signals moving together in a controlled process provide the basis for knowing the correct level of each signal. The natural variation in each signal is captured in the profile, so subtle deviations from the norm can be identified at each time step. A particular signal may be varying within normal limits, yet be moving in a fashion that is contrary to the pattern of movement of other signals with which it is correlated. Conventional set-point controls or operator monitoring will not notice that its behavior is abnormal until a large deviation has occurred. The examples here show that the empirical profile is capable of distinguishing normal excursions of a signal from abnormal ones, with a sensitivity that is very small compared to the normal dynamic range of the signal. Fault detection is thereby more sensitive than conventional process control methods, leading to earlier detection of faulty conditions. Detection may be early enough that down time of the equipment can be scheduled rationally, parts pre-ordered for it and total downtime reduced.

Some strengths of the empirical modeling method are also its weakness. Plant equipment often functions routinely with more variation than desirable for a highly efficient process. Variation due to routine special causes will cause the software model to annunciate alerts routinely. In such cases, alerts can be dismissed as "false alarms" in systems with high variability or suppressed with real-time alert logic. However, in such cases, "retraining" of the model can be performed routinely and in near real-time (Wegerich, et al., 2001) and reliable monitoring continued.

This problem can be illustrated from the data in Fig. 9. The alerts annunciated here are intermittent, indicating a process on the fringe of normality. The data show that the mean HP wheel chamber pressure, for example, is elevated by about 30 units during the period that the voltage was at 1352, compared to the over-voltage period earlier and later. Functionally, then, of the signals in event two, only exciter voltage and CDP – outside of its maintenance period – behaved normally, although the generator produced the power necessary to run the plant during this period.

Application of alert logic when process variation is high may reduce the lead-time before a significant failure somewhat. But this reduction should not represent a

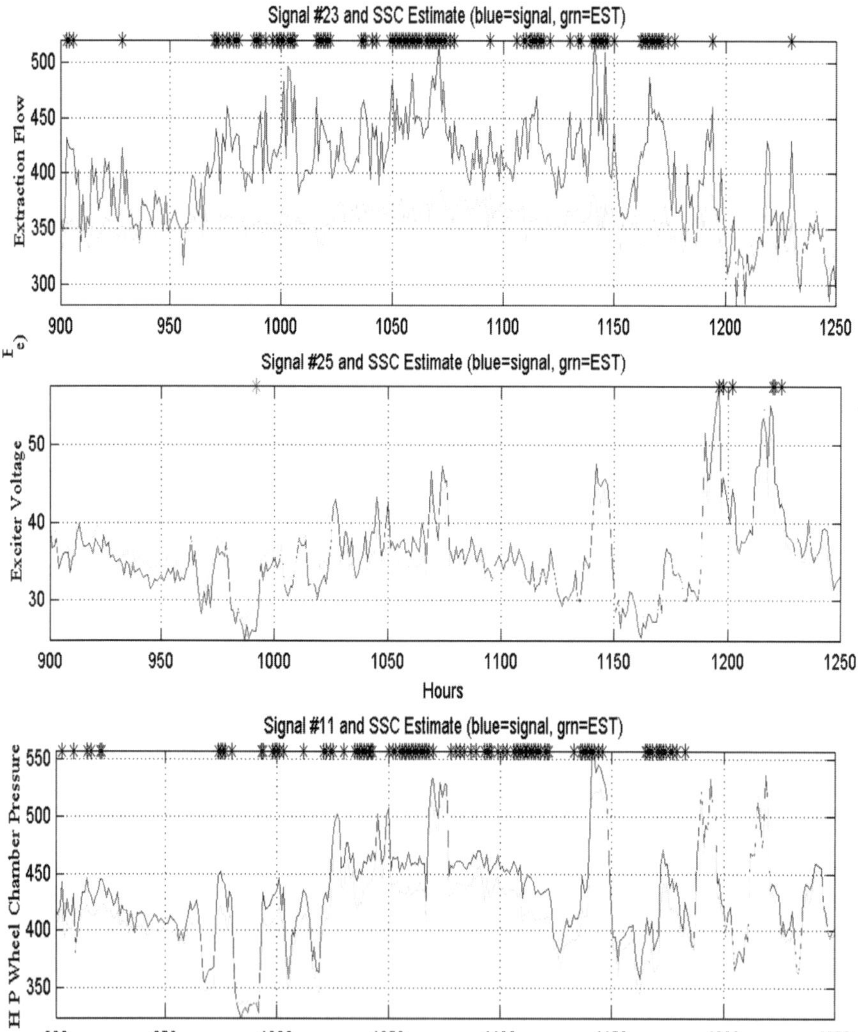

Fig. 9. Extraction flow, HP wheel pressure and exciter voltage show the actual signal levels compared to the model estimate levels. Note the similarity of the HP wheel chamber pressure and exciter voltage signals, and the difference in alert density

significant fraction of the early warning period. The ability of the empirical profiling software to continuously define the normal operating level for an operator is an advantage that remains through process variations with assignable causes. Also, as shown in event two, the patterns of signals that alert simultaneously add a dimension of information to normal process monitoring. Users may become adept at recognizing patterns of alerts and deviations from these patterns that lead to variation-reducing control strategies, as well as early detection of equipment faults.

6 Summary

Two equipment fault events, chosen from normal operating data at the USS GaryWorks co-generation facility in Gary, Indiana, are discussed in this paper to demonstrate the characteristics of an empirical model (the Plant *e*CM) for early detection of equipment failures that produce unplanned maintenance. The examples demonstrate:

The high fidelity, real-time estimation of the data signal by Plant *e*CM is a reliable basis for judging whether processes are at the normal level defined in the profile

The Plant *e*CM annunciates alerts to abnormal behavior on very small deviations from the profile condition, even hours ahead of significant equipment or process deviations in the examples discussed

The annunciation of alerts by the empirical profile focuses attention on particular (groups of) signals in the monitored system that can be showing early indications of equipment failure or process deviations

Acknowledgements

The authors wish to thank USS Gary Works, NiSource Inc., and SmartSignal Corp. for the opportunity to perform the study. Mr. Jesse Cox at USS and Mr. John Kerastas at SmartSignal helped gather and present the information used.

References

1. S. Wegerich, R. Singer, J. Herzog, A. Wilks, "Challenges Facing Equipment Condition Monitoring Systems," Proceedings, International Maintenance and Reliability Conference (MARCON 2001), Gatlinburg, TN, USA (6-9 May 2001)

2. K. C. Gross, R. M. Singer, S. Wegerich, J. P. Herzog, R. Van Alstine, and F. Bockhorst, "Application of a model-based fault detection system to nuclear power plant signals," Proc. Intelligent Power System Application To Power Systems (ISAP 97), Seoul, Korea (6-10 July, 1997) 66-70

10 On-Line Determination of the MTC (Moderator Temperature Coefficient) by Neutron Noise and Gamma-Thermometer Signals

C. Demazière and I. Pázsit

Department of Reactor Physics, Chalmers University of Technology
SE-412 96 Göteborg, Sweden

{demaz, imre}@nephy.chalmers.se

The estimation of the Moderator Temperature Coefficient of reactivity (MTC) by noise analysis is investigated theoretically. It is shown that the main reason of the MTC underestimation by noise analysis that was noticed experimentally previously lies with the heterogeneous structure of the moderator temperature noise throughout the core. The resulting deviation of the reactor response from point-kinetics only accounts for a negligible part. Therefore a new MTC noise estimator is proposed. This estimator relies on the core average temperature noise that has to be evaluated with a proper core weighting function. The coolant temperature noise has to be measured in many points of the reactor so that the core average temperature noise could be approximated. Gamma-Thermometers (GTs) could be used for that purpose since it is demonstrated (both theoretically and via real measurements) that in the frequency range of interest for the MTC estimation they work as thermocouples.

1 Introduction

The Moderator Temperature Coefficient of reactivity (MTC) is an important safety parameter in Pressurized Water Reactors (PWRs). Defined as the variation of reactivity induced by a change in the coolant temperature and divided by the core-averaged temperature change [1], the MTC plays a major role in the feedback mechanism of the reactor and has to be negative so that the reactor is in all circumstances stable[1]. Due to the decrease of the boron content through the fuel

[1] The feedback mechanism in a PWR relies mainly on the Doppler and MTC effects. Therefore, a slightly positive MTC could be permitted if the MTC combined with the Doppler coefficient provides an overall negative reactivity feedback. Due to the economical advantages of high burnup fuel, e.g., longer fuel cycle (fuel which has a positive

Footnote continues on next page

cycle (as a consequence of the decrease of the reactivity excess due to the fuel burnup), the magnitude of the MTC increases. Consequently, the MTC is usually measured at Beginning Of Cycle (BOC) and Hot Zero Power (HZP) to verify that the MTC is negative (therefore preventing from the consequences of a power increase). The MTC has also to be estimated a few months before the expected End Of Cycle (EOC), to verify that the MTC is not too negative (thus preventing from the consequences of a cool-down event). This second measurement is performed at Hot Full Power (HFP) and indicates if the reactor can be operated until its pre-calculated EOC.

Whereas the BOC measurement can be considered as accurate (use of a reactivity meter and homogeneous temperature distribution throughout the core), the EOC measurement has been proven to be less reliable for many different reasons. The most important one is probably the fact that the procedure used to perform the EOC measurement takes some time during which some reactivity changes due to effects other than the MTC take place and which cannot be measured directly [4]. This compromises somehow the experimental technique since the main goal of the measurement is to verify that core calculations predict the EOC MTC accurately. Therefore the experimental technique should rely on as few as possible core calculated parameters. The other major concern for power utilities is the transient induced by the traditional measurement techniques such as the boron dilution method. This transient has to be monitored for 12 to 24 hours, and generally requires operating the reactor at a lower power level.

Another measurement technique was proposed some years ago to monitor the MTC without perturbing the reactor [21, 25]. This method relies on the correlation between the neutron and temperature noise signals. It is known that even if a system is at its steady state, small fluctuations occur around its mean value. These small fluctuations are representative of the dynamics of the system. Therefore monitoring the neutron and temperature noise should allow estimating the MTC while the reactor is at its steady state, thus avoiding the problems of the traditional measurement techniques. Several attempts to monitor the MTC via noise analysis were made in the past [1, 9, 11, 13-17, 19-23, 25-30]. Unfortunately, all of them revealed that the MTC estimated via the noise technique systematically underestimated the actual value of the MTC by a factor two to five. It was even noticed that this calibration factor seems to be constant throughout the fuel cycle and even between different fuel cycles as long as the same pair of neutron and temperature detectors are used.

Recent measurements of the amplitude of the temperature noise at the core exit revealed that the temperature noise seems to be radially non-homogeneous [12]. It was even noticed that the cross-correlation between two temperature detectors located inside the core on a same axial plane was strongly decreasing with the

---

contribution to the MTC), power utilities are trying nowadays to get licensed for a slightly positive MTC at Beginning Of Cycle (BOC).

distance between the two detectors [7]. As will be explained later, the MTC noise estimate was derived in the hypothesis of spatially homogeneous temperature noise throughout the core. Therefore the effect of non-homogeneous temperature noise on the MTC was investigated in the following via an analytical one-dimensional model, which revealed that the discrepancy between the MTC noise estimate and its actual value seemed to be mostly due to the radial heterogeneous structure of the temperature noise throughout the core. A new MTC noise estimator allowing taking this effect into account was thus proposed. This new estimator implicitly assumes that the core average temperature noise could be estimated. This requires being able to measure the temperature noise throughout the core. As will be demonstrated in the following, Gamma-Thermometers (GTs) could offer such a possibility since in the frequency range of interest for the MTC determination via noise analysis they work as temperature detectors.

A nomenclature explaining all the abbreviations used can be found at the end of this chapter.

2 MTC Estimation by Noise Analysis

In this section, the MTC definition and its noise estimator are recalled. Emphasis is put on the hypotheses used to derive the MTC noise estimator. The effect of these hypotheses on the accuracy of the MTC estimation is then quantitatively evaluated via a one-dimensional model and some conclusions are drawn.

2.1 MTC definition

According to the newest American Standard, the Moderator Temperature Coefficient (MTC) of reactivity is defined as the variation of the reactivity $\delta\rho$ induced by a change of the core inlet temperature of the coolant, divided by the coolant average temperature change δT_m^{ave} [3]:

$$\delta\rho(t) = MTC \times \delta T_m^{ave}(t) \tag{1}$$

where t represents the time.

If the temperature fluctuations are assumed to be proportional to the fluctuations of the macroscopic cross-sections via a time- and space-independent coefficient K as follows:

$$\delta T_m(\boldsymbol{r},t) = K \times \delta \Sigma_a(\boldsymbol{r},t) \tag{2}$$

with $\boldsymbol{r}$ representing the position, then the first-order perturbation theory allows calculating the average temperature change as:

$$\delta T_m^{ave}(t) = \frac{\int \delta T_m(\mathbf{r},t)\phi_0^2(\mathbf{r})d\mathbf{r}}{\int \phi_0^2(\mathbf{r})\,d\mathbf{r}} \tag{3}$$

where ϕ_0 is the static flux.

2.2 MTC Noise Estimator

As explained in [4], there are several noise estimators. When the coherence between the measured temperature noise and the measured average temperature noise equals unity, all the noise estimators coincide. The most commonly used noise estimator (the so-called H_1 estimator) is given in the frequency domain by:

$$MTC = \frac{CPSD_{\delta\rho,\delta T_m^{ave}}(\omega)}{APSD_{\delta T_m^{ave}}(\omega)} \tag{4}$$

where the *APSD* and the *CPSD* stand for the Auto-Power Spectral Density and Cross-Power Spectral Density respectively. Even if the right-hand side of Eq. (4) depends on the frequency $f = \omega/2\pi$, this dependence cancels out and the MTC is frequency-independent. The problem is that neither the reactivity noise nor the average temperature noise are measurable quantities.

The reactivity noise can be nevertheless inferred from the following relationship:

$$\delta\rho(\omega) = \frac{1}{G(\omega)}\frac{\delta\phi^{pk}(\mathbf{r},\omega)}{\phi_0(\mathbf{r})} \tag{5}$$

where G is the at-power reactor transfer function and $\delta\phi^{pk}$ is the point-kinetic component of the flux noise. Eq. (5) was derived from the factorization of the flux into an amplitude function P and a shape function Ψ as:

$$\phi(\mathbf{r},t) = P(t)\Psi(\mathbf{r},t) \tag{6}$$

which leads to the following expression of the flux noise in the frequency domain (second-order terms neglected):

$$\delta\phi(r,\omega) = \delta\phi^{pk}(\mathbf{r},\omega) + \delta\psi(\mathbf{r},\omega) \tag{7}$$

where

$$\delta\phi^{pk}(\mathbf{r},\omega) = \phi_0(\mathbf{r})\delta P(\omega) \tag{8}$$

and from the following definition of the at-power reactor transfer function:

$$G(\omega) = \frac{\delta P(\omega)}{\delta \rho(\omega)} \tag{9}$$

There are nevertheless two main problems if one wants to use Eq. (5) for evaluating the reactivity noise. The first one is due to the use of the at-power reactor transfer function, since this transfer function contains the reactivity feedback mechanism and more specifically the MTC. Consequently, this transfer function is unknown. If one considers high frequencies, typically frequencies higher than 0.1Hz, one may assume that the feedback mechanism does not take place due to the relatively large time constant of the power-to-coolant temperature transfer function, i.e., one is interested in frequencies for which the reactor responds before the feedback begins to act. Therefore, the at-power (or closed-loop) reactor transfer function G can be replaced in that case by the open-loop or zero-power reactor transfer function G_0:

$$G(\omega) \approx G_0(\omega) \tag{10}$$

If one also neglects frequencies higher than 1Hz (this upper cut-off frequency has also the advantage of eliminating the effect of the damping of the fluctuations traveling from the neutron to the temperature detectors), the zero-power reactor transfer function further simplifies into:

$$\mathrm{G}_0(\omega) \approx \frac{1}{\beta} \tag{11}$$

where β is the fraction of delayed neutrons which can accurately be estimated by core calculations.

So far, narrowing the frequency region of interest to the frequency band 0.1 to 1.0Hz allowed replacing the reactivity noise $\delta\rho$ by the amplitude function noise δP via Eqs. (9), (10), and (11), i.e., known quantities and parameters without any approximation. As can be seen from Eqs. (7) and (8), the amplitude function noise cannot be expressed as a function of the total flux noise $\delta\phi$ solely, rather as a function of the point-kinetic component of the flux noise $\delta\phi^{pk}$. Unfortunately, this component is not measurable in practice since neutron detectors are sensitive to the total flux noise $\delta\phi$. Therefore Eq. (5) cannot be used to estimate the reactivity noise, unless one approximates the point-kinetic component of the flux noise by the total flux noise. This is equivalent to neglecting the contribution due to the amplitude function noise $\delta\Psi$, i.e., to assuming that the reactor behaves in a point-kinetic way since one has:

$$\begin{aligned}\frac{1}{G(\omega)}\frac{\delta\psi(\boldsymbol{r},\omega)}{\phi_0(\boldsymbol{r})} &= \frac{1}{G(\omega)}\frac{\delta\phi(\boldsymbol{r},\omega)}{\phi_0(\boldsymbol{r})} - \frac{1}{G(\omega)}\frac{\delta\phi^{pk}(\boldsymbol{r},\omega)}{\phi_0(\boldsymbol{r})} \\ &= \frac{1}{G(\omega)}\frac{\delta\phi(\boldsymbol{r},\omega)}{\phi_0(\boldsymbol{r})} - \delta\rho(\omega)\end{aligned} \tag{12}$$

In summary, the reactivity noise can only be approximated by the following expression in the frequency range 0.1 to 1.0Hz:

$$\begin{aligned}\delta\rho^{biased}(\boldsymbol{r},\omega) &= \frac{1}{G_0(\omega)}\frac{\delta\phi(\boldsymbol{r},\omega)}{\phi_0(\boldsymbol{r})} \\ \neq \delta\rho(\omega) &= \frac{1}{G_0(\omega)}\frac{\delta\phi^{pk}(\boldsymbol{r},\omega)}{\phi_0(\boldsymbol{r})}\end{aligned} \tag{13}$$

This reactivity noise is only biased by the deviation of the reactor response from point-kinetics, i.e., by the fact that one uses $\delta\phi$ (measurable) instead of $\delta\phi^{pk}$ (non-measurable). It will be shown below that this bias is not significant.

Regarding the estimation of the temperature noise now, the main problem pertains to the fact that the core average temperature noise should be estimated. Unfortunately, all the commercial PWRs are badly instrumented with respect to temperature detectors. Usually, only core-inlet, core-outlet, and core-exit thermocouples are available. In all the previous experimental investigations, a core-exit thermocouple was used to estimate the MTC by noise analysis [1, 9, 11, 13-17, 19-23, 25-30]. In most cases, the thermocouple was located within the same fuel channel where the neutron noise was measured, but in some cases the two detectors were located in two neighboring fuel channels. Using one core-exit temperature instead of the core average temperature for evaluating the temperature noise is equivalent to assuming that the temperature noise is spatially homogeneous throughout the core.

Another problem related to the temperature noise is the fact that usually the local temperature noise is not measured at the same axial location as the neutron noise. In all the previous experimental work, the temperature was measured at the core-exit. If one assumes that the fluctuations travel upwards unperturbed from the in-core neutron detector to the core-exit thermocouple (no noise source between the two detectors), the only effect that takes place is a damping for frequencies higher than the frequency corresponding to the transit time between the two detectors, i.e., typically 1.0Hz. Consequently, the higher cut-off frequency allows both eliminating the damping effect and simplifying the zero-power reactor transfer function into a very simple expression (see Eq. (11) above.)

In summary, the average temperature can only be approximated by the following expression for frequencies lower than 1.0Hz:

$$\begin{aligned} \delta T_m^{ave,biased}(\boldsymbol{r},\omega) &= \delta T_m(\boldsymbol{r},\omega) \\ \neq \delta T_m^{ave}(\omega) &= \frac{\int \delta T_m^{ave}(\boldsymbol{r},\omega)\phi_0^2(\boldsymbol{r})d\boldsymbol{r}}{\int \phi_0^2(\boldsymbol{r})\,d\boldsymbol{r}} \end{aligned} \quad (14)$$

As a matter of fact, the MTC noise estimators were all derived previously on the assumption that the temperature noise was spatially homogeneous. In that case, the reactor necessarily behaves in a point-kinetic manner [5, 8]. This means that both Eqs. (13) and (14) are fulfilled, i.e., there is no bias in the estimation of the average temperature noise from the local temperature noise and no bias in the estimation of the reactivity noise from the total flux noise.

2.3 Study of the Approximations Used in the Noise Estimation on the MTC

The two previous biases, namely the deviation from point-kinetics of the reactor response and the non-homogeneity of the temperature noise throughout the core, were investigated via an analytical model of a one-dimensional bare reactor with a core radius a. Only the radial direction was taken into account since it is known that the effect of radially inhomogeneous temperature noise sources give stronger effects than the axial ones.

This analytical model relies on the one-group diffusion approximation. This model is described in details in [5, 8]. Only the main characteristics of this model are recalled here. As described by Eq. (1), the temperature noise sources are space-dependent and defined from the fluctuations of the macroscopic absorption cross-section. Furthermore, the noise sources are not given in a deterministic manner, rather via their statistical properties such as the spatial cross-correlation. Their *CPSD* is then directly factorized into an amplitude function σ^2, which corresponds to the space dependence of the noise source strength (or more simply their *APSD*) and an exponential decay function, which simply states that the correlation between two points decreases when the distance between these two points increases. How rapidly this exponential decay function decreases with increasing distances is related to a parameter l, called in the following the correlation length. Regarding the frequency content, the noise sources are assumed to be white noise. Consequently, their *CPSD* and *APSD* are frequency independent. In formula, this can be written as:

$$CPSD_{\delta\Sigma_a}(\boldsymbol{r},\boldsymbol{r}',\omega) = \sigma^2(\hat{\boldsymbol{r}})e^{-\frac{|r-r'|}{l}} \quad (15)$$

with

$$\hat{\boldsymbol{r}} \equiv \frac{\boldsymbol{r}+\boldsymbol{r}'}{2} \tag{16}$$

and

$$\sigma^2(\hat{\boldsymbol{r}})\Big|_{\boldsymbol{r}=\boldsymbol{r}'} = APSD_{\delta\Sigma_a}(\boldsymbol{r},\omega) \tag{17}$$

Many different parameters were investigated in [5, 8]. In the following, a set of parameters corresponding to a realistic case is presented. More specifically, the cross-sections are representative of a PWR core at EOC, the correlation length l was chosen to be equal to 15cm (which seems to correspond roughly to what was experimentally noticed in [7]), and the calculations were performed at a frequency of 1Hz. Regarding the amplitude or shape function, the following one was studied:

$$\sigma(\hat{x}) = \left[1 - \left(\frac{\hat{x}}{a+\delta a}\right)^2\right]^{-1} \tag{18}$$

with $\delta a = a/5$. This shape function somehow expresses the fact that the temperature noise is larger close to the core boundary than at the core center and corresponds therefore to what was experimentally noticed [12].

From Eqs. (1) - (3), it is straightforward to calculate the actual value of the MTC via the first-order perturbation theory and one finds:

$$MTC = -\frac{1}{K\nu\Sigma_{f,0}} \tag{19}$$

where $\nu\Sigma_{f,0}$ represents the static macroscopic fission cross-section multiplied by the average number of neutrons per fission. The MTC can be correctly estimated by the following ideal MTC noise estimator in the frequency range 0.1 to 1.0Hz:

$$H_1^{ideal} = \frac{CPSD_{\delta\rho,\delta T_m^{ave}}(\omega)}{APSD_{\delta T_m^{ave}}(\omega)} = \frac{1}{G_0(\omega)} \frac{CPSD_{\delta\phi^{pk}/\phi_0,\delta T_m^{ave}}(\omega)}{APSD_{\delta T_m^{ave}}(\omega)} \tag{20}$$

As stated before, neither the reactivity noise (or the point-kinetic component of the flux noise) nor the core average temperature noise can be directly estimated.

If one assumes that the core average temperature noise can be measured, one can estimate the discrepancy introduced by the fact that the reactor does not behave in a point-kinetic way. This is accomplished by using in Eq. (4) $\delta\rho^{biased}$ instead of $\delta\rho$ (see Eq. (13)). This leads to the following biased MTC noise estimator:

$$\tilde{H}_1^{biased}(\boldsymbol{r},\omega) = \frac{CPSD_{\delta\rho^{biased},\delta T_m^{ave}}(\boldsymbol{r},\omega)}{APSD_{\delta T_m^{ave}}(\omega)} = \frac{1}{G_0(\omega)} \frac{CPSD_{\delta\phi/\phi_0,\delta T_m^{ave}}(\boldsymbol{r},\omega)}{APSD_{\delta T_m^{ave}}(\omega)} \tag{21}$$

The ratio between this estimator and the actual MTC value allows appreciating the error introduced by the deviation of the reactor response from point-kinetics. The space-dependence of this ratio is plotted in Fig. 1. As can be seen, this MTC noise estimator gives a relatively good estimation of the actual value of the MTC. Consequently, the deviation of the reactor response from point-kinetics does not play a significant role regarding the MTC estimation. One can therefore consider that approximating the reactivity noise $\delta\rho(\omega)$ by $\delta\rho^{biased}(\boldsymbol{r},\omega)$ does not induce any appreciable discrepancy in the MTC estimation.

Using this valid approximation, one can then study the effect of using the local temperature noise instead of the core average temperature noise. This is accomplished by using in Eq. (4) $\delta T_m^{ave,biased}$ instead of δT_m^{ave} (see Eq. (14)). This leads to the following MTC noise estimator:

$$H_1^{biased}(\boldsymbol{r},\omega) = \frac{CPSD_{\delta\rho^{biased},\delta T_m^{ave,biased}}(\boldsymbol{r},\omega)}{APSD_{\delta T_m^{ave,biased}}(\boldsymbol{r},\omega)} = \frac{1}{G_0(\omega)} \frac{CPSD_{\delta\phi/\phi_0,\delta T_m}(\boldsymbol{r},\omega)}{APSD_{\delta T_m}(\boldsymbol{r},\omega)} \tag{22}$$

This noise estimator corresponds to the one that was used so far in all the previous studies and contains the effect of both the deviation from point-kinetics of the reactor response and the approximation of the core average temperature noise by the local temperature noise. The ratio between this noise estimator and the actual MTC value is plotted in Fig. 2. As can be seen, there is a very strong underestimation of the actual value of the MTC, and this underestimation is noticeably space-dependent. Close to the core center, the MTC is roughly underestimated by a factor five, which corresponds to what was noticed in the experimental investigations.

Consequently, it seems that the huge discrepancy between the traditional MTC noise estimator H_1^{biased} and its actual value is due to the fact that the temperature noise is not spatially homogeneous rather than due to the deviation of the reactor behavior from point-kinetics. Consequently, estimating the average temperature noise and using instead the $\tilde{H}_1^{biased}$ noise estimator would allow getting a fairly good estimation of the actual MTC.

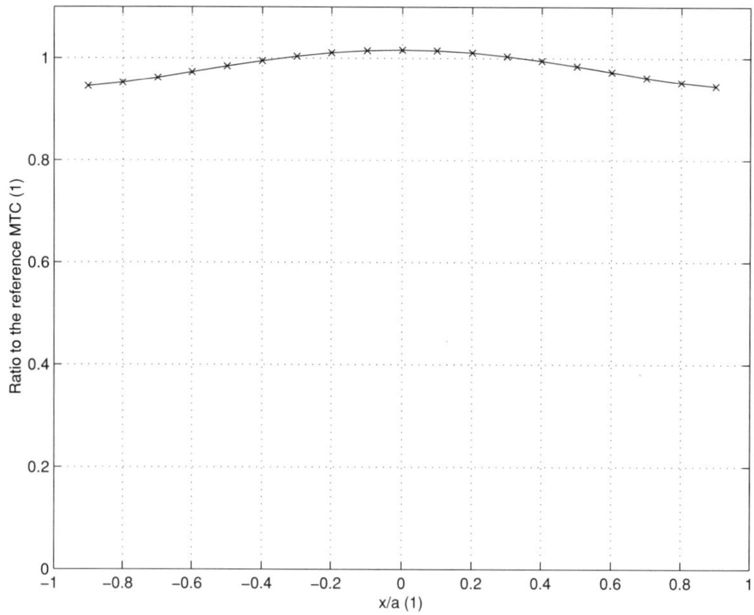

Fig. 1. Discrepancy between the MTC noise estimate $\tilde{H}_1^{biased}$ and its actual value (due to the deviation from point-kinetics of the reactor response solely)

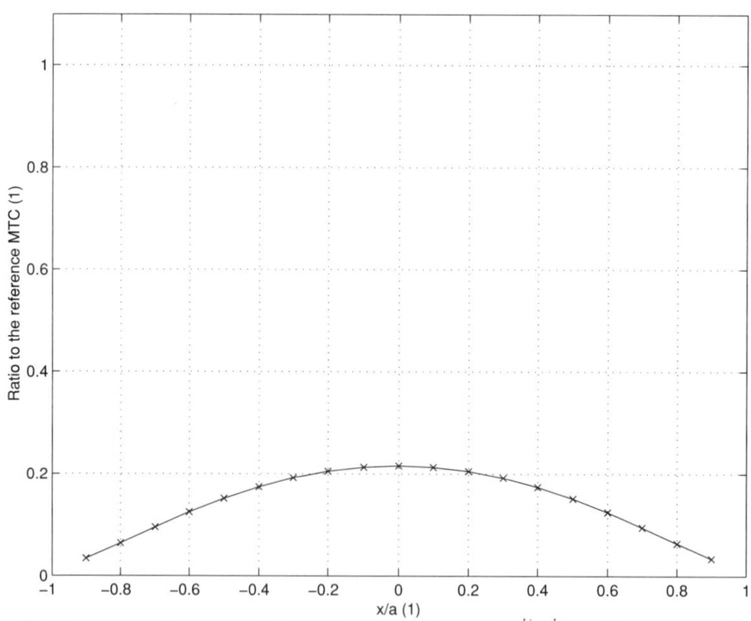

Fig. 2. Discrepancy between the MTC noise estimate H_1^{biased} and its actual value (due to the deviation from point-kinetics of the reactor response *and* the approximation of the average temperature noise by the local one)

3 Measurement of the Temperature Noise via Gamma-Thermometers

Estimating the core average temperature noise can be done by measuring the temperature noise in several points of the reactor core and by using an integration formula so that the integrals in Eq. (3) can be numerically approximated. This requires also knowing the static flux at the corresponding locations. Any static core calculator could be used for that purpose since it is well known that the static flux is nowadays accurately estimated by these codes. As pointed out previously, commercial PWRs are badly instrumented with respect to in-core temperature detectors. Nevertheless the Swedish PWR Ringhals-2 contains 108 permanent Gamma-Thermometers (GTs) installed in the core (12 strings containing each 9 GTs located at different axial levels). It was thus proposed to study the possibility of using GTs for measuring the temperature noise throughout the core. In a different setting, the use of GTs for measuring temperature noise has already been proven in the Halden Reactor Project [10, 28]. Our purpose in this work is to perform the same investigation regarding measurements taken at an operating power plant.

3.1 Description of the Halden Type Gamma-Thermometers

There are several types of GTs. The Ringhals-2 unit is instrumented with the so-called Halden type GTs [2]. In this design, a metallic pin is heated up by the gamma flux mostly (elastic and inelastic collisions) and to a lesser extent by the neutron flux ((n, γ), (n, α), (n, p) reactions, elastic and inelastic collisions). The pin is encapsulated into an outer body or housing, and the thermal insulation between the pin and the outer body is made by some xenon gas. One of the tips of the pin is therefore thermally insulated, whereas the other tip is in direct contact with the body of the GT. A differential chromel/alumel thermocouple measures the temperature drop along the pin with the hot junction located at the insulated tip, and the cold junction in direct contact with the coolant outside the GT body. Fig. 3 gives an overview of the Halden type GT. Even if GTs are designed to measure the gamma flux (or more practically the power), the cold junction of the differential thermocouple is in direct contact with the coolant and could be therefore used for monitoring the moderator temperature fluctuations.

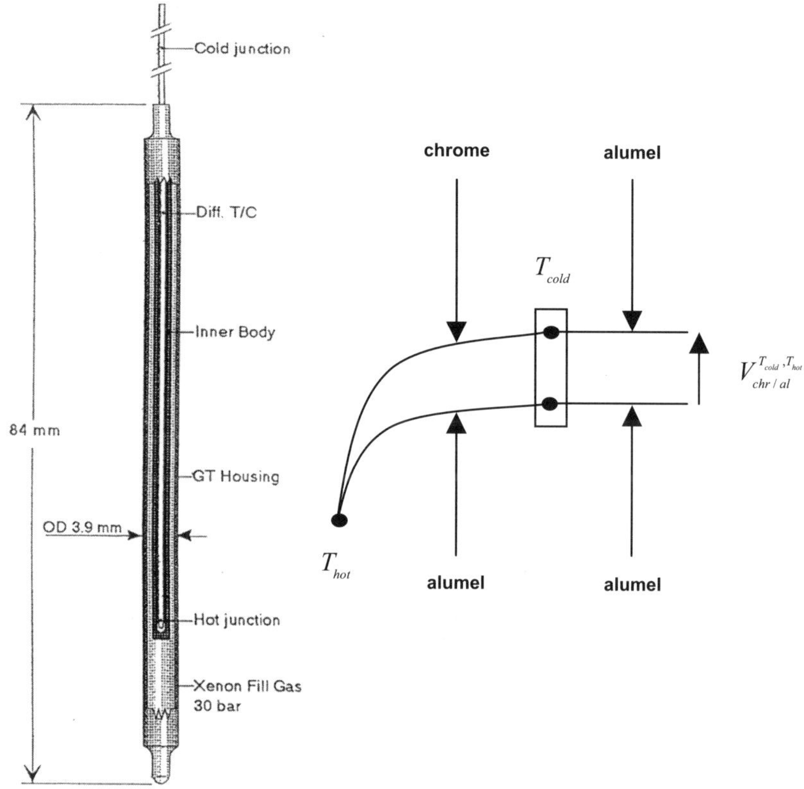

Fig. 3. Sketch of the Halden type GTs (from [10]) on the left, and the corresponding schematic description of the differential thermocouple on the right

3.2 Use of Gamma-Thermometers for Noise Analysis

As pointed out previously, the hot junction of the thermocouple measures the temperature increase due to the heating along the pin with the cold junction acting as a heat sink and which is at the same temperature as the coolant. In case of static measurements, the voltage delivered by the differential thermocouple is therefore directly proportional to the heating Q according to the following formula (in which the heat loss across the gas chamber and the radial conduction across the inner body are neglected):

$$V_{chr/al}^{T_{cold},T_{hot}} = \alpha_{chr/al} \times \left(T_{hot} - T_{cold}\right) = \alpha_{chr/al} \times \frac{QL^2}{2\lambda} \tag{23}$$

where $V_{chr/al}^{T_{cold},T_{hot}}$, $\alpha_{chr/al}$, L, λ, and Q are the voltage, the thermocouple constant (typically approximately 41μV/°C for a K type thermocouple, i.e., chromel/alumel), the separation distance between the hot and cold junctions, the thermal conductivity of the inner body, and the heat flux in the thermometer respectively.

In case of dynamical measurements, because of the different thermal characteristics of the cold and hot junctions, the thermocouple will respond differently, i.e., with different delays to phenomena associated to the hot junction and to phenomena associated to the cold junction. Since the cold junction is located above the GT body and is direct contact with the coolant, it responds very quickly to coolant temperature oscillations (with a thermal time constant typically around 0.1-1.0s). This means that the cold junction acts as a low-pass filter of the coolant temperature noise with a cut-off frequency of a few hertz. On the contrary, the hot junction is thermally insulated within the inner body and has therefore a significantly more sluggish response (with a thermal time constant typically around 10-100s). Consequently, the hot junction acts a low-pass filter of the gamma/neutron flux with a cut-off frequency of a few hundredth of hertz. If one writes the transfer function from the cold and hot junctions temperature noise ($\delta T_{cold} = \delta T_m$ and δT_{hot} respectively) to the GT signal in the frequency domain as follows:

$$\delta V_{chr/al}^{T_{cold},T_{hot}}(\omega) = \frac{\alpha_{chr/al}}{1 + j\tau_{hot}\omega} \times \delta T_{hot}(\omega) - \frac{\alpha_{chr/al}}{1 + j\tau_{cold}\omega} \times \delta T_m(\omega) \tag{24}$$

where the indices τ_{hot} and τ_{cold} refer to the time constants of the hot and cold junctions respectively, then one notices that in the frequency range of interest for the MTC estimation by noise analysis, i.e., typically from 0.1 to 1.0Hz only the signal due to the cold junction remains. The fluctuations due to the gamma/neutron heating are consequently completely filtered out by the hot junction. Only the very low frequencies contain the contribution of both the hot and cold junctions.

3.3 Checking of the GT Ability to Measure the Moderator Temperature Noise

In 2000, during fuel cycle 24, a noise measurement using GTs was performed at the Swedish Ringhals-2 PWR. The Halden type GTs were used for that purpose and three channels (J10, L11, and M03) were instrumented, as shown by the core map of Fig. 4. The detector string J10 is located close to the core center, the string M03 close to the core boundary, and the string L11 occupies an intermediate position. Each string contains nine GTs at different axial levels. For each detector string, all nine different axial levels were measured. The different levels are roughly equally spaced and cover the whole core active height. The numbering of the levels (1-9) is done from the bottom of the core active height to the top of the core active height. The duration of the measurement was almost three hours, and the signals were recorded at a sampling frequency of 8Hz. The time signals were then detrended (if a trend was found), the irregularities in the mean values eliminated by signal processing methods and data analysis was performed in the frequency domain [24]. In order to evaluate the *APSDs* and *CPSDs* of the different signals, Welch's averaged, modified periodogram method was used. The time signals were divided into overlapping sections of 256 points, then windowed by using a Hanning window. The sections were assumed to overlap by 128 points.

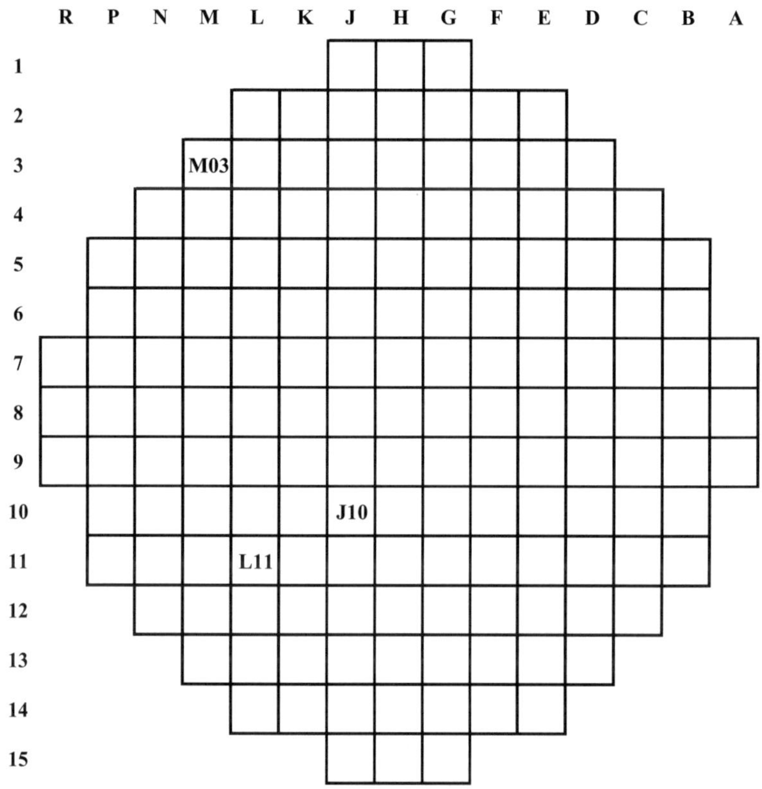

Fig. 4. Radial location of the GTs in Ringhals-2

The ability of the GTs to monitor the temperature fluctuations of the coolant can be first checked qualitatively by cross-correlating two GTs within the same fuel channel. If the GTs are sensitive to the temperature noise, then the temperature fluctuations recorded by the lowermost detector should also be present in the uppermost detector. Due to the separation distance between the two-detectors, the noise recorded by the uppermost one is simply time-shifted compared to the noise recorded by the lowermost one (if one assumes that no noise is added between the two detectors):

$$\delta V^{T_{cold},T_{hot}}_{chr/al,uppermost}(t) = \delta V^{T_{cold},T_{hot}}_{chr/al,lowermost}(t-t_0) \tag{25}$$

where t_0 is the transit time for the temperature fluctuations to travel upwards between the two detectors. In the frequency domain, this reads as:

$$\delta V^{T_{cold},T_{hot}}_{chr/al,uppermost}(\omega) = e^{-i\omega t_0}\,\delta V^{T_{cold},T_{hot}}_{chr/al,lowermost}(\omega) \tag{26}$$

The Wiener-Khinchin theorem allows then writing:

$$CPSD_{\delta V^{T_{cold},T_{hot}}_{chr/al,uppermost},\,\delta V^{T_{cold},T_{hot}}_{chr/al,lowermost}}(\omega) \approx e^{-i\omega t_0}\left|\delta V^{T_{cold},T_{hot}}_{chr/al,lowermost}(\omega)\right|^2 \tag{27}$$

Consequently, in the frequency range where the GTs detect temperature fluctuations, the phase of the *CPSD* between the two detectors should be linear. This can be noticed in Fig. 5, where one possible pair of detectors within the fuel channel L11 was chosen as a representative example. This shows that GTs in the frequency range 0.1 to 1.0Hz are able to detect the moderator temperature fluctuations traveling upwards through the core. GTs were even used successfully to measure flow velocities within the core [7].

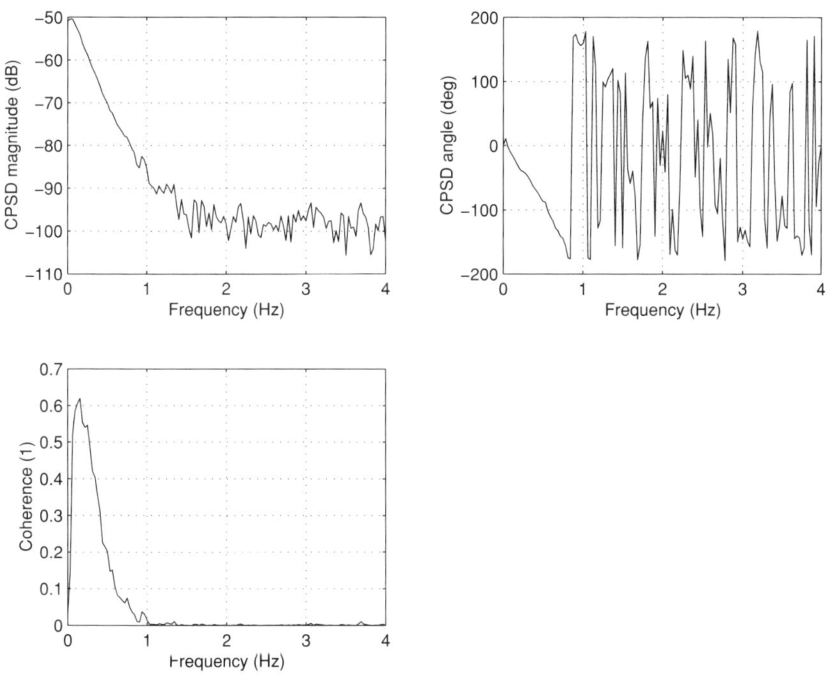

Fig. 5. *CPSD* of the detector string L11, between plane number 4 and plane number 9

Another possibility to check if GTs work as thermocouples in the frequency range of interest for the MTC estimation by noise analysis is to try to numerically estimate the transfer function of the GTs. This can be achieved by the so-called Auto-Regressive Moving Average (ARMA) modeling. If one assumes that the GTs measure a noise *e*, then an ARMA(*n*, -*m*) model consists of trying to write the GTs transfer function as:

$$\sum_{k=0}^{n} a_k \delta V_{chr/al}^{T_{cold},T_{hot}}\left(t-kT\right)=\sum_{l=0}^{m} c_l e\left(t-lT\right) \tag{28}$$

where *n* and *m* are the order of the model, and *T* is the sampling interval. The a_k and c_l coefficients are estimated from the measured output signal $\delta V_{chr/al}^{T_{cold},T_{hot}}$ by using a minimization procedure [18]. The noise *e* is still unknown and assumed to be completely uncorrelated, i.e., white.

Once the coefficients a_k and c_l are estimated, the time constants of the corresponding transfer function can be estimated since any real pole *p* of the transfer function defines a specific system dynamics first-order mode and can therefore be associated to a time constant τ as [31]:

$$\tau=-\frac{T}{\ln p} \tag{29}$$

The corresponding cut-off frequency is thus:

$$f=\frac{1}{2\pi\tau} \tag{30}$$

When comparing Eq. (28) with Eq. (24), it can be noticed that the noise *e* in Eq. (28) is a combination of both the cold and hot junction signals given in Eq. (24). This represents a simplification and an approximation to the actual GT transfer function. Nevertheless, in such a case an ARMA(2,1) model could suffice to estimate the two time constants. Simulations were therefore carried out to verify that the time constants given by the ARMA(2,1) model estimated from simulated data using Eq. (24) and two independent white noise sources for the cold junction and hot junction temperature noise were equal to the two actual time constants, i.e., the one corresponding to the cold junction and the one corresponding to the hot junction.

Due to the hardware/software processing of the signals such as the DC removal and the anti-aliasing filter, an ARMA(2,1) model was found to be in some cases unsatisfactory when applied to measured data. Simulations were then carried out in order to determine what order the ARMA model should be. It was assumed that both the DC removal and the anti-aliasing filter were equivalent to two second-order digital high-pass and low-pass filters respectively with two different cut-off frequencies. Depending on the ratio between the cut-off frequencies of the anti-aliasing filter and of the cold junction, and on the ratio between the cut-off frequencies of the DC remover and the hot junction, several cases can be encountered. The ARMA model can detect the anti-aliasing filter, and/or the cold junction, and/or the DC remover, and/or the hot junction (with *or* having a non-exclusive meaning). This corresponds to the following ARMA models: (1,0), (2,0), (2,1), (3,1), (3,2), (4,2), (4,3), (5,3), (6,4), and (6,5). As before, simulations were carried out in order to see if the ARMA models were able to correctly estimate the time constants, using Eq. (24) and two independent white noise sources for the cold junction and hot junction temperature noise. It was noticed that due to the presence of the DC remover and the anti-aliasing filter, the time constants were sometimes biased, but they still remain in the same range as their actual values.

The criteria for choosing the best ARMA model were thus the following [18]:

- the ARMA model must correspond to the hypotheses for which the model was developed, i.e. if the anti-aliasing filter, and/or the cold junction, and/or the DC remover, and/or the hot junction are assumed to be detected by the model, then their respective zeros/poles locations and values should match realistic cases;
- a zero/pole cancellation means that the model order can be reduced;
- any identical poles/zeros pattern between different ARMA models generally expresses the fact that this specific pattern probably contains the actual system dynamics;
- the fit between the measured data and the data simulated by the ARMA model should be as high as possible;
- the residuals, i.e. the part of the output that the model could not reproduce, should be as low as possible;
- the 99% confidence interval of the zeros and poles should not be too large.

The results of the ARMA modeling are summarized in Table 1, Table 2, and Table 3 for the detector strings J10, L11, and M03 respectively. In some occurrences, it was not possible to detect either the cold junction time constant, or the hot junction time constant, or both. As the simulation revealed, the removal of the DC component and the anti-aliasing filter could explain such a behavior. Other cases show that two time constants can clearly be determined. Even if it was noticed with the test cases that the numerical values of the estimated time constants could have somewhat large uncertainty, it seems that one of the time constant lies in the range 0.1 to 1.0s, and the other time constant lies in the range 2.0 to 20.0s. The first one corresponds therefore to the cold junction, whereas the second one corresponds to the hot junction.

Table 1. Results of the ARMA modeling for the detector string J10

Level	ARMA	Fit [%]	τ_{cold} [s]	f_{cold} [Hz]	τ_{hot} [s]	f_{hot} [Hz]
1	(2,1)	57.6	0.089	1.780	17.113	0.009
2	(2,1)	72.6	-	-	-	-
3	(1,0)	55.4	-	-	14.323	0.011
4	(2,1)	75.7	-	-	-	-
5	(2,1)	58.6	-	-	-	-
6	(3,1)	71.3	0.471	0.338	-	-
7	(3,1)	90.2	0.029	5.413	-	-
8	(5,3)	82.9	0.290	0.549	-	-
9	(5,3)	83.9	0.492	0.324	-	-

Table 2. Results of the ARMA modeling for the detector string L11

Level	ARMA	Fit [%]	τ_{cold} [s]	f_{cold} [Hz]	τ_{hot} [s]	f_{hot} [Hz]
1	(1,0)	44.1	-	-	12.123	0.013
2	(4,2)	58.2	0.224	0.712	11.235	0.014
3	(1,0)	56.7	-	-	22.990	0.007
4	(4,2)	52.1	0.150	1.063	6.070	0.026
5	(4,2)	61.3	0.409	0.389	4.405	0.036
6	(4,2)	56.4	0.309	0.514	2.919	0.055
7	(4,2)	48.6	0.279	0.571	2.626	0.061
8	(4,2)	53.0	0.424	0.375	1.755	0.091
9	(2,1)	54.2	-	-	-	-

Table 3. Results of the ARMA modeling for the detector string M03

Level	ARMA	Fit [%]	τ_{cold} [s]	f_{cold} [Hz]	τ_{hot} [s]	f_{hot} [Hz]
1	(2,1)	57.6	0.091	1.740	23.912	0.007
2	(2,1)	58.7	0.083	1.927	15.245	0.010
3	(2,1)	55.0	0.065	2.433	17.348	0.065
4	(2,1)	58.0	0.048	3.301	20.308	0.008
5	(1,0)	42.2	-	-	7.838	0.020
6	(1,0)	52.0	-	-	17.673	0.009
7	(2,1)	49.8	0.900	0.177	2.103	0.076
8	(2,1)	40.0	0.413	0.386	6.604	0.024
9	(4,2)	52.7	0.281	0.567	3.760	0.042

4 Conclusions

In the past, the bias between the MTC noise estimation and its actual value was considered to be due to the deviation of the reactor response from point-kinetics. A simple one-dimensional analytical model relying on the one-group diffusion approximation revealed that this non point-kinetic response of the reactor accounts for only a negligible part in the MTC underestimation. Most of the discrepancy comes from the fact that the temperature noise is not spatially homogeneous throughout the core. A new MTC noise estimator was therefore proposed. This estimator still relies on the point-kinetic response of the reactor, but uses a core average temperature noise. This core average temperature noise is estimated by using the square of the static flux as a weighting function. By doing so, the discrepancy between this MTC noise estimator and its actual value is very little, and in practice only due to the still remaining deviation of the reactor response from point-kinetics.

In order to use this new MTC noise estimator, one should be able to measure the local temperature noise in many points within the core, so that the average temperature noise could be numerically evaluated. Gamma-Thermometers (GTs) could be used for that purpose since in the frequency range of interest for the MTC determination by noise analysis, typically 0.1 to 1.0Hz, they work as thermocouples. The main reason for this is the large thermal time constant of the GT hot junction. This large time constant completely filters out the signal induced by the gamma/neutron noise, and only the signal associated due to the cold junction remains. This cold junction is in direct contact with the coolant and therefore monitors the moderator temperature noise. This was proven both qualitatively and quantitatively. The qualitative analysis relied on the linear phase behavior of the *CPSD* between two GTs located within the same fuel channel but at different axial elevations, this linear slope being characteristic of coolant temperature fluctuations traveling upwards and unperturbed. The quantitative analysis was carried out by ARMA modeling of the GTs transfer functions. Two time constants were in most cases found, one lying in the range 2.0 to 20.0s and representative of the hot junction, and another one lying in the range 0.1 to 1.0s and representative of the cold junction.

It is intended to test this new MTC noise estimator in the Swedish Ringhals-2 PWR, which is instrumented with 108 GTs throughout the core. The neutron noise would be monitored by an in-core neutron detector synchronized with the GTs signals. The ratio between the local temperature noise (close to the core center) and the average temperature noise is expected to give the calibration factor that was noticed experimentally previously, i.e., in the range of one half to one fifth. This means that in principle only the GTs signals should allow checking that the heterogeneous structure of the temperature noise throughout the core is responsible for the MTC underestimation.

Acknowledgements

We acknowledge Ringhals Vattenfall AB and more specifically tekn. lic. Tell Andersson for the gamma-thermometer measurements and their support regarding the MTC investigations. This work was supported by a research grant from Ringhals Vattenfall AB, research contract 502270-003, the Swedish Centre of Nuclear Technology (SKC), and a research grant from the French "Commissariat à l'Energie Atomique" (CEA/DRN/DER), research contract V. 315 8001.

References

1. Aguilar O, Por G (1987) Monitoring temperature reactivity coefficient by noise method in a NPP at full power. Ann. Nucl. Energy 14: 521-526
2. Andersson T (1994) On-line monitoring of thermal margins at Ringhals 2. Licentiate thesis, Chalmers University of Technology, Göteborg, Sweden
3. ANSI (1997) Calculation and measurement of the moderator temperature coefficient of reactivity for water moderated power reactors, an American National Standard. American Nuclear Society, ANSI/ANS-19.11-1997
4. Demazière C (2000) Development of a non-intrusive method for the determination of the moderator temperature coefficient of reactivity (MTC). CTH-RF-157, Chalmers University of Technology, Göteborg, Sweden
5. Demazière C, Pázsit I (2000) Theory of neutron noise induced by spatially randomly distributed noise sources. In: PHYSOR2000 conference, ANS International Topical Meeting on Advances in Reactor Physics and Mathematics and Computation into the Next Millennium, May 7-12, 2000, Pittsburgh, PA, USA
6. Demazière C, Pázsit I, Pór G (2000) Estimation of the moderator temperature coefficient (Analysis of an MTC measurement using boron dilution method). In: Embedded Topical Meeting #1: International Topical Meeting on Nuclear Plant Instrumentation, Controls, and Human-Machine Interface Technologies (NPIC&HMIT 2000), November 13-16, 2000, Washington, DC, USA
7. Demazière C, Arzhanov V, Pázsit I (2000) Final report on the research project Ringhals diagnostics and monitoring, Stage 5. CTH-RF-156/RR7, Chalmers University of Technology, Göteborg, Sweden
8. Demazière C, Pázsit I (2002) Theoretical investigation of the MTC noise estimate in 1-D homogeneous systems. Ann. Nucl. Energy 29: 75-100
9. García Cuesta J, Blázquez J (1995) PWR moderator temperature coefficient measurement using autoregressive modelling. In: SMORN VII, June 19-23, 1995, Avignon, France, pp 249-255
10. Haaland A, Devold H, Tosi V (1991) In-pile performance of miniaturised Gamma-Thermometers. HWR-313, OECD Halden Reactor Project, Halden, Norway
11. Herr JD, Thomas JR Jr. (1991) Noise analysis for monitoring the moderator temperature coefficient of pressurized water reactors: II. Experimental. Nuc. Sci. Eng. 108: 341-346
12. Karlsson J K-H (2000) Personal communication. GSE Power Systems AB, Nyköping, Sweden

13. Kostic LJ, Runkel J, Stegemann D (1988) Thermohydraulics surveillance of pressurized water reactors by experimental and theoretical investigations of the low frequency noise field. Prog. Nuc. Energy 21: 421-430
14. Kostic L, Husemann A, Runkel J, Stegemann D, Kahlstatt P (1991) Estimation of PWR moderator temperature coefficient as a function of neutron noise amplitude. In: SMORN VI, May 19-24, 1991, Gatlinburg, TN, USA, pp 35.01-35.12
15. Kostic L (1997) Monitoring of the temperature reactivity coefficient at the PWR nuclear power plant. Ann. Nucl. Energy 24: 55-64
16. Laggiard E, Runkel J (1997) Evaluation of the moderator temperature coefficient of reactivity in a PWR by means of noise analysis. Ann. Nucl. Energy 24: 411-417
17. Laggiard E, Runkel J (1999) Noise analysis estimation of the moderator temperature coefficient for a PWR fuel cycle. Ann. Nucl. Energy 26: 149-156
18. Ljung L (2000) System identification toolbox for use with Matlab, User's guide, Version 5. The Math Works, Inc., Natick, MA, USA
19. Oguma R, Lorenzen J, Bergdahl B-G, Liao B (1995) Study of noise analysis method for estimation of moderator temperature coefficient in a PWR. In: SMORN VII, June 19-23, 1995, Avignon, France, pp 32-40
20. Oguma R, Bergdahl B-G, Liao B, Lorenzen J (1995) Development of moderator temperature coefficient monitoring technique using noise analysis in PWRs, Result of noise analysis for measurement on 95-12-07. EuroSim report ES-95-29
21. Pór G, Izsák E, Valkó J (1985) Some results of noise measurements in a PWR NPP. Prog. Nuc. Energy 15: 387-393
22. Por G, Jozsa I (1995) Estimation of the temperature reactivity coefficient in operating nuclear power plant. In: SMORN VII, June 19-23, 1995, Avignon, France, pp 41-47
23. Shieh DJ, Upadhyaya BR and Sweeney FJ (1987) Application of noise analysis technique for monitoring the moderator temperature coefficient of reactivity in pressurized water reactors. Nuc. Sci. Eng. 95: 14-21
24. The Math Works (2000) Signal Processing Toolbox for use with Matlab, User's guide, Version 5. The Math Works, Inc., Natick, MA, USA
25. Thie JA (1977) Neutron noise sources in PWRs. Prog. Nuc. Energy 1: 283-292
26. Thomas JR Jr, Herr JD, Wood DS (1991) Noise analysis method for monitoring the moderator temperature coefficient of pressurized water reactors: I. Theory. Nuc. Sci. Eng. 108: 331-340
27. Thomas JR Jr, Clem AW (1991) PWR moderator coefficient via noise analysis: Time series methods. In: SMORN VI, May 19-24, 1991, Gatlinburg, TN, USA, pp 34.01-34.12
28. Tosi V, Haaland A (1993) The use of the gamma thermometer as a multipurpose sensor. HWR-357, OECD Halden Reactor Project, Halden, Norway
29. Türkcan E (1982). Review of Borssele PWR noise experiments, analysis and instrumentation. Prog. Nuc. Energy 9: 437-452
30. Upadhyaya BR, Shieh DJ, Sweeney FJ, Glöcker O (1988) Analysis of in-core dynamics in pressurized water reactors with application to parameter monitoring. Prog. Nuc. Energy 21: 261-269
31. Wu SM, Hsu MC, Chow MC (1979) The determination of time constants of reactor pressure and temperature sensors: The dynamic data system method. Nuc. Sci. Eng. 72: 84-96

Nomenclature

APSD	Auto-Power Spectral Density
ARMA	Auto-Regressive Moving Average
BOC	Beginning Of Cycle
CPSD	Cross-Power Spectral Density
EOC	End Of Cycle
GT	Gamma-Thermometer
HFP	Hot Full Power
HZP	Hot Zero Power
MTC	Moderator Temperature Coefficient
PWR	Pressurized Water Reactor

11 Detecting Impacting of BWR Instrument Tubes by Wavelet Analysis

V. Arzhanov* and I. Pázsit

Department of Reactor Physics, Chalmers University of Technology
Gibraltargatan 3, S-412 96 Göteborg, Sweden

arj@nephy.chalmers.se, imre@nephy.chalmers.se

*Permanent address: Keldysh Institute of Applied Mathematics, Moscow, Russia

Detection of impacting of detector tubes, also called instrument strings, has been a matter of interest both in Swedish and foreign BWRs. Although detection of the vibrations is relatively simple, the discovery and quantifying of the severity of impacting is far more complicated. There exists no single method that gives absolute results without the need for calibration or comparison with reference measurements.

Most known methods that were frequently applied in the past require comparison with a reference, i.e., impacting-free state, such as the broadening of the peak, decreasing of the decay ratio, or distortion of the probability distribution function. We describe a calibration-free method to detect and quantify impacting of instrument tubes in BWR reactors. The method is tested on measurements, taken by GSE Power Systems AB at the Swedish BWR Oskarshamn-2 during three fuel cycles.

1 Introduction

Discovering detector tube vibrations, and especially detecting impacting, has been of interest since relatively long. There are several ways of identifying and quantifying impacting from the signals of the vibrating detectors. These are based on the distortion of the phase between two detectors, widening of the peak in the auto power spectral density (APSD) function or increasing of the decay ratio (DR) associated with the vibration peak, distortion of the probability distribution function (PDF), and finally wavelet analysis.

Most of the methods are not absolute, rather relative and need access to data from the same string before impacting. Widening of a peak, decreasing of the decay ratio, distortion of the PDF are all, as the terminology discloses, methods that require a comparison to the impacting-free data in order to detect impacting.

Wavelet analysis is one of the few methods, if not the only, which to a large degree is absolute or calibration-free. It is based on a suggestion of Thie [1] that each impacting of a detector tube against the wall of a fuel assembly will induce short, damped oscillations of the fuel assembly itself, which then will contribute to the detector signal. High-frequency damped oscillations of the fuel assembly will manifest themselves as spikes, and the task of detecting impacting is then reduced to the task of detecting spikes in the signal. Such detection can be performed with wavelet analysis. Nevertheless it has to be emphasised that the method is based on a hypothesis, i.e. the high-frequency vibration of the fuel assembly on impacting, which is difficult to verify experimentally. However, there exists some evidence that the hypothesis is true [5].

We were given access to a large number of measurement data, taken by GSE power systems at Oskarshamn-2 between 1991 and 1994 through an agreement between GSE, SKI and Chalmers. A vibration monitoring system VIBMON [2, 3], installed by GSE at Oskarshamn, was used for data acquisition. In most cases there is also information available on whether or not impacting occurred, and if so, with what severity, through inspection of detector tube damage, which was performed during refuelling. Such a set of data is invaluable for the test of an independent method.

2 Description of the Problem

Unit 2 at the Oskarshamn nuclear power plant suffered excessive instrument (or guide) tube vibrations in the early 1990's. The guide tube vibrations reportedly caused wear and damage to neighbouring fuel boxes. In two cases the vibrations lead to the occurrence of big holes in adjacent fuel boxes [3]. Fig. 1 gives a visual explanation for the problem. The guide tube, which is of about 1.8 cm in diameter and 380 cm in length, contains several neutron detectors, also called Local Power Range Monitors (LPRM). The guide tubes are located in the space between four fuel boxes. Since the long thin tube is fixed only at its ends, it performs free vibrations when placed into the streaming water flowing around the string. The resonance frequency is typically 2 to 3 Hz.

Fig. 2 shows the core cross section of Oskarshamn-2 BWR reactor that has totally 24 in-core instrument guide tubes. They are numbered from left to right and from top to bottom. Also Fig. 2 indicates two holes found in 1992 and 1994 during inspections.

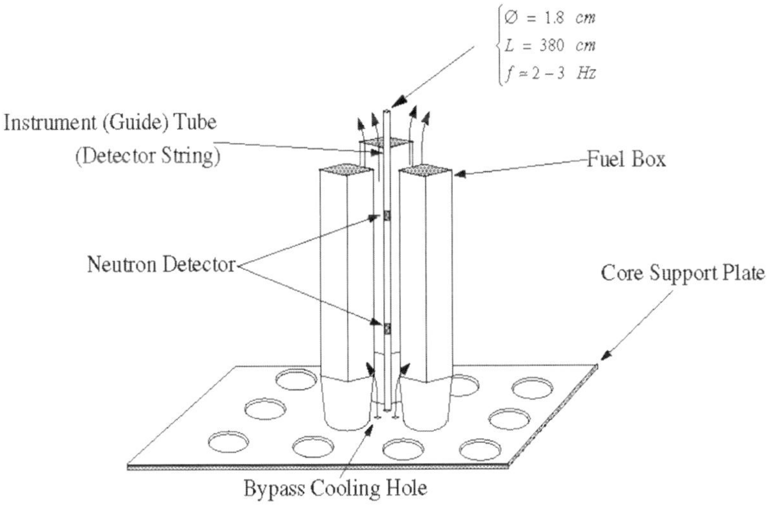

Fig. 1. An instrument tube between fuel boxes in the core

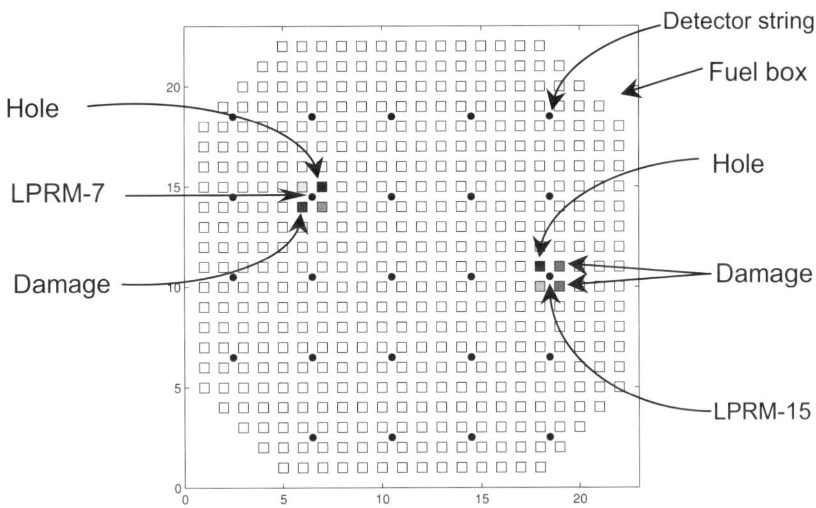

Fig. 2. The core map of Oskarshamn 2

The vibrations history is summarised in Fig. 3, which displays all the 16 measurements with dates ranging from 1991-Oct-21 when the guide tube vibration was first identified as a problem until 1994-Nov-29, by which time the problem was resolved. This time period covers fuel cycles 17 (C-17) to fuel cycle 20 (C-20). In addition, three visual inspections, designated as RA2-92, RA2-93, and RA2-94 are indicated together with the most significant findings. More information is found in [2, 3]

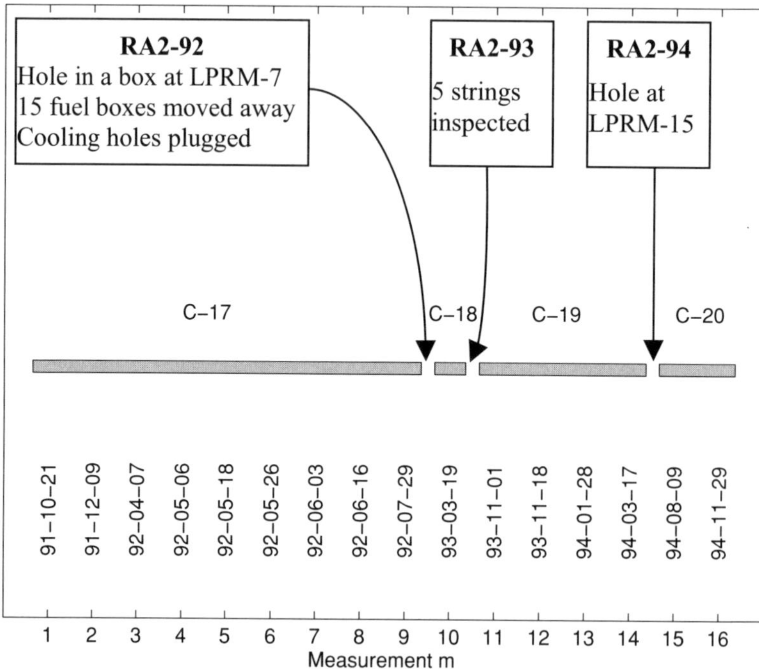

Fig. 3. Measurement history

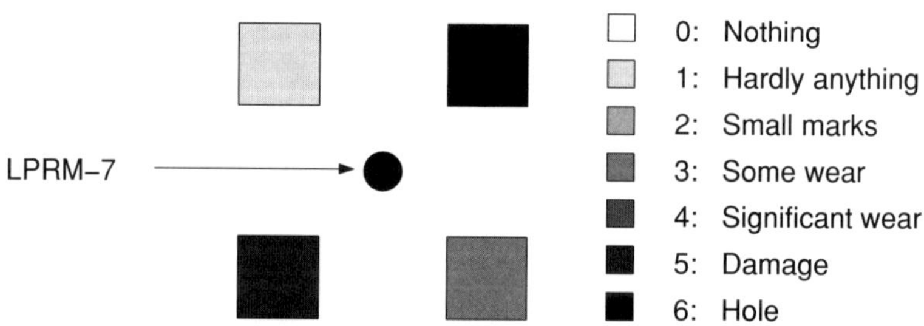

Fig. 4. Experts' damage grading of visual inspection

Fig. 4 shows a part of the damage map in the neighbourhood of detector string 7 together with the wear scale graded 0 to 6. It should be noted that the grading was done by visual inspection thus being not completely objective.

To compress the information, all wear marks corresponding to the surrounding fuel boxes were summed up to yield a total damage grade associated with a particular instrument tube. Fig. 5 displays the distribution of the total damage over the core map and the LPRM index respectively.

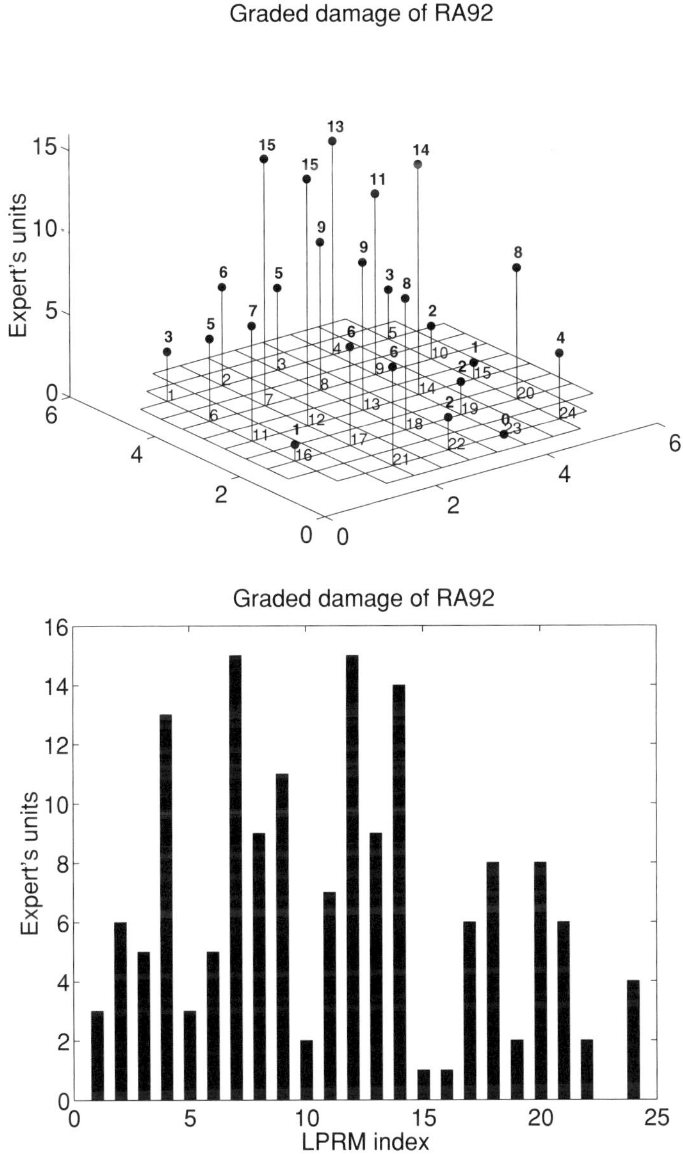

Fig. 5. Damage distribution of inspection RA2 92

3 Physical Model to Detect Vibrations

The physical quantity that allows us to identify mechanical vibrations of a neutron detector is the neutron flux gradient at the detector position. In case of a non-zero gradient, the vibrating detector moves between positions with different neutron flux thus giving an output signal proportional to the mechanical oscillations. Of course, in reality the detector vibrations are not monochromatic. In practice therefore one applies spectral analysis to the detector signal to find the peak in the APSD function corresponding to the mechanical vibrations. The whole scheme is illustrated in Fig. 6.

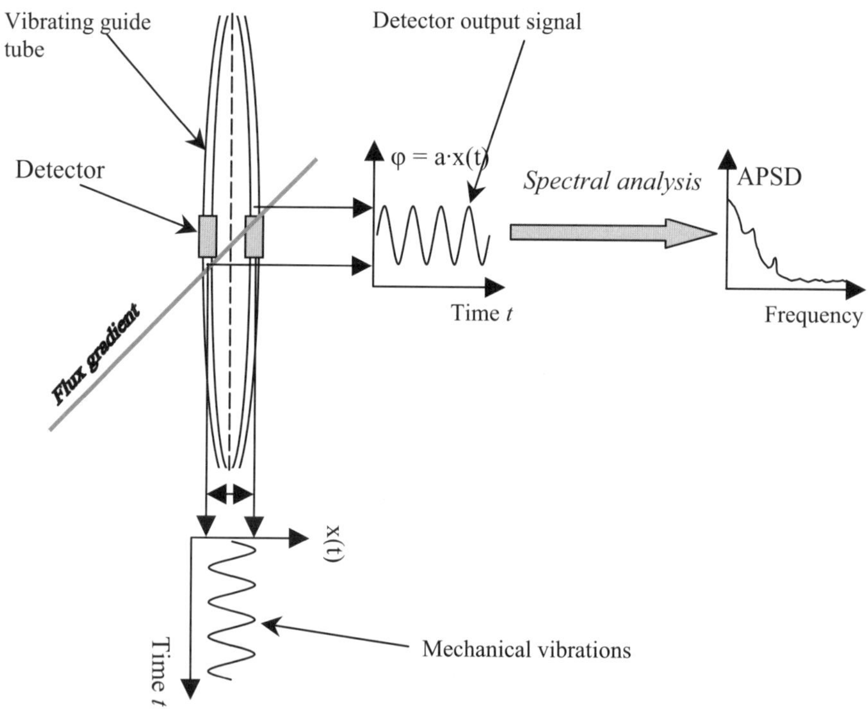

Fig. 6. Physical model to detect vibrations

It is worthwhile to notice the following two points. First, the existence of the flux gradient is necessary for obtaining oscillations in the LPRM signals. Otherwise there is no possibility of detecting the vibrations. Nevertheless, in practice the flux gradient is seldom close to zero. Second, usually we do not know the flux gradient and that prevents us from deducing the amplitude of mechanical vibrations via spectral analysis.

4 Physical Model to Detect Impacting

The physical model of impacting is based on a suggestion of Thie [1] that each impacting of a detector tube against the wall of a fuel assembly will induce short, damped oscillations of the fuel assembly itself, which will contribute to the detector signal. The situation is illustrated in Fig. 7. According to the model we assume the detector signal to consist of:

- Stationary global neutron noise, *N(t);*
- Neutron noise coming from the mechanical vibrations of the detector, *S(t),* which is additionally supposed to be stationary;
- Neutron noise induced by impacting, *T(t),* which is additionally supposed to be transient.

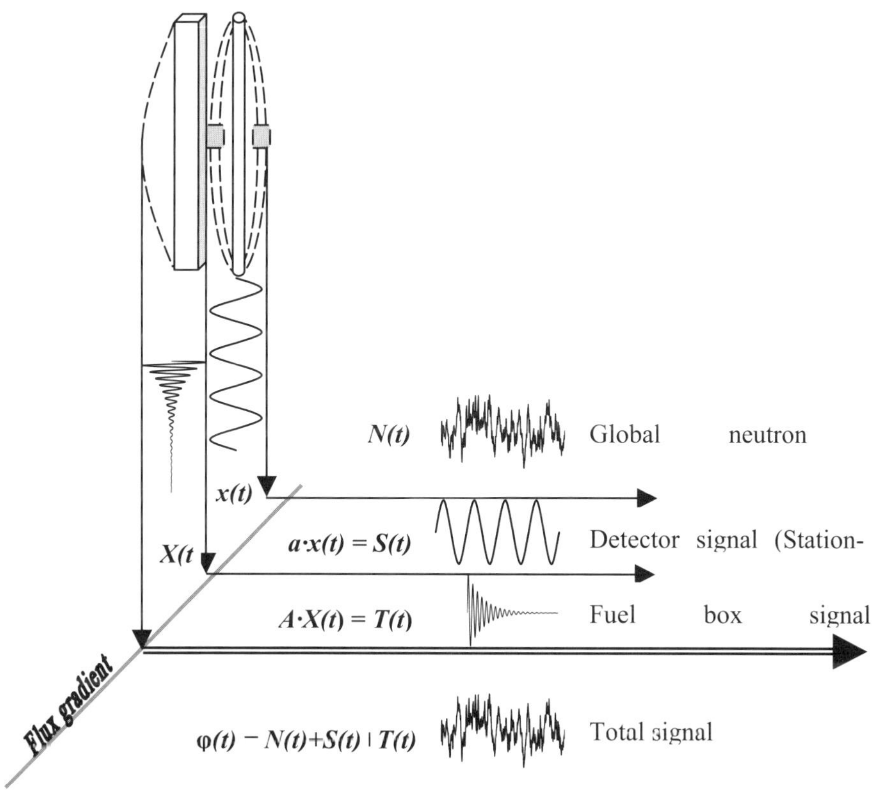

Fig. 7. Physical model to detect impacting

Summing up the three components we arrive at the representation for the detector signal

$$\phi(t) = N(t) + S(t) + T(t) \tag{1}$$

5 Basic Definitions in Wavelet Analysis

According to the model, presented briefly above, the neutron detector signal consists of a stationary process, which is perturbed by addition of a non-stationary process on impacting. Wavelets, in contrast to Fourier analysis, have reportedly proved to be a powerful tool in dealing with non-stationary data [4]. Very briefly, wavelet analysis consists of breaking up a signal into shifted and scaled versions of the original (mother) wavelet. More specifically, an arbitrary signal, $S(t)$, is cast into a double series

$$S(t) = \sum_{n,j} s_{n,j} \psi_{n,j}(t) \tag{2}$$

where the wavelet basis, $\{\psi_{n,j}(t)\}$, is constructed by using solely the original function $\psi(t)$

$$\psi_{n,j}(t) = \frac{1}{\sqrt{2^j}} \psi\left(\frac{t - 2^j n}{2^j}\right) \tag{3}$$

One often speaks of approximations and details. The approximations are the high-scale, low frequency components of the signal. The details are the low-scale, high frequency components. At the most basic level, this looks as follows

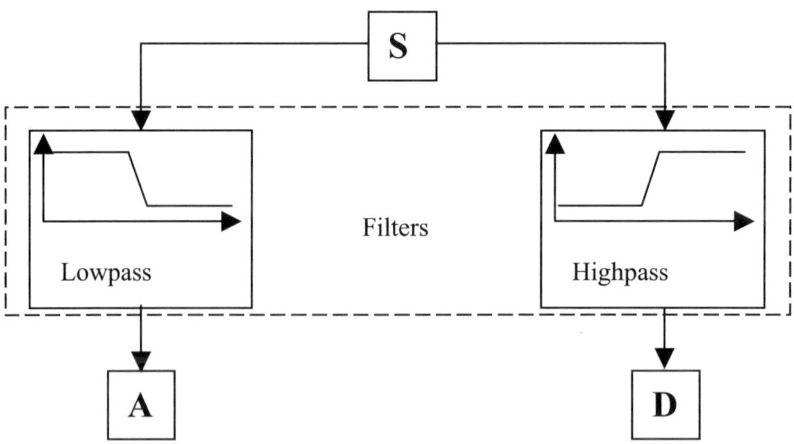

Fig. 8. Basic step in wavelet decomposition

This decomposition can be written as $S(t) = A(t) + D(t)$. A very simple and crude, but sometimes quite efficient de-noising method consists of neglecting details completely by setting $D(t) = 0$. A more advanced method is to set a threshold τ and disregard all details that are under this barrier by setting $D(t)$ to zero

$$\hat{D}(t) = \begin{cases} 0 & |D(t)| \le \tau \\ D(t) & |D(t)| > \tau \end{cases} \tag{4}$$

In general, wavelet analysis gives a more flexible way of de-noising the signal by recursively repeating the basic decomposition into approximations and details as shown in Fig. 9.

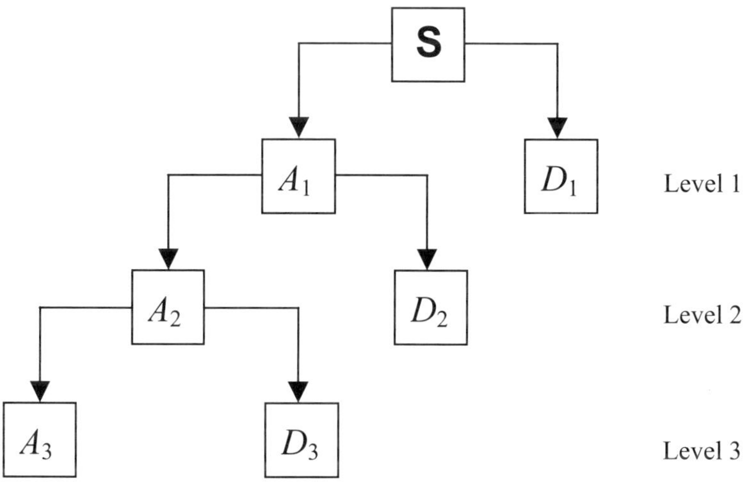

Fig. 9. Multi-level wavelet decomposition

The most general technique to filter the signal is to set different thresholds at different levels and afterwards to reconstruct the signal as

$$\hat{S}(t) = A_n(t) + \hat{D}_n(t) + \ldots + \hat{D}_1(t)$$

6 One-Dimensional Model to Simulate Impacting

Although the vibration in the X-Y plane is two-dimensional, no qualitative difference was found between 1-D and 2-D simulations [5]. Because of this fact, a simplified one-dimensional simulation has been used in the current work to study the possibility to detect impacting of a detector string against surrounding fuels boxes, as shown in Fig. 10.

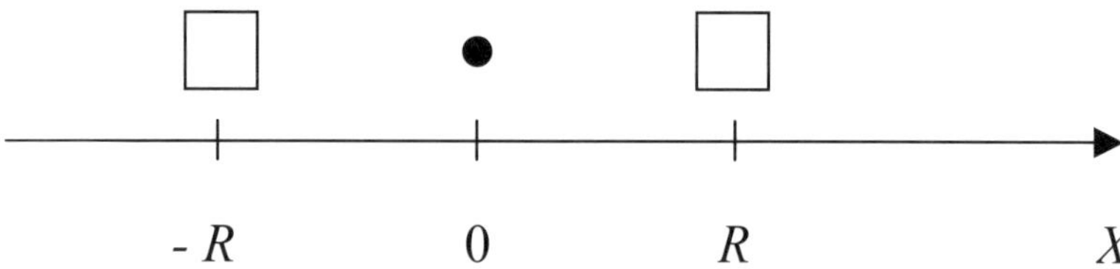

Fig. 10. One-dimensional model to detect impacting

Following [6], we assume that the damped oscillations of a detector guide tube may be described by the equation

$$\ddot{x}(t) + 2\theta\dot{x}(t) + \omega_0^2 x(t) = f(t) \tag{5}$$

with $\langle x(t) \rangle = 0$ being the equilibrium position, θ standing for the damping factor, and *f(t)* being a random driving force. It was found reasonable [6, 7] to model the stochastic force *f(t)* by a discrete series of impulses arriving at regular times $t_n = n \cdot \Delta t$

$$f(t) = F_c \sum_{n=-\infty}^{+\infty} r_n \delta(t - t_n) \tag{6}$$

where $\{r_n\}$ is a normal random variable with mean value $\mu_r = 0$ and standard deviation $\sigma_r = 1$. The parameter F_c, also called the "force coefficient," describes the strength of the driving force. The pulse repetition frequency must be chosen much higher than that of the oscillator, i.e.

$$\frac{1}{\Delta t} >> \frac{\omega_0}{2\pi}$$

Impacting is simulated by confining the detector string motion within *(-R,+R)*. Whenever $|x(t)|$, in the course of simulation, exceeds *R*, the velocity $\dot{x}(t)$ is reversed, i.e., $\dot{x}(t) \to -\dot{x}(t)$, which models elastic reflection from an infinite mass without energy loss. According to our model we assume that the neutron noise due to the mechanical vibration, *S(t),* is linearly related to the displacement *x(t),* i.e. $S(t) = a \cdot x(t)$.

The model, used in the current work, also involves the fuel box vibration that obeys the equation

$$\ddot{X}(t) + 2\Theta\dot{X}(t) + \Omega_0^2 X(t) = F(t) \tag{7}$$

where the force $F(t)$ is induced by impacting. At each impact, the detector tube transfers a certain impulse, $2m|\dot{x}(t)|$, to the fuel rod. Thus, if the box/tube mass ratio is $M/m = k$, then, at impacting, one has $|\dot{X}| = 2\dot{x}/(k+1)$. This will lead to the representation

$$F(t) = \sum_i 2\frac{M}{k+1} \cdot \delta(t-\tau_i)|\dot{x}(\tau_i)| = c\sum_i \delta(t-\tau_i)|\dot{x}_i| \tag{8}$$

with τ_l being the moments of impacting.

Once again, we assume the neutron noise caused by impacting to be proportional to the displacement $X(t)$, i.e., $T(t) = A \cdot X(t)$. Finally, the detector signal reads as

$$\phi(t) = N(t) + a \cdot x(t) + A \cdot X(t) \tag{9}$$

7 Results with Simulated Data

Three simulated signals are shown in Fig. 11. The stationary component of the detector signal, $a \cdot x(t)$, is shown in Fig. 11a, the transient component, $T(t) = A \cdot X(t)$, is given in Fig. 11b, and finally, the third plot, Fig. 11c, displays the total detector signal $\phi(t)$. It should be noted here that the amplitude of the fuel box signal, $T(t)$, is 2 orders of magnitude smaller as compared to the detector tube signal, $a \cdot x(t)$, and thus it is completely invisible in the total signal, $\phi(t)$.

Various wavelets available in MATLAB Wavelet Toolbox [8] have been tried to identify impact events from the simulated detector signal, $\phi(t)$. A typical example is presented in Fig. 12, the first row of which displays the first four members of the Daubechies wavelet family with *db1* being nothing else but the Haar function. For the sake of comparison the last row reproduces impact events that are known in the course of simulation. Finally, rows D1, D2, and D3 display details of the first three levels evaluated by the corresponding wavelets.

The other wavelet functions tried in this research give a similar picture, some of them being more preferable while others being less suitable to detect impacting. In a sense, *db4* was found to be the best for our purposes.

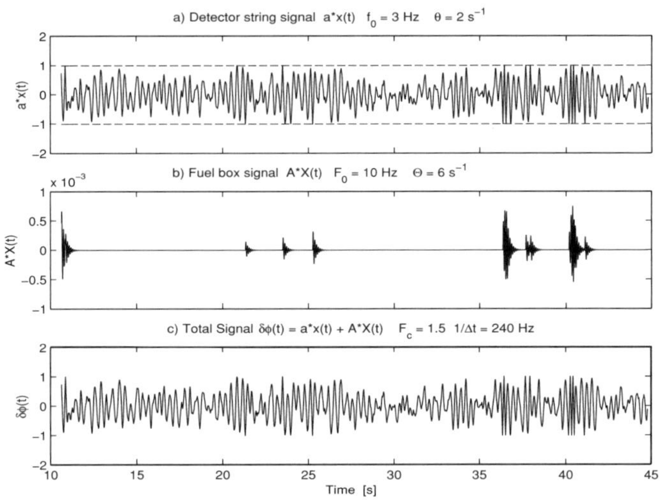

Fig. 11. Simulated signals

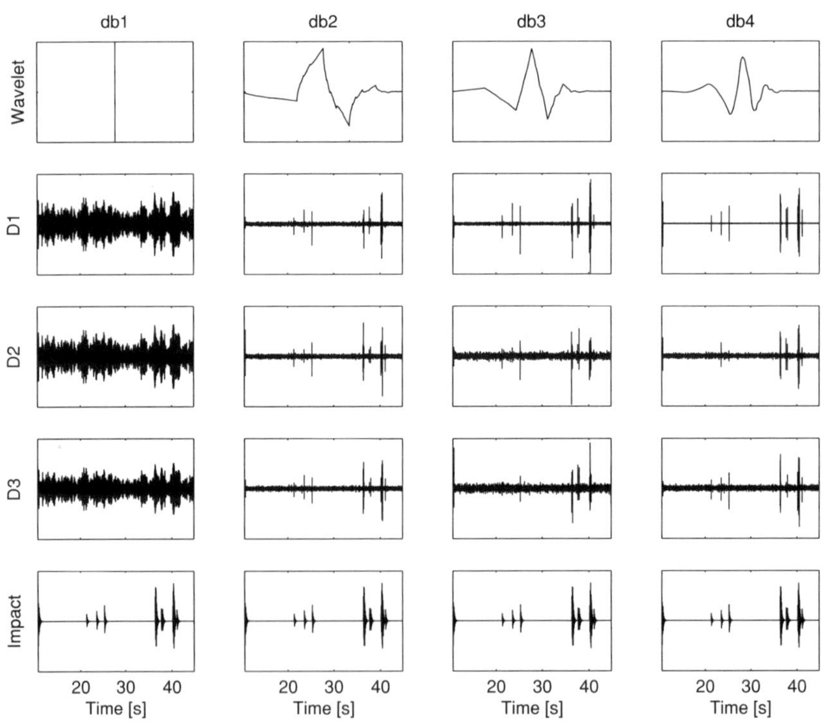

Fig. 12. Examples of wavelet analysis using different wavelets

8 Results with Real Data

To obtain reliable results we need to evaluate the global neutron noise $N(t)$ in the detector signal $\phi(t)$

$$\phi(t) = N(t) + S(t) + T(t)$$

Let f_0 and F_0 be eigenfrequencies of the detector string and the fuel box respectively. By assumption, $N(t)$ is white noise and $F_0 > f_0$. Because of this, we have

$$\phi_f(t) \equiv HP_f\{\phi(t)\} = HP_f\{N+S+T\} = HP_f\{N\} + HP_f\{S+T\} \approx HP_f\{N\} \quad (10)$$

if we choose $f > F_0$, where HP_f is a high-pass filter with a cut-off frequency f. This allows us to evaluate

$$\sigma_N \equiv Var\left[N(t)\right] \approx Var\left[\phi_f(t)\right]$$

and hence to set up the threshold in (4) $\tau = \sigma_N$.

In order to quantify the severity of impacting, we define an impacting rate (*IR*) as the number of spikes per unit time of observation survived the wavelet threshold procedure (4). Typical examples of wavelet-filtered signals, together with the *IR* values, are given in Fig. 13.

Detector string 8 is known to be strongly impacting during cycle 17 (measurement 7) and less impacting during cycle 19 (measurement 11). Two factors have contributed to this change. The first was an improvement in construction, which led to a decrease of the turbulence of the coolant. The second is the decreased coolant flow, from 7.6 to 7.0 ton/s. This experimental evidence agrees well with the piece of wavelet analysis presented in Fig. 13, which shows the *IR* indexes evaluated for the upper sensor of string 8 (LPRM 8.1) for measurement 7, $IR = 0.875$, and for measurement 11, $IR = 0.140$.

Fig. 14 summarises the results of the wavelet analysis performed for all the measurements. Each individual subplot of Fig. 14 displays a distribution of the IR index over all 24 LPRM signals. The 25-th bar gives the water flow for comparison reason. The *IR* indicator is given relative to the maximal value over all signals and all measurements. This maximal value was found to be $IR = 2.57$ for detector tube 15 in measurement 13.

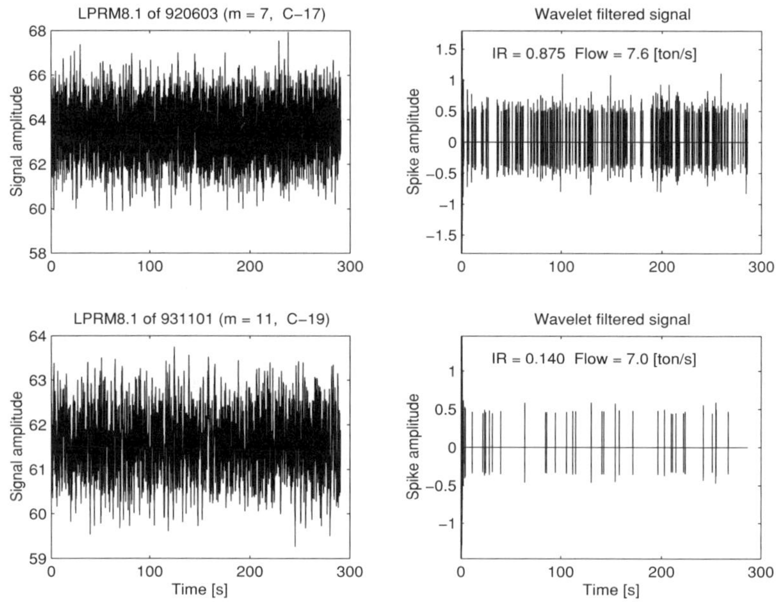

Fig. 13. Wavelet analysis as applied to real signals

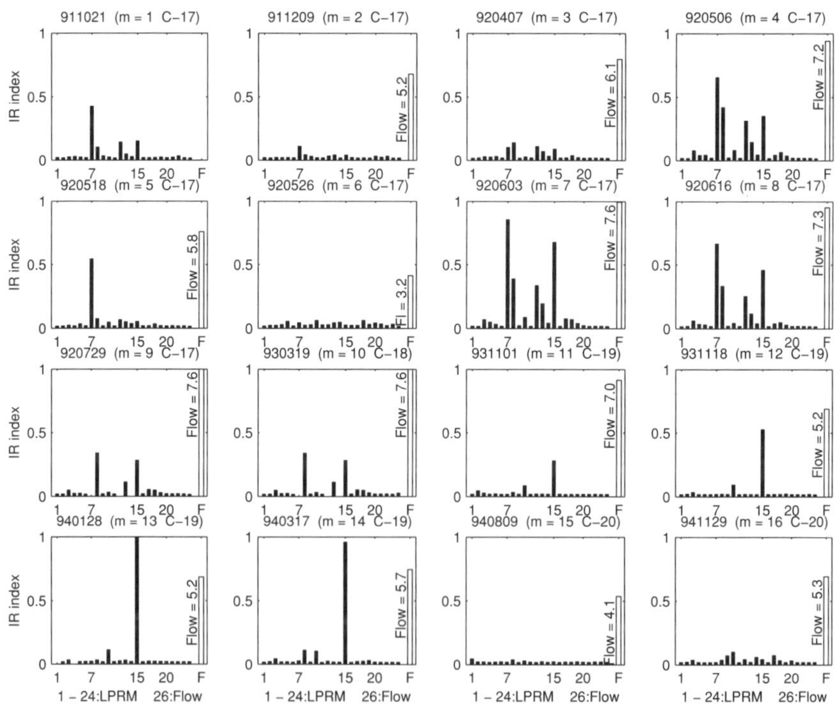

Fig. 14. Results of wavelet analysis for all measurements

As one can see in Fig. 14, there is a strong dependence of the impacting rate on the water flow that fully corresponds to our expectations. For example, water flow of 6.1 ton/s causes small impacting in measurement 3 as contrast to measurement 4 with water flow of 7.2 ton/s.

The first construction improvement was made between cycles 18 and 19. As a result we can see a qualitative change in the behaviour of *IR*, measurements 7 to 10 as compared to measurements 11 to 14. The second improvement was made between cycles 19 and 20, once again this resulted in a dramatic change of the *IR* distribution, measurements 11 to 14 as contrast to measurements 15 and 16. Moreover, tube 7 had the largest impacting rate before the first change in construction was made. This fully conforms to the visual inspection RA2-92 that revealed a big hole in the neighbourhood of instrumentation tube 7. In the period of time between the first and second improvements, cycle 19, string 15 was found to show the most intense impacting, which perfectly agrees with the inspection RA2-94 that discovered another large hole in a fuel box next to tube 15.

To take into account the cumulative nature of the damage due to impacting we introduce a cumulative impacting rate index as a sum of corresponding indices during the time of observation. Fig. 15 presents the distribution of the cumulative *IR* evaluated for cycle 17 as compared to the total damage reported in the inspection RA2-92.

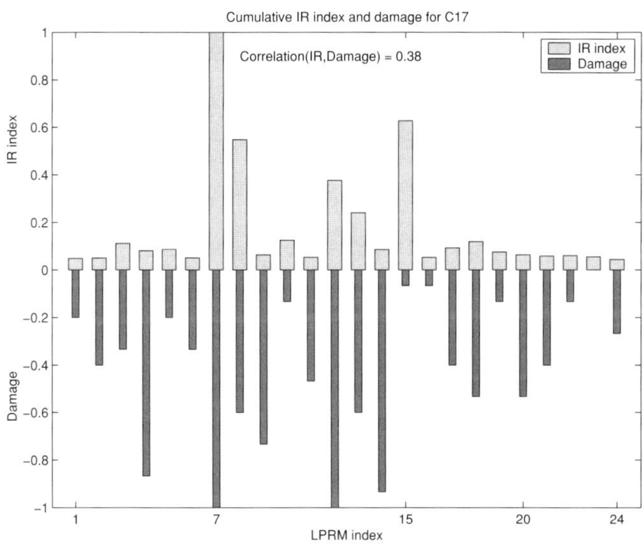

Fig. 15. Cumulative impacting rate

The correlation of this indicator with the reported damage was found to be 0.38 that is quite good as compared to other indicators. The comparison is given in Fig. 16 that shows the *IR* index as the third best correlated indicator.

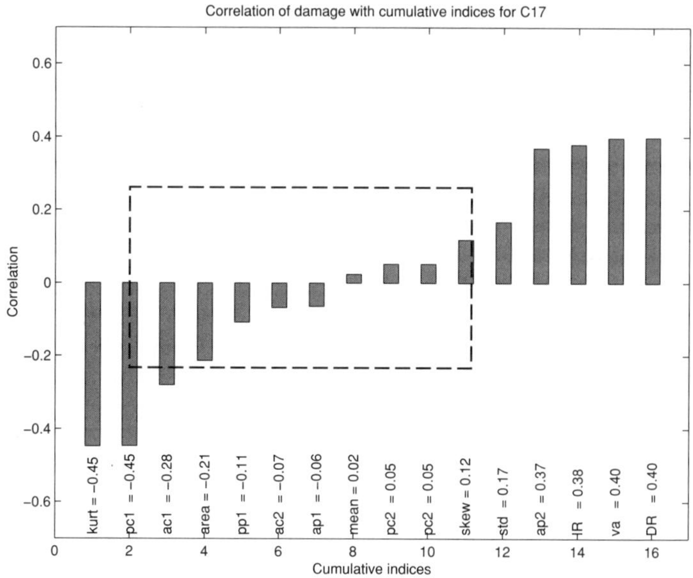

Fig. 16. Wavelet analysis as compared to other impacting indicators

Indicators displayed in Fig. 16 are as follows:

- *mean* - mean value;
- *std* - standard deviation;
- *kurt* - kurtosis;
- *skew* - skewness;
- *DR* - decay ratio;
- *pp1* - power (amplitude) of the first peak (peak 1) in APSD;
- *ap1* - area under peak 1 in APSD;
- *pp2* - power (amplitude) of the second peak (peak 2) in APSD;
- *ap2* - area under peak 2 in APSD;
- *pc1* - power (amplitude) of the first peak in the coherence function between the signals of the upper and lower detectors;
- *ac1* - area under the first peak 1 in coherence;
- *pc2* - power (amplitude) of the second peak in the coherence function;
- *ac2* - area under peak 2 in coherence;
- *va* – apparent vibration amplitude;
- *IR* - impacting rate.

The apparent vibration amplitude, *va*, is calculated by the formula proposed in [2]:

$$va = 3.28 \cdot \sqrt{\left\langle HP_f(S_1) \cdot HP_f(S_2) \right\rangle}$$

where S_1 and S_2 are signals of the upper and lower detectors, HP_f is a high pass filter with a cut-off frequency, $f = 4\ Hz$.

It is interesting to note that, firstly, kurtosis and peak 1 in the correlation function between the upper and lower signals were found to be strongly anti-correlated with the reported damage; secondly, the indicators embraced in the dashed rectangle have very little to say about impacting; whereas thirdly, *IR* together with *DR*, *va*, and *ap2* are mostly correlated with the reported damage.
It should be noted however, that the impacting rate is the only indicator that practically needs no calibration prior to the use. Let us consider, for example, the best correlated indicator, *DR*, which was calculated by, first, fitting an estimate of the auto correlation function of the signal $x(t)$

$$C_{xx}(\tau) \equiv E\{[x(t)-\mu]\cdot[x(t+\tau)-\mu]\} \qquad \mu \equiv E\{x(t)\}$$

to the parametric function

$$C(\tau) = \alpha e^{-\lambda\cdot\tau}\cos(2\pi\nu\cdot\tau+\phi)+\beta e^{-2\pi\eta\cdot\tau}$$

and, second, defining *DR* as

$$DR \equiv e^{-\lambda/\nu}$$

Table 1. Comparison between *IR* and *DR* indicators in measurement 4

	Non-impacting tube, LPRM 16 *Damage 1 out of 15*	*Impacting tube, LPRM 12* *Damage 15 out of 15*
Upper detector	$\frac{IR = 0.017}{DR = 0.299}$	$\frac{IR = 1.039}{DR = 6.6\cdot 10^8}$
Lower detector	$\frac{IR = 0.017}{DR = 0.354}$	$\frac{IR = 0.652}{DR = 0.366}$

It is seen from Table 1 that one has to know in advance the *DR* value corresponding to a non-impacting tube, whereas *IR* is estimated close to zero in this situation. As a matter of fact, it should be noted that sometimes the estimation of *DR* becomes unstable, such as the unrealistic value of $6.6\cdot 10^8$ in Table 1.

9 Conclusion

We itemise the conclusion as follows.

a) The large measurement set, together with knowledge of occurrence and severity of impacting, gave an outstanding opportunity to test the method of detecting impacts by wavelet methods;

b) For heavy impacting, the wavelet analysis was found to be very effective of identifying and quantifying impacting without the need of calibration to measurements with known impacting or without impacting;

c) For mild impacting, the performance of the wavelet method is not perfect, although fully comparable with that of other methods;

d) Further development and tests will be made with various types of wavelets as well as further improvement of the threshold procedure and the way to estimate the cumulative impacting rate is planned.

References

1. Thie JA (1981) "Power Reactor Noise," American Nuclear Society, USA
2. Bergdahl B-G, Oguma R (1996) "Investigation of In-Core Instrument Guide Tube Vibrations at Oskarshamn BWR Unit 2 Based on Noise Analysis," SKI Report 96:37
3. Karlsson JK-H, Bergdahl B-G, Oguma R (2001) "The guide tube vibrations at Oskarshamn-2 1991-1994." GSE Power Systems AB, GSE-01-006
4. Mallat S (1998) "A Wavelet Tour of Signal Processing," Academic Press, San Diego
5. Rácz A, Pázsit I (1998) "Detection of detector string impacting with wavelet techniques," Annals of Nuclear Energy 25: 387-400
6. Pázsit I, Antonopoulos-Domis M, Glöckler O (1984) "Stochastic aspects of two-dimensional vibration diagnostics," Progress of Nuclear Energy 14: 165-196
7. Pázsit I, Glöckler O (1994) "BWR Instrument Tube Vibrations: Interpretation of Measurements and Simulation," Annals of Nuclear Energy 21: 759 - 786
8. Misiti M, Misiti Y, Openheim G, Poggi J-M (2000) "Wavelet Toolbox for Use with MATLAB," User's Guide, Version 2, The Math Works, Inc.

12 Development of Advanced Core Noise Monitoring System for a Boiling Water Reactor

Michitsugu Mori*, Shigeru Kanemoto**, Mitsuhiro Enomoto**, Shigeo Ebata**

*)Tokyo Electric Power Company
4-1, Egasaki-cho, Tsurumi-ku, Yokohama, 230-8510 Japan,

mori-mcy@rd.tepco.co.jp

**)Toshiba Corporation
8, Shinsugita-cho, Isogo-ku, Yokohama, 235-8523 Japan,

shigru.kanemoto@toshiba.co.jp

For an efficient nuclear power plant operation, advanced core design is very important. However, advances in design may result in approaching unexplored areas where unforeseen events are increasingly common. The purpose of this paper is to describe a BWR core noise monitoring system for addressing core anomaly problems in future advanced core operation. To monitor in-core status from a limited number of signals, we introduce various new algorithms to compensate for a lack of information. We demonstrate that up-to-date signal processing technologies, such as independent component analysis, nonlinear principal component regression, factor analysis and neural network, are effective tools for core monitoring. Also, the combination of the first principle or design model and the empirical fitting model is applied to estimate unobserved in-core state variables. Several concrete examples, such as stability monitoring, reactivity coefficient estimation, core flow estimation and in-core signal validation, are shown based on long-term monitoring experiences in real BWR plants. Through these examples, we demonstrate that the core noise monitoring system is a general platform for providing a variety of monitoring tools to meet the requirements of an advanced core operation strategy.

1 Introduction

In view of the scaling back of the construction of new plants, an efficient nuclear power plant operation has become increasingly important in recent years. Hence, a number of countermeasures are discussed such as plant life extension, up-rating, prolongation of fuel cycle, shortening outage period or online maintenance. Among them, reactor core design for an effective fuel usage is widely recognized as being the most realistic countermeasure and is the focus on intensive development efforts. In particular, MOX fuel usage is desired in Japanese plants from the viewpoint of fuel cycle optimization.

Considering these circumstances, core integrity monitoring is emerging as an important issue. Introduction of new core and fuel design may have side effects that reduce the safety margin even if the design is validated carefully, since the core behavior is dominated by complex physics based on nuclear, thermal hydraulics, chemical and mechanical phenomena. It is reasonable to expect noise signal analysis to be an effective tool for monitoring such unanticipated and undesired side effects brought about by advanced core design and operation, since the noise signals are rich in information on in-core complex dynamic behavior. Also, noise analysis is expected to be a cost-effective method since it utilizes existing instrumentation signals.

The purpose of the present paper is to develop the core noise monitoring system for a boiling water reactor (BWR) by utilizing the latest signal processing technology. In recent years, a number of new signal processing technologies are proposed in various fields concerning the complex system. They include technologies for brain wave signal analysis, sound and image analysis, and financial analysis. In the present system, we utilize various methods, such as principal component analysis, factor analysis, independent component analysis, and neural network, in addition to the conventional Fourier transformation and the autoregressive modeling analysis. Also, we utilize the combination of the first principle or design model and the empirical model to estimate unobserved parameters in core dynamics. We define this combination model as the gray model. The use of these various new algorithms is essential for accurate, robust and flexible monitoring of reactor core status from a limited amount of instrumentation.

These new methods are applied to monitor BWR core dynamic behavior. The core and regional stability monitoring is one of the important objects of the present system due to the recent core power density and size increase or MOX fuel usage. Also, the void and temperature reactivity coefficients should be evaluated more correctly. The sensor validation is also an important target in core dynamics monitoring. As an example, the result of the core flow signal validation is shown in the present paper using the gray model based on a thermal hydraulic model and in-core neutron noise analysis. The present paper shows the concept of the core noise monitoring system and its algorithms through concrete application examples based on real plant data .

2 Outline of the Core Noise Monitoring System

Advanced core design is directed to achieving higher burn up of fuel, MOX fuel usage, prolongation of fuel cycle, up-rating of core thermal output, and so forth.

These strategies may affect core dynamics behavior since the core will approach to an operating area previously not experienced. The direct effects are reactor physics parameter change, such as void or temperature reactivity coefficients. They also affect core nuclear thermal hydraulic stability behavior. Also, increase of power density and core size induce complicated stability phenomena such as regional neutron flux oscillation. From the viewpoint of thermal hydraulics, it will be important to monitor a flow and energy balance to confirm operational safety margin changes due to higher burn up and long-term operation. Also, it is important to perform sensor validation using an analytic redundancy technique before applying various diagnosis algorithms. Mechanical oscillation monitoring will be important to prevent unanticipated phenomena, since advances in core design will approach unexplored areas where unforeseen events might occur. These monitoring objects of the present system are summarized in Table 1 together with monitoring algorithms.

Table 1. Monitoring items and methods in the System

Area	Monitoring Items	Used Signals	Methods
Reactor physics parameter	Void reactivity coefficient	APRM, Core flow	AR model & Transfer function
	Temperature reactivity coefficient	SRNM Temperature	Factor analysis
Reactor stability	Core & Regional Stability	LPRMs	AR model, Independent Component Analysis
	Fuel Channel Stability	LPRMs	AR model
Thermal Hydraulics	Flow balance change	Pressure drop. Core flow, Jet pump flow	Mass & momentum balance model
	Heat balance change	Core flow, Pressure drop, Temperature	Mass & energy balance model
	Core Flow Rate	LPRMs, Core Pressure drop Temperature	Drift flux model & correlation function
	General sensor signal validation	Flow, Pressure, Temperature, Water level etc	Principal Component Regression GMDH, Neural Network
Mechanical	Instrument tube & fuel channel vibration	LPRMs	Power Spectral Density & coherence function

Another feature of the present system is provision of cost-effective monitoring tools to satisfy the above monitoring objects by analyzing fluctuating component signals from existing instrumentation. Here, we utilize state of the art technologies, such as neural network modeling, independent component analysis in addition to the conventional signal processing technologies. Also, we utilize the first principle or design model to compensate for a lack of information due to a limited amount of in-core instrumentation and to estimate unobserved core states. Here, to increase estimation accuracy, we combine the first principle model with the empirical fitting (black-box) models, since the first principle model does not have an adaptation capability to simulate an actual plant status with a high accuracy. This combination model is considered to be a gray model since it has the good features of both models; that is, the generalization or robust capability of the first principle (white-box) model and the adaptive or accurate capability of the fitting (black-box) model. The core flow estimation in Table 1 is a good example for utilization of the gray model.

The present system is required to provide the latest monitoring algorithms, user-friendly interfaces to handle the high-level monitoring functions, and flexibility for future extension of new monitoring functions. To meet these requirements, we adopt Visual Basic language for the user interface, DLL library based on C and FORTRAN languages to construct signal processing algorithms, and MATLAB software to utilize the latest algorithms developed in other engineering areas. This combination is very powerful for constructing the system at a low cost and to provide a flexible extension capability. We can also easily utilize the past assets of various program source codes for plant design or monitoring system development by adopting DLL libraries.

Figures 1 and 2 show the appearance and system configuration of the core noise monitoring system, which was set up in an actual BWR plant. The system consists of a personal computer, an A/D converter and LAN, which was connected with the process computer and the plant diagnosis system. The core performance calculation results from the process computer and various time scaled raw data are stored periodically and offered as an open structure database for various kinds of monitoring functions. Here, the EWS based data server is used to assure the security of the plant control network. This system has been in operation from the start up test from 1996 until now in a joint study being conducted by TEPCO and Toshiba. In the joint study, various kinds of monitoring algorithms have been developed, and, some of them were verified by this on-line system and long-term real plant data. Furthermore, they were applied to other types of BWR plants to verify their generality using off-line acquired data. In the following section, several examples of the core noise monitoring functions and their verification results are shown.

Fig. 1. System appearance

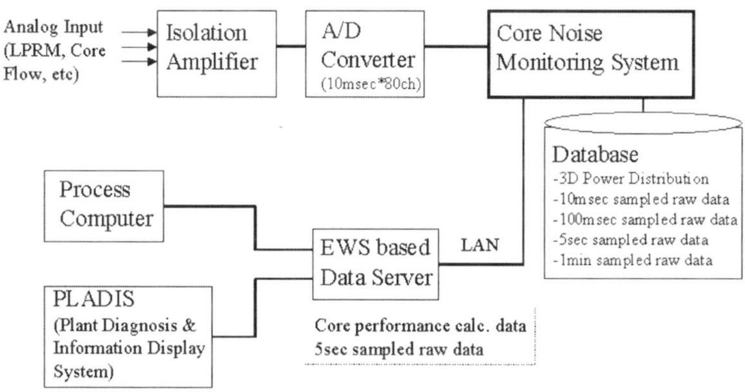

Fig. 2. System configuration

3 Examples of Individual Core Noise Monitoring Functions

We will show typical core noise monitoring sub-systems. In order to extract maximum information from a limited number of observed signals, we introduce various new algorithms. Here, we explain why the new algorithms are necessary and how they work in actual plant monitoring.

3.1 Stability Monitoring Sub-System

Core stability monitoring is one of important monitoring items in BWR plants. Due to coupling effects of neutron diffusion and two-phase flow dynamics, a variety of oscillation modes may be observed. In the current reactor core design, three kinds of instability modes are considered, that is, a core global oscillation mode called core stability, a local fuel channel flow oscillation mode called channel stability, and an out of phase power oscillation mode called regional stability.

The present sub-system provides a monitoring tool for these three kinds of stability mode. For the core and channel stability monitoring, we applied the conventional autoregressive (AR) modeling method to evaluate a stability index (decay ratio) using APRM and LPRM noise signals (Kanemoto et al. 1988). The AR model is expressed by

$$x(t) = \sum_{m} a(m)x(t-m) + \varepsilon(t) \tag{1}$$

AR coefficients can be obtained by the recursive or batch least squares algorithms. Then, the decay ratio is estimated from the impulse response function calculated from the AR model.

For the regional stability monitoring, we proposed a regional power range monitor (RPRM) signal(Takeuchi et al. 1999, Yamashita 1997), which is defined as a weighted sum of LPRM signals as follows:

$$RPRM_k(t) = \int W_k(r) LPRM(r,t) \tag{2}$$

Here, weighting factor $W_k(r)$ is obtained by an eigenmode of a neutron diffusion equation, which is approximated by the Bessel function. The present system adopts two kinds of weighting factors that consider azimuthally orthogonal oscillation modes. The decay ratios for regional stability are estimated from the RPRM signals.

An example of the stability monitoring subsystem display is shown in Figure 3. Here, oscillation waveforms for APRM, RPRMs and LPRMs are shown together with decay ratios, since waveform is important for grasping a complicated oscillation mode intuitively. Furthermore, a Lissajous figure of two RPRMs is displayed to identify a regional oscillation direction. The system also provides other kinds of display such as power spectral density and decay ratio trends.

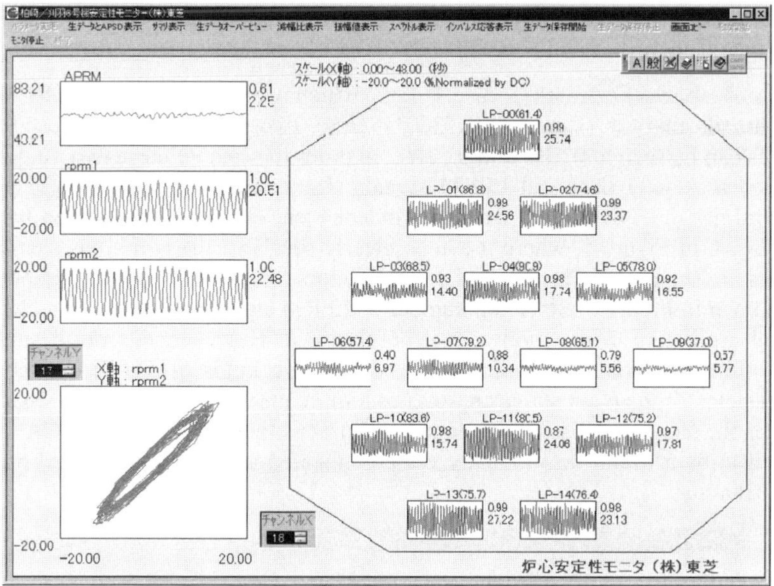

Fig. 3. An example of stability monitor display

In addition to the decay ratio estimation, we applied a new method called independent component analysis (ICA), which has been studied in the areas of brain wave analysis and sound analysis (Jutten & Herault 1991). ICA is a method of time series data analysis to obtain independent component axis. Here, the meaning of "independent" is different form that of "uncorrelated" which is used in principal component analysis (PCA). Among the several different ICA algorithms, we utilize the following algorithm called the deflation algorithm (Thawonmas 2000):

$$
\begin{aligned}
& x(t) = As(t) + \varepsilon(t), \qquad y_i(t) = w^T x(t) \\
& I = \frac{K_4(y_i)}{K_2(y_i)^2} = \frac{E\left[(w^T x)^4\right] - 3E\left[(w^T x)^2\right]^2}{E\left[(w^T x)^2\right]^2} \Rightarrow \min \\
& w = w + \mu\left(E\left[y_i x\right] - \frac{E\left[y_i^2\right]}{E\left[y_i^4\right]} E\left[y_i^3 x\right]\right) \\
& x_n = x_{n-1} - \frac{E\left[x_{n-1} x_{n-1}{}^T\right] w}{E\left[y^2\right]}
\end{aligned}
\tag{3}
$$

Here, the i-th scalar estimate $y_i(t)$ for independent component vector s(t) is obtained from linear transformation of raw data vector x(t) using a weight vector w. The weight vector, w, is obtained by the recursive algorithm so as to minimize the 4-th order moment, that is, the kurtosis factor. This feature is quite distinct from those of the conventional PCA method, which minimizes the second moment correlation factor. Due to this feature, ICA can distinguish a delicate independent eigenmode from mixed and complicated oscillation modes.

Figure 4 shows Forsmark-2 case.4 data, which were used in the international benchmark analysis (Verdu et al. 2001). Here, two slightly different oscillation modes can be seen at 0.5Hz and 0.52Hz. In order to separate these two modes, we applied ICA to 7 observed LPRM signals (Kanemoto et al. 2001). The power spectral densities calculated from two independent components obtained by ICA are shown in Figure 5, where it can be clearly seen that the two kinds oscillation modes are separated. The signs of weight factors of two independent components are shown in Figure 6. Since the number of LPRM signals is not many, a clear regional oscillation distribution cannot be observed. However, this result suggests two different spatial distribution modes cause two kinds of different oscillation frequencies. If the plant status approaches a more unstable state, it is reasonable to expect the dominant oscillation mode to attract other modes. Thus, ICA is expected to be a useful tool to analyze a complicated and delicate regional oscillation behavior.

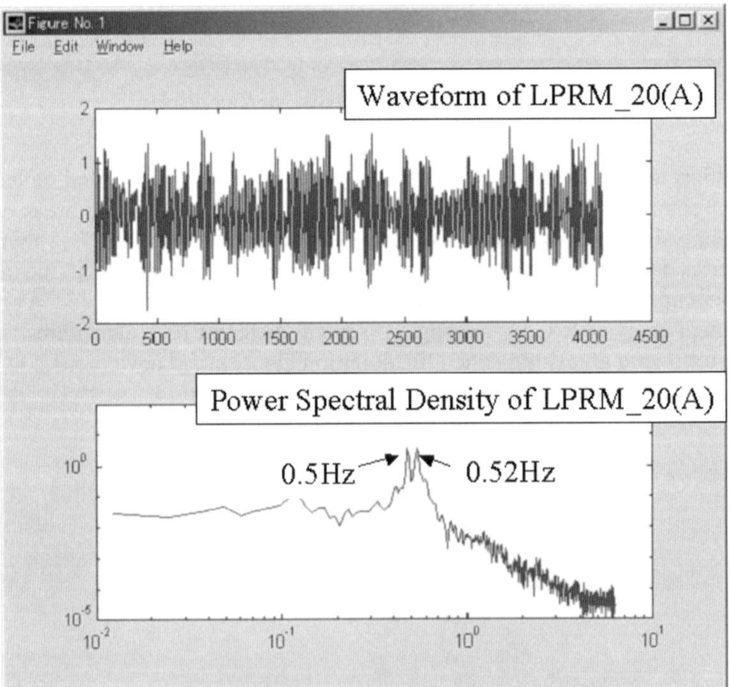

Fig. 4. Raw data and APSD

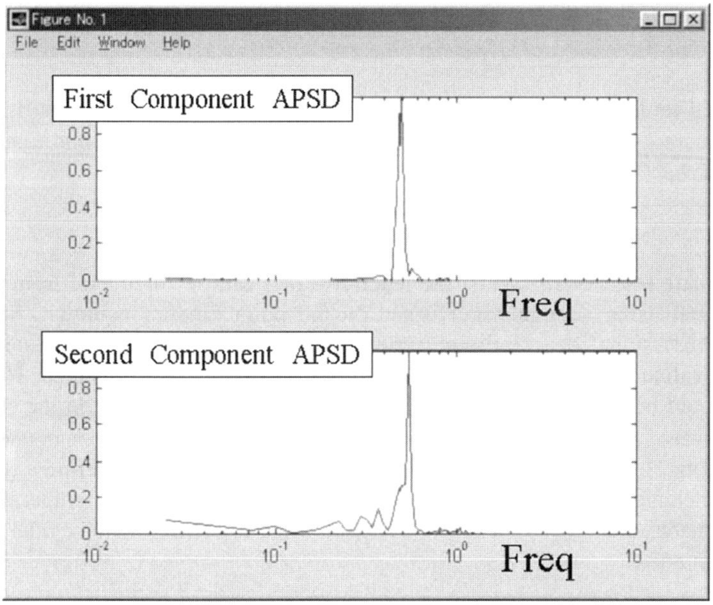

Fig. 5. APSD of two independent components

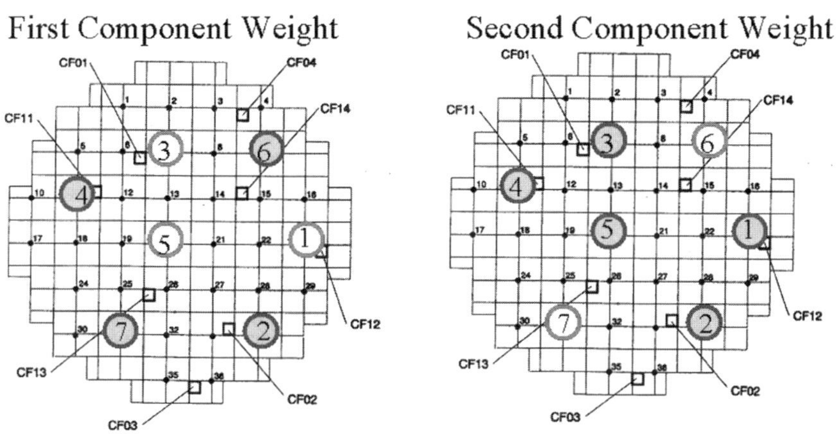

Fig. 6. Weighting parameter obtained by ICA

3.2 Reactivity Coefficient Monitoring Sub-system

Void and temperature reactivity coefficients are important parameters that affect reactor core dynamics. These coefficients are defined in a linear approximation as follows:

$$\rho(t) = K_\alpha(\alpha(t) - \alpha_0) + K_d(T(t) - T_0) \tag{4}$$

To estimate these coefficients, the reactivity ρ(t) can be calculated from neutron flux signal using Kalman filter based on the point kinetics equation. Since the temperature signal can be directly measured, the temperature coefficient can be easily evaluated from the ratio of reactivity and temperature change. However, care should be taken regarding the inclusion of observation noise in the signal. A simple least squares fitting between ρ(t) and T(t) yields a bias error in coefficient estimation. Hence, we introduced the factor analysis where we define a real temperature change as the factor f(t) distinguished from the observed temperature signal (Tamaoki et al. 2000). That is, we assume the following factor model and observation noise:

$$\rho(t) = af(t) + \varepsilon(t), \qquad T(t) = bf(t) + \eta(t) \tag{5}$$

Based on a standard algorithm of factor analysis, coefficients a and b are obtained. The temperature coefficient can be evaluated from the ratio a/b. In Table 2, results from three kinds of methods based on simulation data are compared. The simulation data, which are shown in Figure 7, are created to approximate actual reactivity and temperature change data as well as observation noise. In Table 2, it can be seen that the factor analysis gives the best accuracy from the viewpoints of both bias error and scattering error. The simple least squares fitting yields bias error due to observation noise included in the temperature signal. The coefficient can be also obtained from the low frequency gain of transfer function between reactivity and temperature signals. In this case, due to the averaging effect in transfer function calculation, the bias error becomes small. However, the scattering is larger than in the factor analysis. From theoretical considerations, it can be proven that the factor analysis gives the best results for minimizing the bias and scattering error for the same number of data points.

Table 2. Factor analysis results

Method	Factor analysis	Transfer function	Least squares fitting	True value
Results (¢/deg)	-1.581 ± 0.032	-1.587 ± 0.072	-1.044 ± 0.023	-1.6

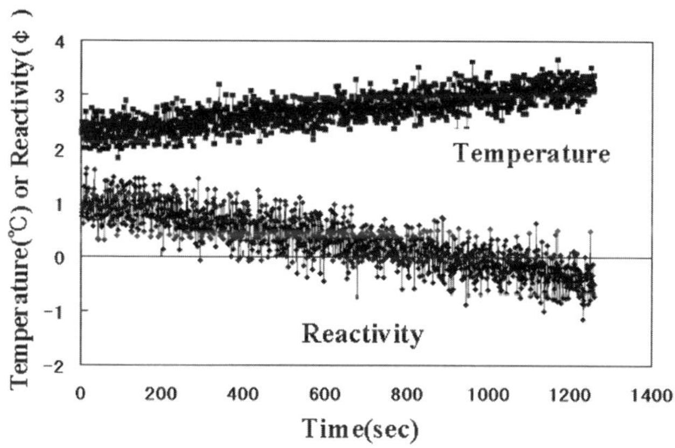

Fig. 7. Simulated data for factor analysis results

Regarding the void reactivity coefficient estimation, the problem is complicated since the void fraction cannot be directly observed. Hence, it is necessary to estimate the void fraction indirectly. Although Kalman filter is one of the promising methods for this purpose, we adopt an indirect estimation method based on the transfer function obtained from core flow and APRM noise signals. Since the dominant noise source is core flow fluctuation in BWR, the transfer function from core flow to APRM has the most reliable information on the void reactivity coefficient through fuel box inlet flow and void fraction dynamics. However, care should be taken regarding the inclusion of observation noise in the input core flow signal. The calculated transfer function has a bias gain due to this observation noise as follows:

$$Calkulated\ TF = \frac{CPSD\,(APRM\,,CF\,)}{APSD\,(CF\,)} = \frac{CPSD\,(APRM\,,CF\,)}{APSD\,(realCF\,) + APSD\,(observatio\ nNoise\,)} \tag{6}$$

Here, CF and realCF mean an observed core flow signal and a real core flow. In the present system, the transfer function is calculated using the ARMA model. Then, typical shape features of the transfer function, such as average gain, peak gain and damping factor, are evaluated (Tamaoki et al. 2001). By preparing the relation map between these feature parameters and the void reactivity coefficient based on the simulation using plant dynamics design code, we can estimate the void reactivity coefficient through the transfer function and the relation map. Typical application results in an actual BWR plant are shown in Figure 8. Here, the results are compared with ones from core performance calculation. The good agreement suggests the effectiveness of the present method. However, further improvement of the present method is desired, since the preparation of the relation map based on the design code for all types of BWR is not so easy.

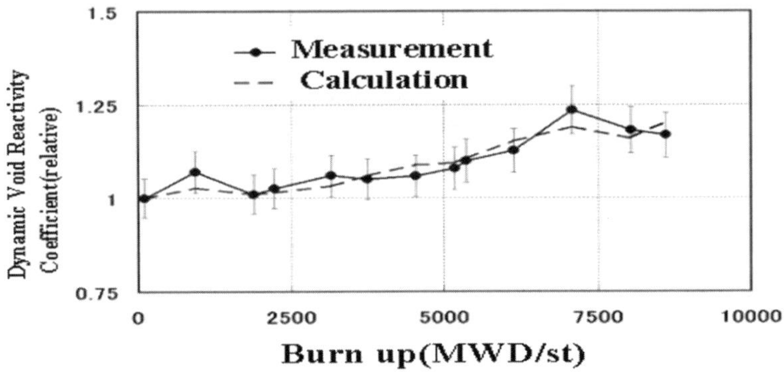

Fig. 8. Void reactivity coefficient estimation results

3.3 Core Flow Estimation Sub-System

The core flow signal is important for assuring fuel burn up calculation or thermal margin evaluation. The current core flow measurement is based on pressure head measurement of jet pumps or internal pumps. Other methods are available; the core flow can be estimated by enthalpy balance or core pressure drop balance based on the current in-core instrumentation. However, there are essential unknown factors such as pressure loss due to crud of fuel rods, or carry under enthalpy. Hence, another method is desired to calibrate the conventional core flow instrumentation signal. The neutron noise method, which utilizes a void speed obtained from LPRM noise correlation between axially located detectors, has been developed (Mori et al. 1996, 1997).

This method is based on the drift flux model of two-phase flow and the correlation analysis to measure void velocity from LPRM noise. The drift flux model is expressed as follows:

$$J = (1-\alpha)v_f + \alpha v_g, \qquad v_g = C_0 J + V_{gj} \tag{7}$$

Here, volumetric flux of mixture J is calculated from void fraction α and velocity of liquid and vapor, v_f and v_g. Vapor velocity is obtained from an empirical void correlation factor C_0 and drift velocity V_{gj}. Once the power distribution, inlet mass flow rate and inlet enthalpy are given, vapor and liquid velocity can be calculated by Eq.(7). On the other hand, if one of the velocities is measured from the LPRM noise correlation, the inlet mass flow rate of fuel channel can be inversely estimated. There are many studies on this method (Stekelenburg & Hagen 1993). Here, the problems that must be solved in order to assure the estimation accuracy of this method can be summarized as follows:

(1) How to define the velocity in the drift flux model corresponding to the measured time delay from LPRM noise signals. Here, it is important to take account of the fact that LPRM noise signals are sensitive to both vapor and liquid velocity.

(2) How to weight four surrounding fuel channel velocities to compare with a single measurement velocity. Here, considerations should be placed on the fact that the sensitivity area of LPRM detectors are narrow and easily affected by microscopic power distribution changes due to control rod position and fuel enrichment.

To solve these problems, we introduce two kinds of empirical fitting models. The combination of drift flux model and fitting model is considered to be the gray model as mentioned above.

At first, we introduce the concept where a mixed transit time corresponds to a measured transit time between LPRM detectors. The mixed transit time between LPRM B and C is defined as follows:

$$\tau_{BC} = w(\alpha_B)\tau_{g,BC} + (1 - w(\alpha_B))\tau_{f,BC}, \qquad w(\alpha) = 1 - \alpha^m \tag{8}$$

Here, the transit time of liquid and vapor is obtained from integration of each velocity along vertical axis.

The point of this assumption is that the measured transit time should be affected by both vapor and liquid velocities in correspondence to the void fraction at an upstream detector position. The weighting factor is obtained in the form of Eq.(8) from the empirical relationship measured in a real plant. Figure 9 shows this relationship. Here, the fitting parameter m is obtained as 1.7. The structure of the weighting factor is based on the physical assumption that LPRM noise signals are sensitive to both vapor and liquid velocity. In the low void fraction area, the LPRM signals should be more sensitive to vapor behavior. On the other hand, in the high void fraction area where the void fraction is larger than the liquid fraction, they should be rather sensitive to liquid behavior.

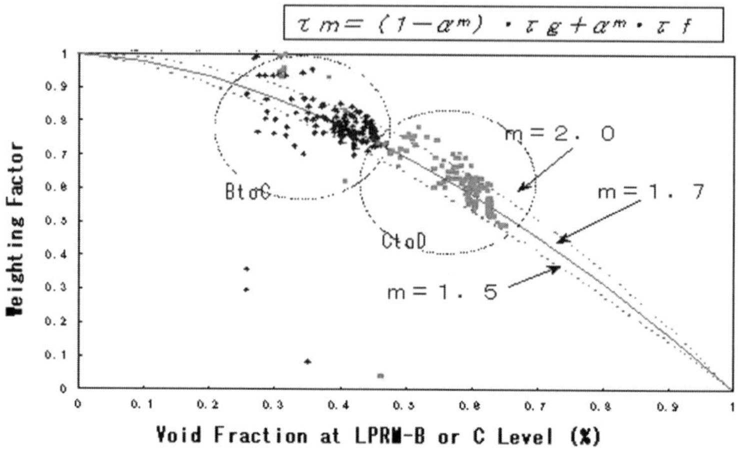

Fig. 9. Weighting factor correlation for mixed transit time

To solve the second problem, we adopt the following weighting model for calculating the transit time from four surrounding fuel channel transit time:

$$\tau_{BC} = \sum_{i=1}^{4} \frac{\alpha(i,B)^q}{V_g(i,B)^q} \tau_{BC}(i), \tag{9}$$

Here, the weighting factor is defined as a function of void fraction and void velocity at the upstream detector position. The p and q are empirical adjusting parameters, and are set to be 0.5 in the present system. Here, the model structure is based on the assumption that LPRM noise signals are sensitive to larger void fraction and smaller void velocity.

Also, it is necessary to introduce the correction factor which considers microscopic power distribution change. Since the field of view of thermal neutron detector is small, the measured transit time is strongly affected by the fuel channel corner void velocity neighboring LPRM detectors (Stekelenburg & Van der Hagen 1993) Hence, we should consider the horizontal microscopic power distribution in a fuel channel. This distribution is affected by the control rod position near the LPRM detector as well as the sub-channel fuel rod enrichment and burn up history. Since it is difficult to evaluate these effects precisely, we introduce the empirical correction factor depending on the power level of detector position as follows:

$$CF_{BC} = f(DC_B) \tag{10}$$

Here, the correction factor is obtained as a function of DC level of observed LPRM signals at the upstream detector position.

On the basis of the above considerations, we apply the present core flow methods in an actual BWR plant during long period and various operating conditions. Figure 10 shows an example of long-term core flow estimation results at the rated power condition. Although the power distribution and control rod position are changed due to fuel burn up, the estimation results well agree with the direct core flow measurements. The good agreement with the conventional core flow instrumentation output shows the above empirical fitting model contributes to the improvement of core flow estimation accuracy. The combination of first principle model and empirical fitting model gives both generality and adaptability, thereby contributing to highly accurate monitoring. We summarize the present algorithm, which is installed in the online core flow estimation system in Figure 11.

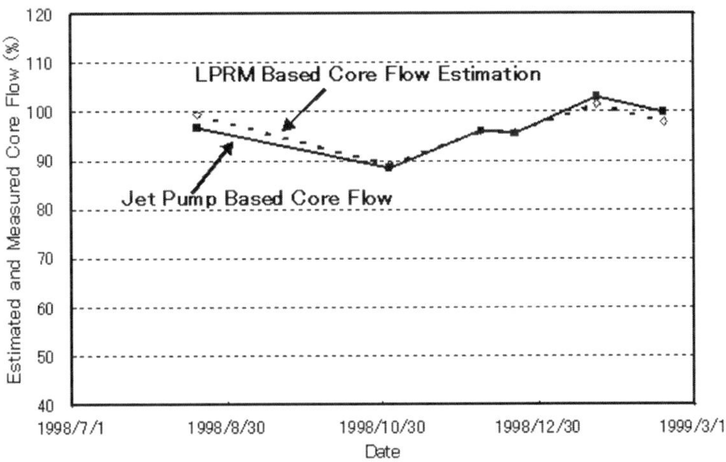

Fig. 10. Long-term estimation trend for core flow in a actual BWR plant

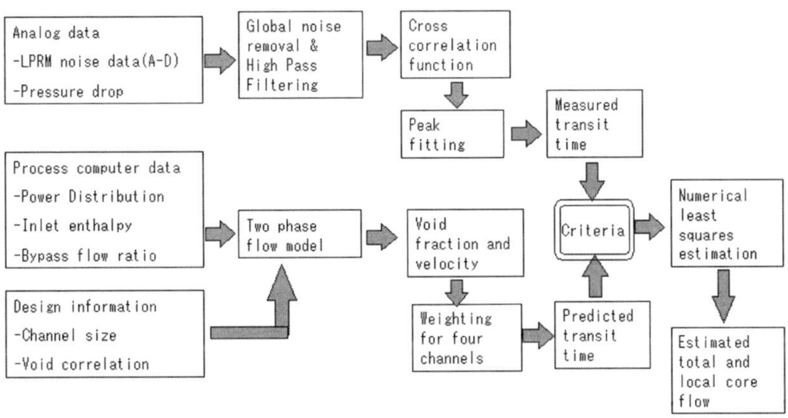

Fig. 11. On-line core flow estimation algorithm

3.4 Sensor Signal Validation Sub-system

In-core sensor signal validation is very important since a small number of sensors perform the key parameter monitoring to assure the core safety margin. Regarding the specific parameter monitoring such as core flow sensor or heat balance, the first principle model or the gray model will be useful due to their universal capability. On the other hand, a general-purpose algorithm is also desired when quantitative physical relations between monitored signals are not clear. These general algorithms using auto associative neural network (AANN), principal component analysis (PCA), partial least squares (PLS) fitting or support vector regression (SVR) have been extensively studied (Uhrig 2000, Rasmussen et al. 2000). An essential feature of these algorithms is a nonlinear fitting model combined with projection to a small number of freedom state spaces. The nonlinear model contributes to higher prediction accuracy, and the projection to small freedom space increases generality and robustness of the model. Since these algorithms can be easily developed by the MATLAB software, we examined several kinds of combination algorithms for the sensor validation based on the nonlinear model and the projection. The five layer AANN proposed by Uhrig has all essential features of the sensor validation. However, AANN does not have a unique expression for the prediction model. This point leads to lack of transparency in a practical application stage. Hence, we adopted the principal component regression combined with the nonlinear fitting model similar to the GMDH model in the output layer, that is, a nonlinear principal component regression model (NLPCR). This model is similar to the three layer AANN as shown in Figure 12. Hence, this model has the above-mentioned two important features, that is, generality and accuracy.

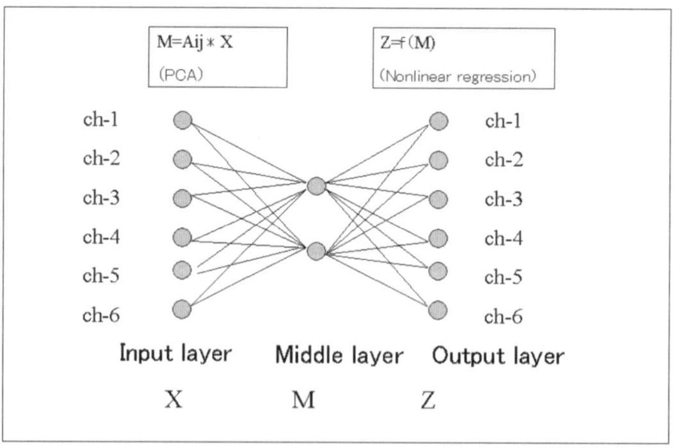

Fig. 12. Nonlinear principal component regression model

An example of application to typical plant signal validation is shown in Figure 13. Here, 3 nodes are used in the middle layer and second-order fitting model is used in the output layer regression. The basic parameters are fitted using the first cycle start up data (240 points with 1 hr interval), and, the prediction results in the third cycle start-up data (71 points with 1 hr interval) are shown in Figure 13. Each data is normalized from –1 to 1 based on the first cycle data. Prediction errors in Figure 13 are shown in the scale from –0.1 to 0.1. Here, a small drift of a main steam header pressure signal can be seen. The prediction capability for different fuel cycles is very important. The present results show good capability of the NLPCR model from the viewpoints of generality and accuracy.

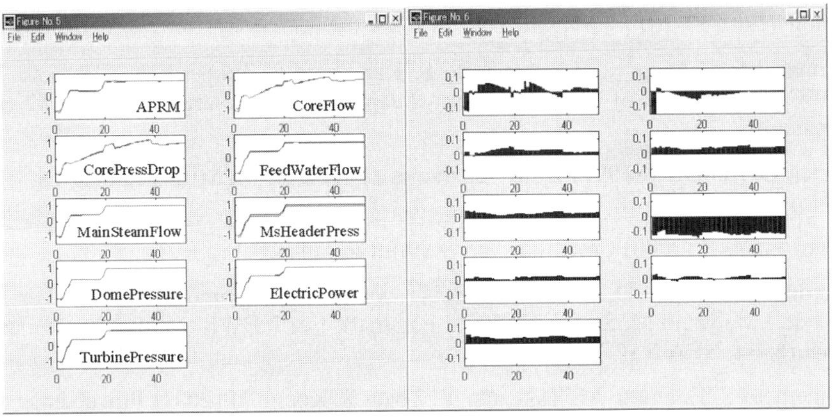

Fig. 13. Comparison of predicted and measured signal trends, and prediction error of the NLPCR model

4 Summary

The core noise monitoring system is a general platform for providing a variety of monitoring tools to meet the requirements of advanced core operation strategy. Here, we described the concept of the system and several application examples based on long-term monitoring experiences in real BWR plants. Depending on the concrete monitoring requirements, adequate algorithms from up-to-date technologies should be combined. Also, past knowledge about plant design should be extensively utilized. The most important requirement is to provide monitoring tools that have general and accurate prediction capability to satisfy practical needs. The flexible and open structure of the system is expected to result in the extension of application areas of nuclear plant integrity monitoring.

References

Kanemoto S, Enomoto M, Ando Y, Namba H, Ebata S, Takagi A, Hattori T, Kitamura N, Yoshizawa A (1988) Development of an on-line reactor stability monitoring system in a boiling water reactor, Progress in Nuclear Energy (SMORN-5) 21:745-752

Takeuchi Y, Takigawa Y, Miyamoto S (1999) A Study on Regional Stability Analysis Methodology with a One-Point Neutron Kinetics Model, Nuclaer Technology 128:257-267

Yamashita K, Anegawa T, Miyamoto S, Kanemoto S, Takeuchi Y, Enomoto M (1997) Proceedings of annual meeting of Japan Atomic Energy Society, E52 (in Japanese)

Jutten C, Herault J (1991) Blind separation of sources, Signal Processing 24-1:1-10

Thawonmas R (2000) Computer Today 99-9:(in Japanese)

Verdu G, Ginestar D, Munoz-Cobo J,L, Navarro J, Palomo M.J, Lansaker P, Conde J.M, Recio M, Sartori E (2001) Forsmark 1 & 2 BWR stability benchmark final report, NEA/NSC/DOC(2001)2

Kanemoto S, Enomoto M, Takeuchi Y, Ebata S, Konno H (2001) Proceedings of annual meeting of Japan Atomic Energy Society, F51 (in Japanese)

Tamaoki T, Enomoto M, Kanemoto S, Ebata S, Chuman K, Hirose J (2000) Proceedings of fall meeting of Japan Atomic Energy Society, R51 (in Japanese)

Tamaoki T, Enomoto M, Ebata S, Sugawara M, Hara T (2001) Proceedings of fall meeting of Japan Atomic Energy Society, H32 (in Japanese)

Mori M, Kanemoto S, Enomoto M, Ebata S (1996) On-line estimation of local and total core flow rates by neutron noise analysis in BWR, Proceedings of INCORE-96, Mito-shi, JAPAN:145-155

Mori M, Kanemoto S, Enomoto M, Ebata S (1997) Proceeding of Japan Atomic Society 39-9:788-(in Japanese)

Stekelenburg AJC, van der Hagen THJJ (1993) Two phase flow monitoring by analysis of in-core neutron detector noise analysis, Ann. Nucl.Energy 20:611-117

Uhrig RE (2000) Reviewing various types of intelligent systems for application to nuclear plants, International Topical Meeting on NPIC & HMIT 2000, Washington DC, November: 109-117

Rasmussen B, Hines JW, Uhrig RE (2000) A novel approach to process modeling for instrument surveillance and calibration verification, International Topical Meeting on NPIC & HMIT 2000, Washington DC, November: 890-900

13 Diagnosis of Measuring Systems Using Cluster Analysis Applied to Hydrostatic Water Level Measurement

A. Traichel, V. Schneider, W. Kästner, R. Hampel

Institute of Process Technique, Process Automation and Measuring Technique (IPM) at the University of Applied Sciences Zittau/Görlitz
Theodor-Körner-Allee 16, D-02763 Zittau, Germany

ipm@hs-zigr.de

For a continuous monitoring, control, and diagnosis of high transient technical processes in safety-related systems as Nuclear Technology it is necessary to provide the best knowledge about the actual state of process. Especially the supervision of water level within pressure vessels with boiled water-steam mixture (pressuriser, steam generator, reactor pressure vessel) during operational and accidental transient processes is very important. A diagnosis system is proposed, which is subdivided in two components, a part for fault detection and a part for fault identification. The diagnosis component for fault detection using cluster analysis is described. The procedure of cluster analysis is demonstrated. The results of fault diagnosis using new validity functions are presented for a Blow Down Experiment at our pressuriser test facility.

1 Introduction

For a continuous monitoring, control, and diagnosis of high transient technical processes in safety-related systems as Nuclear Technology it is necessary to provide the best knowledge about the actual state of process. Especially the supervision of water level within pressure vessels with water-steam mixture (pressuriser, steam generator, reactor pressure vessel) during operational and accidental transient processes is very important [1]. For the determination of water level hydrostatic measuring systems are almost exclusively applied, because they stand the test over many decades in conventional and nuclear plants [2]. The hydrostatic measuring systems are normally installed n-fold redundant in different measuring ranges. The indication errors or malfunctions, released by faults, are often situated within the measuring range, so that they are hardly identified by the operator. How experiences show, synchronous occurrence of indication errors and disturbances could have an escalating effect on the accident behaviour.

Consequentially the aim of development of Diagnosis system for hydrostatic water level measurement is deduced, on the basis of intelligent signal processing, by utilization of existing hardware redundancies. This research work is integrated in the concept, which is demonstrated in Fig. 1. The circles present an iterative process, included former projects, which results in our concept of Diagnosis system.

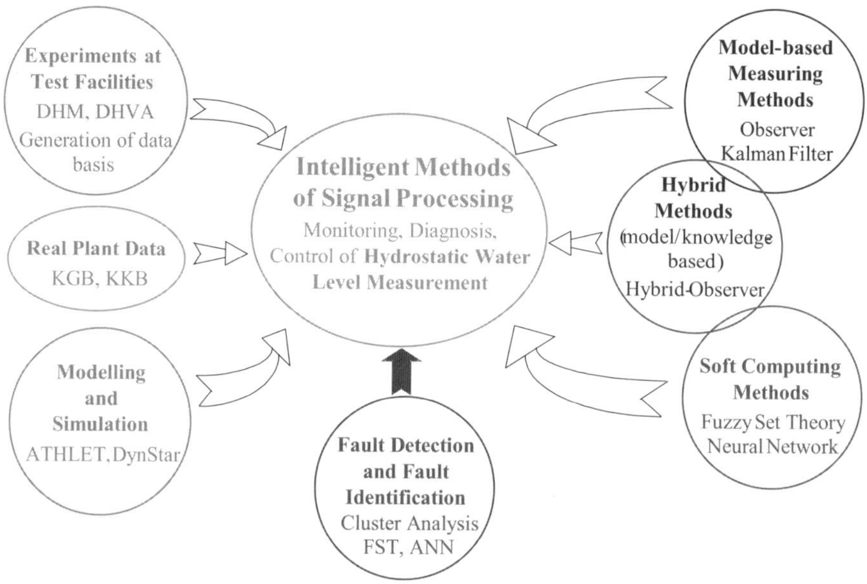

Fig. 1. Conception of the Monitoring and Diagnosis system

This paper deals with the new part of fault detection of hydrostatic water level measurement by cluster analysis methods. The procedure of a cluster analysis is demonstrated. The results of fault diagnosis are demonstrated as for one Blow Down Experiment at our pressuriser test facility.

2 Hydrostatic Water Level Measuring Systems

Fig. 2 presents a scheme of water level measurement at pressuriser test facility with stand pipes and the other components. Hydrostatic measuring systems are n-fold redundant with different measuring ranges (wide and narrow range). The basic principle of its measuring principle is described in [3]. The definition of water level within the pressure vessel corresponding Fig. 2 is as follows:

⇒ collapsed level – measureable value, presents the water inventory,

⇒ mixture level – high transient non-measurable value concerning usual hydrostatic water level measurement [2].

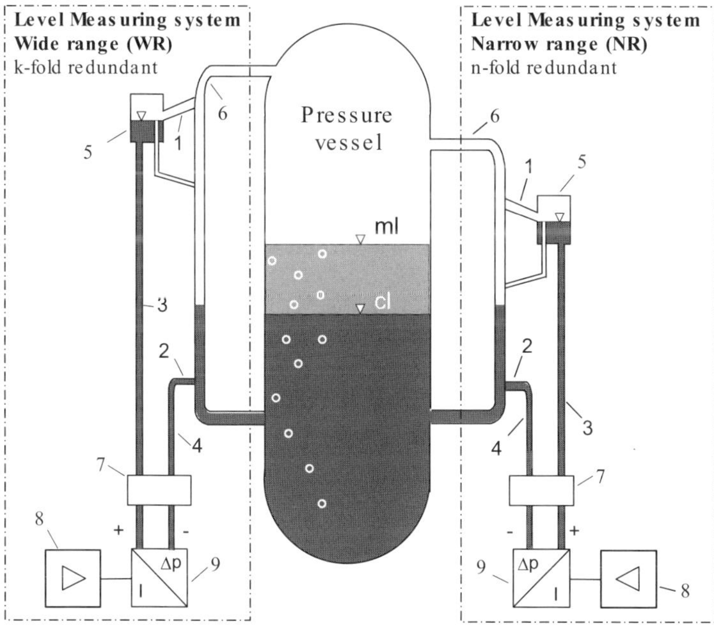

1 - upper fitting
2 - lower fitting
3 - reference pipe
4 - minus chamber
5 - comparison chamber
6 - stand pipe
7 - valve block
8 - power supply system
9 - differential pressure transmitter

Fig. 2. Components of hydrostatic redundant water level measuring system with stand pipe at the test facility DHVA-SWR and definition of water level

The knowledge about the mixture level is safety-relevant in nuclear power plant. During operational and accidental transient processes in measurement occur:

⇒ indication errors due to e.g.,

- deviation from working point (pressure, temperature, steam content) of operating medium,
- steam leak in pressure vessel,
- change of original water level in reference pipe of measuring system (by evaporation of volume water, degassing of non-condensable gases),
- dynamic pressure differences near by pipe fittings,

⇒ malfunction due to e.g.,

- loss of attenuator,
- leak in reference pipe,
- leakiness at valve block,
- disturbances in power supply system.

The **thesis** for diagnosis is, that the deviation of redundant measuring systems from each other will be used.

The **aim of diagnosis** is the determination of fault kind and location. The implementation takes place in the following way:

⇒ detection of indication errors and malfunction using the deviation between measured values **(fault detection)**, and

⇒ identification of fault kind and location **(fault identification)**.

So the proposed **Diagnosis system** is subdivided in two components, a part for fault detection and a part for fault identification, like it is shown in Fig. 3.

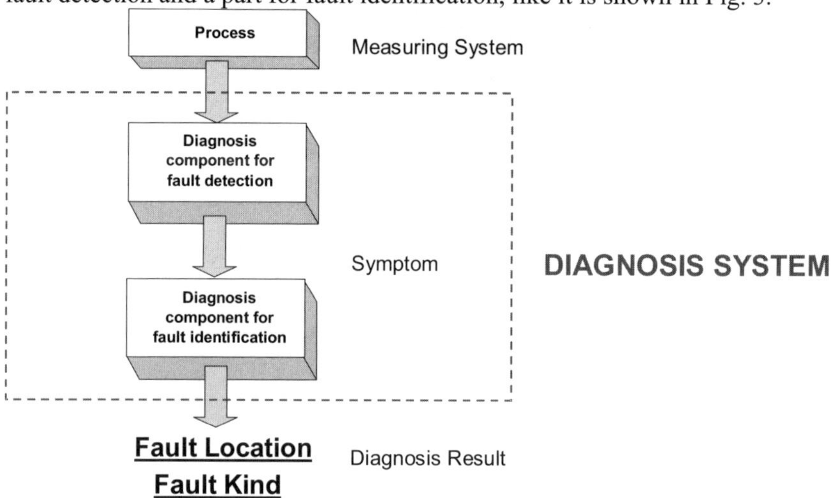

Fig. 3. Structure of Diagnosis system with two components

With the help of fault detection deviations between redundant measuring systems will be identified. Measuring systems with nearly similar measured values will be combined in groups. Significantly deviating measuring signals will be subdivided into separate groups. The resulting classes establish the basic point for the fault identification. This one localizes at first the faulty group (determination of fault location). Afterwards the fault kind will be determined.

For the detection of deviation between measured values classification methods, like cluster analysis are suited. The next part gives an overview about the methodology of cluster analysis. Furthermore it deals with the determination of exact number of classes of a given amount of objects.

Thereto different developed so called validity functions will be presented and verificated by a Blow Down Experiment at the described test facility (investigation object). The last part of the paper gives a summary about the research results.

3 Fault Detection by Cluster Analysis

3.1 Clustering Method

Cluster analysis is an unsupervised classification method, because the characteristics of clusters at the beginning are unknown. The aim of cluster analysis is the division of n-element-groups in k groups or clusters. The set of generated clusters is labeled as partition. By [4] cluster analysis contains generally the following steps, like it is demonstrated in Fig. 4:

⇒ Data preparation,

⇒ Determination of cluster tendency,

⇒ Configuration of data with the help of clustering method, in place single linkage algorithm,

⇒ Evaluation of calculated partitions with validity functions.

The results of fault detection by cluster analysis are the favourite number of classes, cluster configurations and cluster centres at sample time.

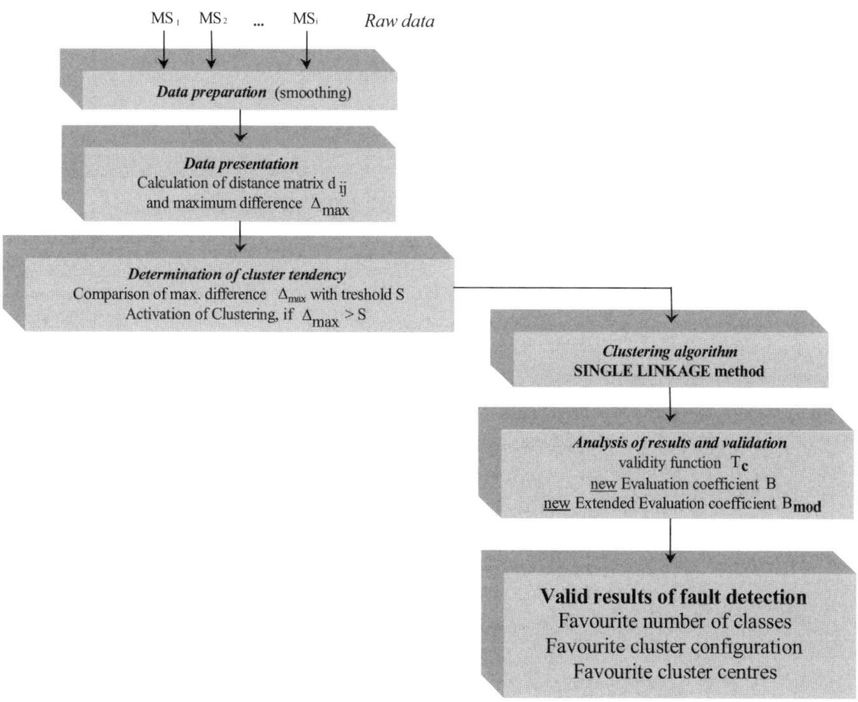

Fig. 4. Structure of Diagnosis component for fault detection at hydrostatic measuring systems

As input data for cluster analysis the redundant measuring signals will be used, which will be calibrated and smoothed within the data preparation. With the help of cluster tendency will be checked, if the data are distributed in groups or randomly. For this special case it is enough to calculate significant deviation between measured values. For this one we use the maximum difference within pairs of measured values Δ_{max} and the distance matrix d_{ij}. The cluster tendency will be determined by the comparison of maximum difference Δ_{max} with a process depending threshold S (Eq. 1):

$$\Delta_{max} = \max_{\forall i,j} d_{ij} > S \tag{1}$$

The clustering will be activated, if the maximum difference Δ_{max} exceeds the threshold S. Above the threshold the data split up in more groups, below they are merging in one group.

The practically used Single Linkage clustering method is an transformation of distance matrix d_{ij} into a sequence of partitions. The Single Linkage algorithm establishs the kernel of the cluster analysis. Fig. 5 shows the procedure of this algorithm, which is a hierarchical method. Stepwise, beginning with the finest partition, the cluster with the smallest distance will be fused. Each object is uniquely dedicated to one group.

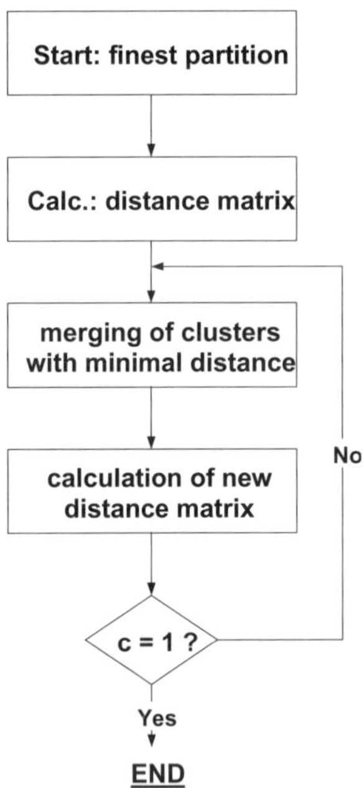

Fig. 5. Procedure of single linkage clustering method

Fig. 6 represents for example a simple data structure for clustering, where the clusters, the centres and the distances are shown. The cluster centres are calculated by Eq. (2):

$$v_k = \frac{1}{n_k} \sum_{i \in C_k} x_i \tag{2}$$

Eq. (3) gives the cluster distance:

$$D(C_k, C_l) = \min_{i \in C_k, j \in C_l} d_{ij} \tag{3}$$

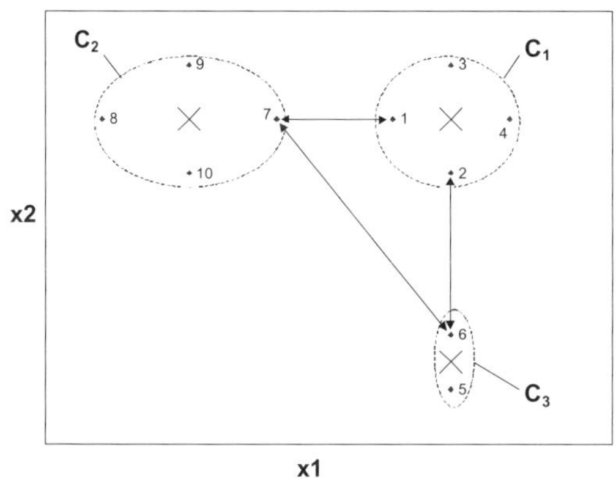

Fig. 6. Example of a data structure (x_1, x_2) with 3 clusters (C_i), cluster centres (×, v_k) and cluster distances (arrows)

The hierarchy of classes in form of dendrogram resulting by single linkage algorithm for 10 elements is presented in Fig. 7. The intersection graph h_l demonstrates the selection of the 10 elements in three classes.

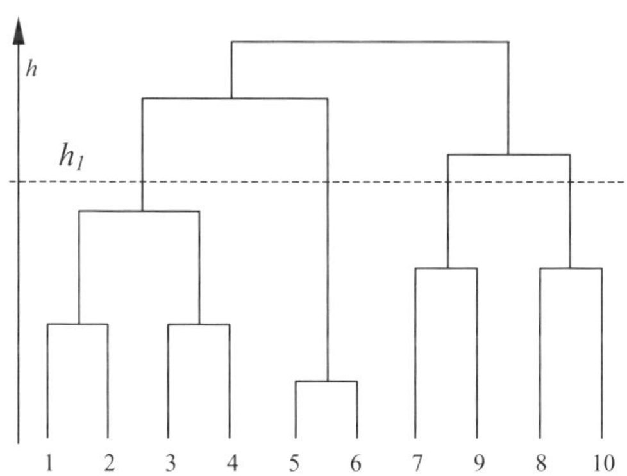

Fig. 7. Dendrogram generated by single linkage algorithm (hierarchy of classes)

3.2 Validity Functions

As aforementioned there are no information about the number of clusters in the considered data set at the beginning of cluster analysis. The cluster algorithm yields several cluster configurations. For determination of optimal number usually the criterion of validity functions is used, which the quality of the partition are reproducing by only one real value [5]. Validity functions will be calculated for all possible number of classes and in terms of minima and maxima evaluated. The common character of all validity function is the fact, that the number of classes belong to the calculated extremum is the searched number of clusters.

For the special task the modified Xie-Beni Index was developed. In [5] a good partition is described as a cluster subdivision with compact and well separated clusters. As parameter for the compactness π Eq. 4 will be used:

$$\pi = \frac{1}{c}\sum_{i=1}^{n}\sum_{k=1}^{c} u_{ik}^{m} \left\| x_i - v_k \right\|^2 \tag{4}$$

where the symbols are:

u_{ik} - membership of object x_i to cluster k
v_k - prototype of k-th cluster
x_i - object
m - weighting exponent
c - number of classes

Compactness is interpreted as a parameter for weighted distribution of elements x_i around the class centres v_k. For compact clusters results a small total distribution.

For the evaluation of separation sep , the value for good partition of built clusters, the minimum distance between class centres will be calculated. Well parted classes are characterized by high separation.

$$\text{sep} = \min_{k \neq l} \left\| v_l - v_k \right\|^2 \tag{5}$$

From this two parameters results the validity function Modified Xie-Beni Index T_c (Eq. 6):

$$T_c = \frac{\pi}{\text{sep}} \xrightarrow[2 \leq c \leq n-1]{} \min \tag{6}$$

The minimum value represents the right number of classes.

For the improvement of subjective intuition of output of modified Xie-Beni Index the results will be scaled to the interval [0, 1]. The new evaluation criterion should show by its maximum the best partition. This requirement will be realized by standardisation and negation of Index T_c (Eq. 7). The result is the standardised validity function. It is signified as Evaluation coefficient B_c (Eq. 7):

$$B_c = 1 - \frac{T_c}{\max\limits_{\forall c} T_c} \qquad \text{with } B_c \in [0,1] \tag{7}$$

The maximum B_c marks the best partition. The worst partition is marked by $B_c \equiv 0$.

3.3 Diagnosis Results for One Blow Down Experiment

The test and verification of diagnosis system component for fault detection takes place by means of experimental data of our test facility DHVA-SWR (IPM). There were carried out a lot of Blow Down Experiments with different boundary conditions.

The collapsed level is represented by five redundant measuring signals, which are indicating nearly the same values during normal operational regime (Fig. 8). Indication errors of water level measurement systems occur by Blow Down (BD). The reason of this indication error is the changing of the level in the reference pipe of the water level measurement system, due to evaporation of water inventory of the comparison vessel and degassing of noncondensable solved gases (SG) in the reference pipe.

In the reference pipe arises a two-phase flow due to depressurization, which has the energy for the generation of gas bubbles (degassing processes). An equivalent water discharge appears at the comparison vessel in combination with gas release. This discharge leads to significant spurious indications of the water level in pressure vessels [1].

The procedure of the Blow Down Experiments will be briefly explained [1]:

⇒ Achieving a steady state ($p = 7.0$ MPa),

⇒ Opening of leak (starting point of Blow Down),

⇒ Feed in of water from preheater ($p = 1{,}5$ MPa; $\vartheta \cong 200$ °C),

⇒ Closing of leak (end of Blow Down),

⇒ Achieving new steady state.

The water inventory of reference pipe for a narrow range measuring system was completely enriched with solved gases according to initial concentration of 75%. The oversaturation factor for the water air mixture, which is necessary for the generation of gas bubbles, is about 10. That means, that for an initial saturation pressure of 7 MPa the degassing process begins at 0,525 MPa during the depressurisation for a gas enrichment of 75%.

Fig. 8 shows this procedure and degassing process for the above explained experiment. The indication error, the deviation due to degassing, has to detect by the clustering algorithm.

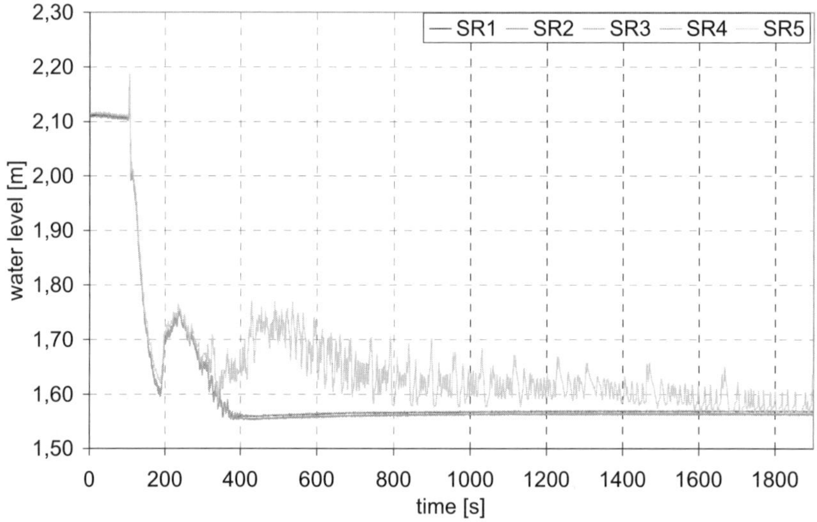

Fig. 8. Deviation of one redundant measuring signal (narrow range level) due to degassing (Blow Down Experiment)

Fig. 9 demonstrates the activation of clustering, depending on threshold S, by using the maximum difference Δ_{max} as parameter for cluster tendency.

The clustering starts, if the threshold S = 3,5 cm is exceeded. This chosen process considered depending threshold the deviation of measuring signals due to noise. The clustering begins nearly 310 s (simulation time), which corresponds very well to the undershooting of dissolution pressure and consequential indication error.

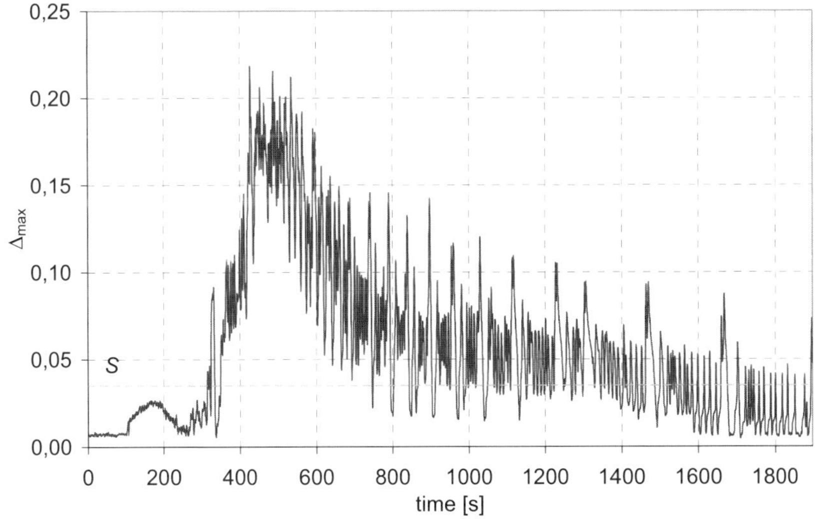

Fig. 9. Cluster tendency (maximum difference) for water level values with threshold S = 3,5 cm (Blow Down Experiment)

Figs. 10 and 11 present the modified Xie-Beni Index T_c and hence derivated Evaluation coefficient B_c , which are used for determination of favourite number of classes. In accordance with the existing circumstances partition of two classes will be preferred over wide ranges.

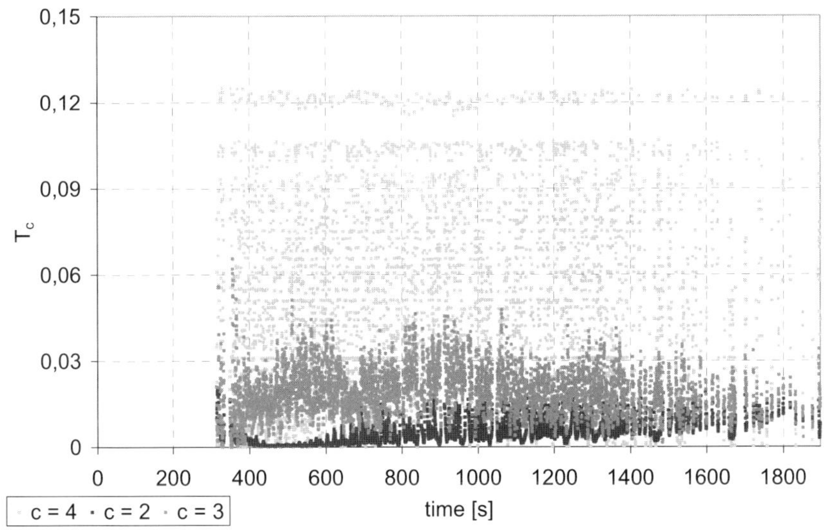

Fig. 10. Modified Xie-Beni Index T_c for c = 2, 3, 4 classes (Blow Down Experiment). Favourite number of classes is characterized by minimum

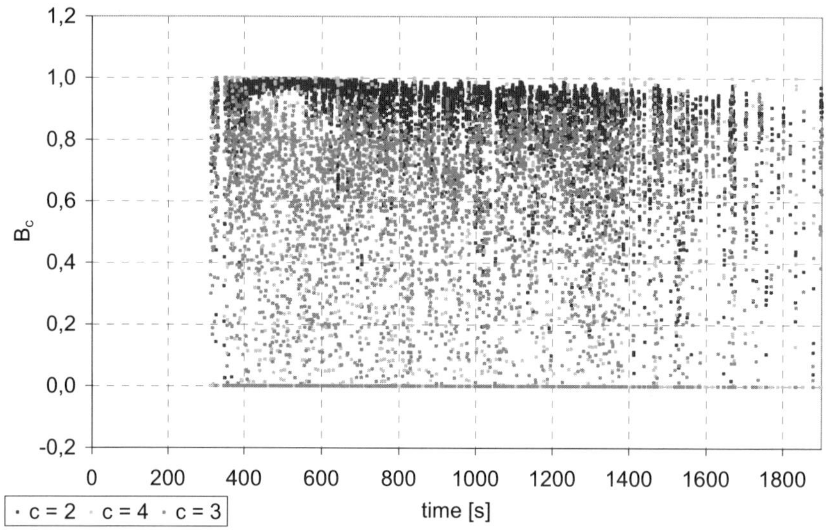

Fig. 11. Evaluation coefficient B (standardised validity function) for c = 2, 3, 4 classes (Blow Down Experiment). Favourite number of classes is characterized by maximum

Admittedly the Evaluation coefficient shows wild fluctuation for partitions with three or four classes, which results in an overvaluation of this partition. Founded are these in the very minimal distance between the four normal measuring signals.

To obviate the falsification of diagnosis results, possibilities for enhancement of robustness of evaluation coefficient are needed. That one is the developed extended evaluation coefficient.

3.4 New Extended Evaluation Coefficient

In Figs. 12 and 13 are contrasted the class centres for partitions with two and four classes for the above mentioned example. The centres for two classes are better separated than for four classes.

Thence the following, on separation based, punish function can be defined (Eq. 8):

$$\beta = \begin{cases} 0 & \text{if } \sqrt{\text{sep}} \le S_{CD} \\ 1 & \text{else} \end{cases} \tag{8}$$

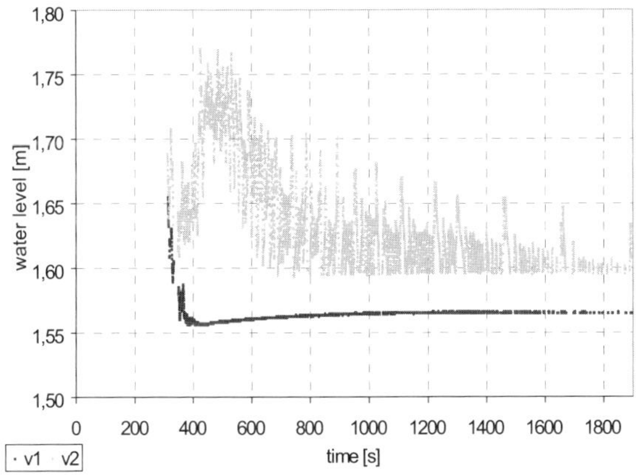

Fig. 12. Cluster centres for two cluster configuration (Blow Down Experiment)

The distance of cluster centres has a big influence on the robustness of validity function.

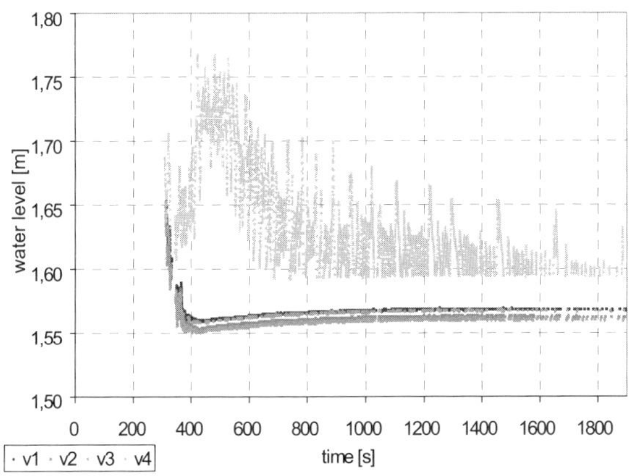

Fig. 13. Cluster centres for four cluster configuration (Blow Down Experiment)

Thereby the threshold S_{CD} has to be dimensioned on particular process. With Eq. (8) the new extended evaluation coefficient B_{mod} can be determined:

$$B_{mod} = \beta\, B_c \tag{9}$$

The maximum of B_{mod} characterizes the best partition. This validity function discards partitions, which class centres are to be closed, like for the four classes solution. Because the minimum class number for non-trivial solution is given by two, the punish function β for partitions with two classes is per definition 1. For solutions with more than two classes Eq. 8 in connection with Eq. (9) is valid.

Fig. 14 presents the parameter $\sqrt{sep}$ for partitions with c = 2, 3, 4 classes determined by threshold S_{CD} = 1 cm for the above Blow Down Experiment.

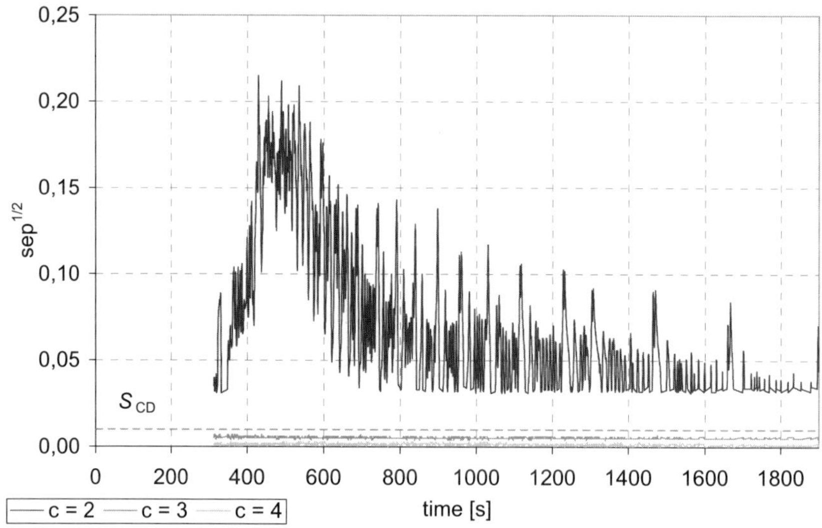

Fig. 14. Parameter $\sqrt{sep}$ for partitions with c = 2, 3, 4 classes determined by threshold S_{CD} = 1 cm (Blow Down Experiment)

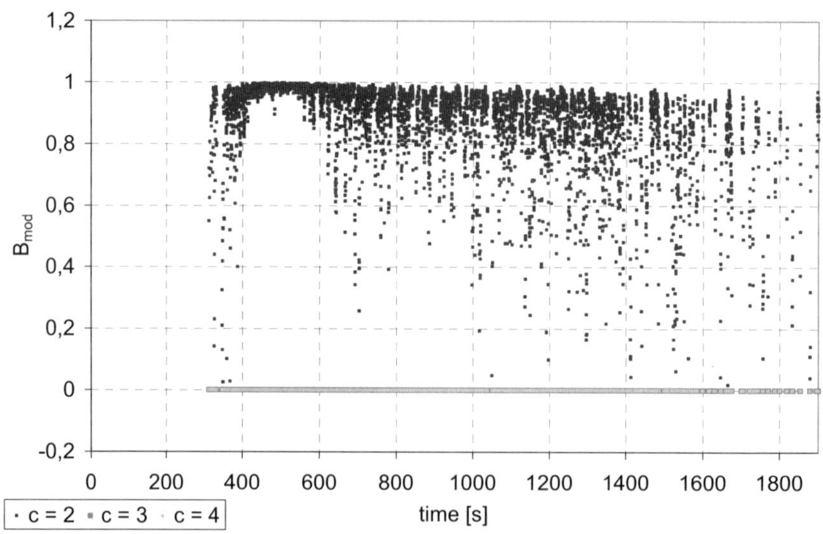

Fig. 15. Extended evaluation coefficient B_{mod} for c =2, 3, 4 classes (Blow Down experiment)

The above Fig. 15 shows the extended evaluation coefficient B_{mod}. It is clearly to be seen, that a high improvement of diagnosis result is achieved as against the other validity functions. Two classes are in evidence always prefered.

4 Summary

As results of investigations it is to be realized that:

⇒ Clustering algorithms are practicable for fault detection at hydrostatic water level measuring systems.

⇒ Verification and validation of diagnosis component for fault detection using cluster methods by generated data basis (experiments at test facility and real data) succeeds and allows the detection of deviation between measured values.

⇒ With new evaluation and extended evaluation coefficient (B, B_{mod}) an advance regarding validation of cluster results is achieved.

⇒ The used clustering method classifies measuring systems with similar measured values in the right way.

⇒ The results of fault detection are the optimal number of classes, respective partition and cluster centres.

⇒ This results will be used as input for the rule-based system for fault identification.

This research work was supported by the Federal Ministry of Germany (BMWi) within the reactor safety research under project No. 150 1204.

References

[1] Hampel R, Kästner W (2001) *Wissensbasierte Systeme zur Diagnose der Funktionsfähigkeit und Fehlerdiagnose für hydrostatische Füllstandsmeßsysteme - Band I und III.* Technischer Bericht zum BMBF-Projekt 150 12 04, HS Zittau/Görlitz (FH)

[2] Traichel A, et. al. (2000) *Verification of Simulation Results of Mixture Level Transients and Evaporation Processes in pressure vessels using Needle-shaped probes.* Workshop Measuring Technique for two-phase transient and stationary flow, Rossendorf, Germany

[3] Hampel R, Kästner W, Fenske A, Vandreier B, Schefter, S (1999) *Meß- und Automatisierungstechnik zur Störfallbeherrschung - Methoden der Signal-verarbeitung, Simulation und Verifikation.* Abschlußbericht zum BMBF-Projekt 150 10 15, HTWS Zittau/Görlitz (FH)

[4] Backer E (1995) *Computer-assisted reasoning in cluster analysis.* Prentice Hall, New York, London

[5] Höppner F, Klawonn F, Kruse R (1997) Fuzzy-Clusteranalyse: *Verfahren für die Bilderkennung, Klassifikation und Datenanalyse.* Vieweg, Braunschweig

[6] Beni G, Xie LX (1991) A Validity Measure for Fuzzy Clustering, *IEEE Transactions on Pattern Analysis and Machine Intelligence* 13 8: pp. 841-847

14 A Hybrid Fuzzy-Fractal Approach for Time Series Analysis and Prediction and Its Applications to Plant Monitoring

Oscar Castillo and Patricia Melin

Dept. of Computer Science, Tijuana Institute of Technology, P.O. Box 4207, Chula Vista CA 91909 USA

{ocastillo, pmelin}@tectijuana.mx

We describe in this paper a new hybrid fuzzy-fractal approach for plant monitoring. We use the concept of the fractal dimension to measure the complexity of a time series of observed data from the plant. We also use fuzzy logic to represent expert knowledge on monitoring the process in the plant. In the hybrid fuzzy-fractal approach a set of fuzzy if-then rules are used to classify different conditions of the plant. The fractal dimension is used as input linguistic variable in the fuzzy system to improve the accuracy in the classification. An implementation of the proposed approach is shown to describe in more detail the method.

1 Introduction

Diagnostic systems are used to monitor the behavior of a process and to identify certain pre-defined patterns that are associated with well-known problems [10]. These problems, once identified, imply suggestions for specific solutions. Most diagnostic systems are in the form of a rule-based expert system: a set of rules is used to describe certain patterns. Observed data are collected and used to evaluate these rules. If the rules are logically satisfied, the pattern is identified, and the problem associated with that pattern is suggested. In general, the diagnostic systems are used for consultation rather than replacement of human expert.

Most current plant monitoring systems only check a few variables against individual upper and lower limits, and start an audible alarm should each variable move out of its predefined range [9]. Other more complicated systems normally involve more sensors that provide more data but still follow the same pattern of independently checking individual sets of data against some upper and lower limits. The warning alarm from these systems only carries a meaning that there is something wrong with the process in the plant.

In this paper a new fuzzy-fractal approach for plant monitoring is proposed. The concept of the fractal dimension is used to measure the complexity of the time series of relevant variables for the process [3]. A set of fuzzy rules is used to represent the knowledge for monitoring the process [11]. In the fuzzy rules, the fractal dimension is used as a linguistic variable to help in recognizing specific patterns in the measured data. The fuzzy-fractal approach has been applied before in problems of financial time series prediction [4, 5, 7, 8] and for other types of problems [6, 14], but now it is proposed to the monitoring of plants.

This paper is structured as follows. First, the problem of plant monitoring and diagnosis is described in more detail. Second, some basic concepts of fractal theory are given. Third, the problem of fuzzy modeling for monitoring and diagnosis is described in more detail. Then, the fuzzy-fractal approach for plant monitoring is proposed and explained. Finally, some experimental results are given and future work is proposed.

2 Monitoring and Diagnosis

Monitoring means checking or regulating the performance of a machine, a process, or a system [10]. Diagnosis, on the other hand, means deciding the nature and the cause of a diseased condition of a machine, a process, or a system by examining the symptoms. In other words, monitoring is detecting suspect symptoms, whereas diagnosis is determining the cause of the symptoms.

The importance of monitoring and diagnosis of plant processes now is widely recognized because it results in increased productivity, improved product quality and decreased production cost. As a result, in the past decade, a large number of research and development projects have been carried and many monitoring and diagnosis methods have been developed [10, 11]. The commonly used monitoring and diagnosis methods include modeling-based methods, pattern recognition methods, fuzzy systems methods, knowledge-based systems methods, and artificial neural networks. It is interesting to note that even though these methods are rather different, they share a very similar structure as shown in Fig. 1.

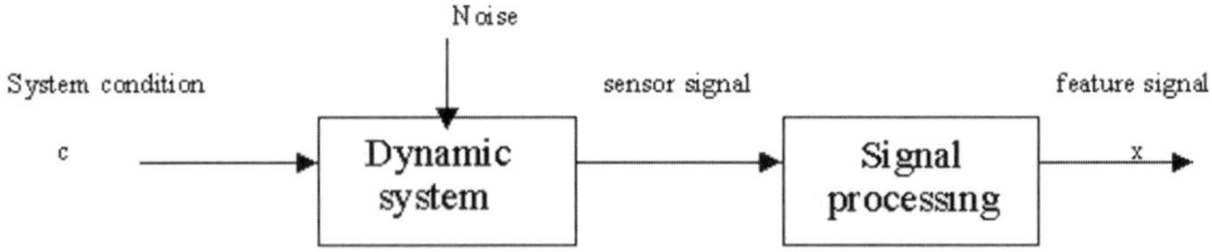

Fig. 1. Plant monitoring and diagnosis

The "health" of a machine, a process, or an engineering system (which will be referred to as system condition and denoted by $c \in \{c_1, c_2, ..., c_m\}$) can be considered as the "input," the system working conditions and noises (including system noise and sampling noise) can be considered as the "noise," and the sensor signals are the "outputs" from the system. Typically, the sensor signals are processed by a computer, after which the signals are transformed into a set of features called feature signals, denoted as $\mathbf{x} = \{x_1, x_2, ..., x_n\}$. In general, the systems conditions are predefined, such as normal, critical, etc. On the other hand, the features may be the mean of a temperature signal, the variance of a displacement signal, etc. Sensing and signal processing are very important to the success of plant monitoring and diagnosis [9].

More formally, the goal of monitoring is to use the feature signals, x, to determine whether the plant is in an acceptable condition (a subset of $\{c_1, c_2, ..., c_m\}$). On the other hand, the objective of diagnosis is to use the feature signals, x, to determine the system condition, $c \in \{c_1, c_{2,...,} c_m\}$. No matter how monitoring and diagnosis methods may differ, monitoring and diagnosis always consist of two phases: training and decision making. Training is to establish a relationship between the feature signals and the systems conditions. Without losing generality, this relationship can be represented as

$$\mathbf{x} = F(c). \tag{1}$$

It should be pointed out that F(c) represents a fuzzy system, a neural network or another method that could be used to obtain this relationship. In fact, it is the form of the relationship that determines the methods of monitoring and diagnosis, as well as the performance of the methods. The relationship F(c) is established based on training samples, denoted by $\mathbf{x}_1, \mathbf{x}_2, ..., \mathbf{x}_k, ..., \mathbf{x}_N$, where the system condition for each training sample is known and denoted as $c(\mathbf{x}_k)$]

After the relationship is established, when a new sample is given (from an unknown system condition), its corresponding condition is estimated based on the inverse relationship

$$c = F^{-1}(\mathbf{x}). \tag{2}$$

This is called decision-making, or classifying. Whereas it is not likely that the training samples will cover all possible cases, decision making often involves reasoning or inferencing.

In particular, when a fuzzy system is used, the relationship is given by a set of rules as shown in Fig. 2. The input to the fuzzy system is the feature signal and the output of the fuzzy system is the estimated plant condition(s) [i.e., $\mathbf{z} = (z_1, z_2, ..., z_m)$ is an estimate of $\mathbf{c} = (c_1, c_2, ..., c_m)$]. In other words, the fuzzy system models the inverse relationship between the system conditions and the feature signals.

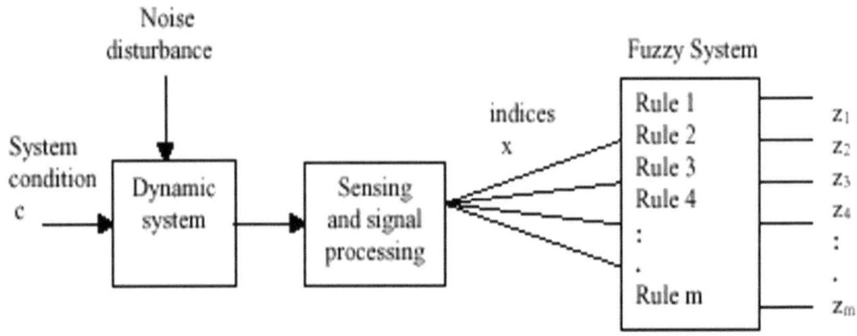

Fig. 2. Fuzzy system for plant monitoring and diagnosis

3 Fractal Dimension of a Geometrical Object

Recently, considerable progress has been made in understanding the complexity of an object through the application of fractal concepts [13] and dynamic scaling theory. For example, financial time series show scaled properties suggesting a fractal structure [2, 5, 7]. The fractal dimension of a geometrical object can be defined as follows:

$$d = \lim_{r \to 0} [\ln N(r)] / [\ln(1/r)] \tag{3}$$

where N(r) is the number of boxes covering the object and r is the size of the box. An approximation to the fractal dimension can be obtained by counting the number of boxes covering the boundary of the object for different r sizes and then performing a logarithmic regression to obtain d (box counting algorithm). In Fig. 3, we illustrate the box counting algorithm for a hypothetical curve C. Counting the number of boxes for different sizes of r and performing a logarithmic linear regression, we can estimate the box dimension of a geometrical object with the following equation:

$$\ln N(r) = \ln\beta - d \ln r \tag{4}$$

this algorithm is illustrated in Figure 4.

The fractal dimension can be used to characterize an arbitrary object. The reason for this is that the fractal dimension measures the geometrical complexity of objects. In this case, a time series can be classified by using the numeric value of the fractal dimension (d is between 1 and 2 because we are on the plane xy). The reasoning behind this classification scheme is that when the boundary is smooth the fractal dimension of the object will be close to one. On the other hand, when the boundary is rougher the fractal dimension will be close to a value of two.

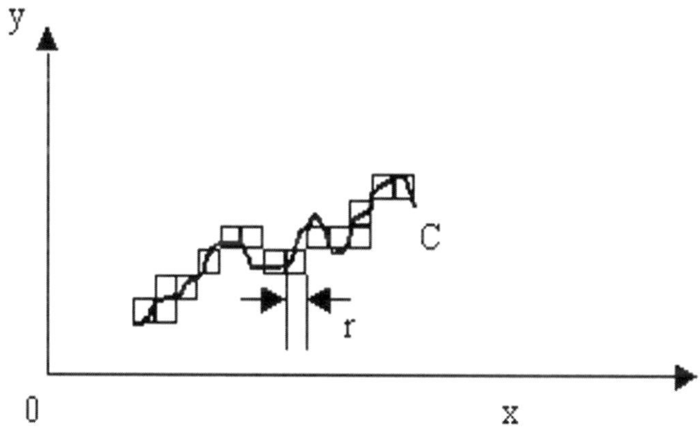

Fig. 3. Box counting algorithm for a curve C

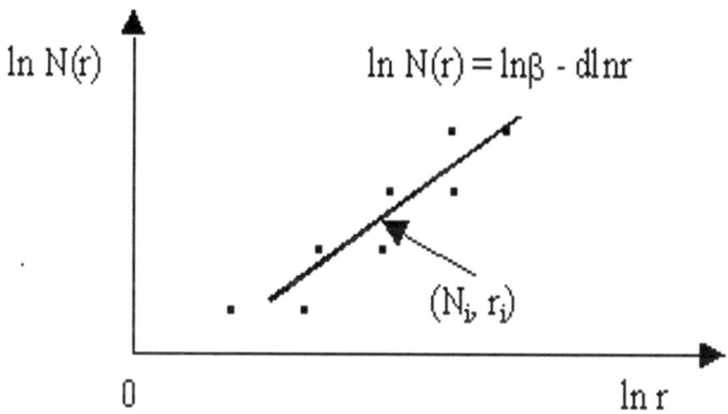

Fig. 4. Logarithmic regression to find the fractal dimension

4 Fuzzy Estimation of the Fractal Dimension

The fractal dimension of a geometrical object is a crisp numerical value measuring the geometrical complexity of the object. However, in practice it is difficult to assign a unique numerical value to an object. It is more appropriate to assign a range of numerical values in which there exists a membership degree for this object. For this reason, we will assign to an object O a fuzzy set μ_o, which measures the membership degree for that object.

Lets consider that the object O is in the plane xy, then a suitable membership function is a generalized bell function:

$$\mu_o = 1 / [1 + | (x-c) / a |^{2b}] \quad (5)$$

where a, b and c are the parameters of the membership function. Of course other types of membership functions could be used depending on the characteristics of the application.

By using the concept of a fuzzy set [16] we are in fact generalizing the mathematical concept of the fractal dimension. In fact, our definition of the fuzzy fractal dimension is as follows.

Definition 1: Let O be an arbitrary geometrical object in the plane xy. Then the fuzzy fractal dimension is the pair: (d_o, μ_o)
where d_o is the numerical value of the fractal dimension calculated by the box counting algorithm, and μ_o is the membership function for the object.

With this new definition we can account for the uncertainty in the estimation of the fractal dimension of an object. Also, this new definition enables easier pattern recognition for objects, because it is not necessary to match an exact numerical value to recognize a particular object [1, 15].

5 Plant Monitoring Using the Fuzzy-Fractal Approach

In this section, we show how to implement a fuzzy rule-based expert monitoring system with two basic sensors: temperature, and pressure. We also use as input the fuzzy fractal dimension of the time series of the measured variables. Individual sensors can identify three linguistic values (normal, high, and low). The three inputs can be combined to give 9 different scenarios. With the perfectly normal case (where all three input variables have normal values), there are additionally 9 more cases where combinations of abnormal readings can be observed.

Let x_1 be the temperature, x_2 the pressure, x_3 the fuzzy fractal dimension, and y the diagnostic statement. Let L_i, N_i, and H_i, represent respectively the three sets of low range, normal range, and high range for input data x_i, where i = 1, 2, or 3. Furthermore, let C_1, C_2,..., C_9 be the individual scenarios that could happen for each combination of the different data sets. The fuzzy rules have the general form:

$$R^{(0)}\text{: IF } x_1 \text{ is } N_1 \text{ AND } x_2 \text{ is } N_2 \text{ AND } x_3 \text{ is } N_3 \text{ THEN } y \text{ is } C_1 \qquad (6)$$

... ...

$$R^{(i)}\text{: IF } x_1 \text{ is } V_1 \text{ AND } x_2 \text{ is } V_2 \text{ AND } x_3 \text{ is } V_3 \text{ THEN } y \text{ is } C_i$$

... ...

$$R^{(26)}\text{: IF } x_1 \text{ is } H_1 \text{ AND } x_2 \text{ is } H_2 \text{ AND } x_3 \text{ is } H_3 \text{ THEN } y \text{ is } C_9$$

In this case, V_i represents L_i, H_i, or N_i, depending on the condition for the plant. Experts have to provide their knowledge in plant monitoring to label the individual cases Ci for i = 1, 2,..., 9. Also, the membership functions for the linguistic values of variables have to be defined according to historical data of the problem and expert knowledge.

We can use the Fuzzy Logic Toolbox of the MATLAB programming language to implement the fuzzy monitoring system described above. In this case, we need to specify the particular fuzzy rules and the corresponding membership functions for the problem. We show below a sample implementation of a health monitoring system using the MATLAB language. First, we show in Fig. 5 the general architecture of the fuzzy monitoring system. In this figure, we can see the input linguistic variables (temperature, pressure, and fractal dimension) and the output linguistic variable (condition of the plant) of the fuzzy monitoring system. Of course, in this case the fractal dimension is estimated using the box counting algorithm, which was implemented also as a computer program in MATLAB.

In Fig. 6 the implementation of the fuzzy rule base in MATLAB is shown. The actual 27 rules were defined according to expert knowledge on the process. In Fig. 7, the Gaussian membership functions for the output variable (condition of the plant) are shown. In Fig. 8 the non-linear surface for the problem of plant monitoring is shown. Finally, the fuzzy rule viewer of MATLAB is shown in Fig. 9. In this case, the fuzzy rule viewer is used to calculate the output condition of the plant for specific values of the input variables.

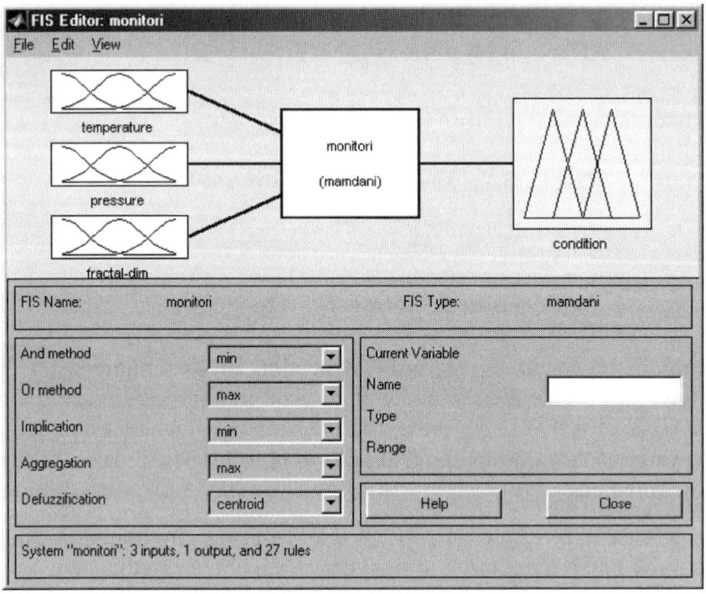

Fig. 5. Architecture of the fuzzy system for plant monitoring

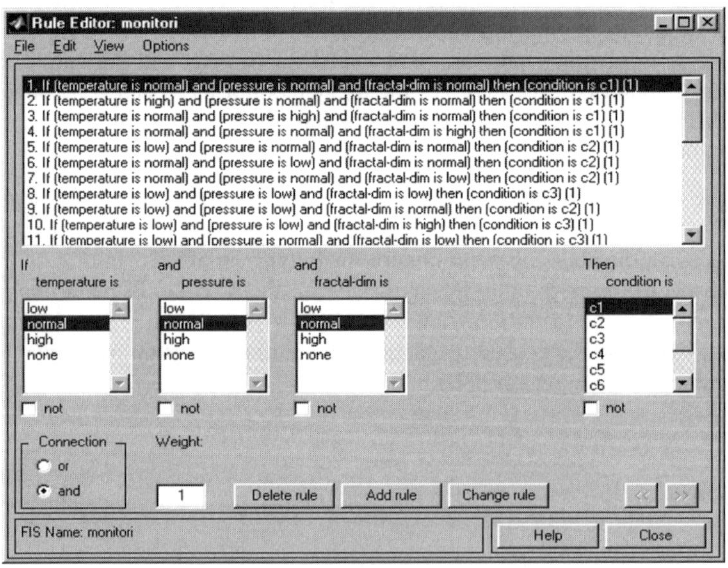

Fig. 6. Fuzzy rule base for plant monitoring

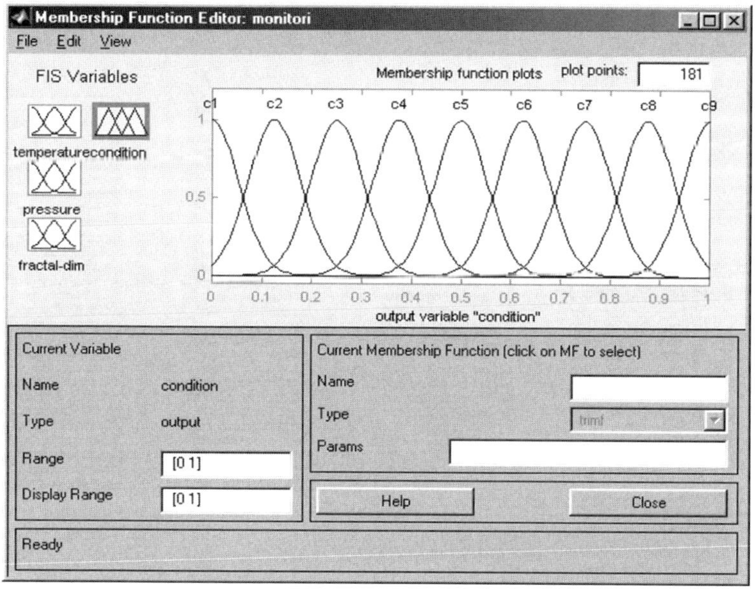

Fig. 7. Membership functions of the output variable

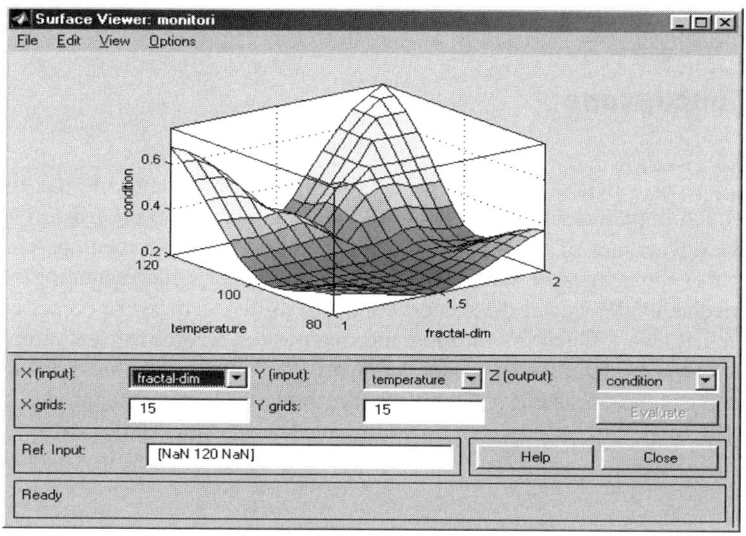

Fig. 8. Non-linear surface for plant monitoring

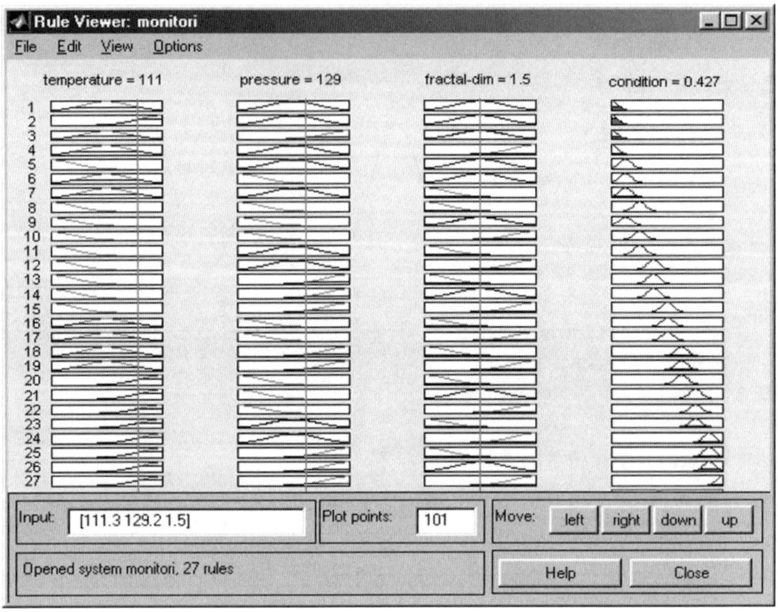

Fig. 9. Fuzzy rule viewer for specific values of the input variables

6 Conclusions

In this paper a hybrid fuzzy-fractal approach for plant monitoring has been proposed. An implementation in MATLAB has been shown, to describe in more detail the advantages of the new approach. The hybrid fuzzy-fractal approach combines the advantages of fuzzy logic (expert knowledge representation) with the advantages of the fractal dimension concept (ability to measure object complexity), to achieve efficient monitoring and diagnostics. A problem yet to be considered, is how to automatically learn (or adapt) the membership functions and rules of the fuzzy system using real data for the problem. A genetic approach could be used to evolve the fuzzy system (including the fractal dimension).

Acknowledgments

We would like to thank the financial support given to this research project by COSNET. We would also like to thank the Department of Computer Science of Tijuana Institute of Technology for the time and resources given to this project.

References

1. Bezdek JC (1981) Pattern Recognition with Fuzzy Objective Function Algorithms, Plenum Press.
2. Brock WA, Hsieh DA, LeBaron B (1991) "Nonlinear Dynamics, Chaos and Instability," MIT Press, Cambridge, MA, USA.
3. Castillo O, Melin P (1994) Developing a New Method for the Identification of Microorganisms for the Food Industry using the Fractal Dimension, *Journal of Fractals*, **2**, No. 3, pp. 457-460.
4. Castillo O, Melin P (1996) "Automated Mathematical Modelling for Financial Time Series Prediction using Fuzzy Logic, Dynamical System Theory and Fractal Theory," Proceedings of CIFEr'96, IEEE Press, New York, NY, USA, pp. 120-126.
5. Castillo O, Melin P (1998) "A New Fuzzy-Genetic Approach for the Simulation and Forecasting of International Trade Non-Linear Dynamics," Proceedings of CIFEr'98, IEEE Press, New York, NY, USA, pp. 189-196.
6. Castillo O, Melin P (1998) "A New Fuzzy-Fractal-Genetic Method for Automated Mathematical Modelling and Simulation of Robotic Dynamic Systems," Proceedings of FUZZ'98, IEEE Press, Alaska, USA, Vol. 2, pp. 1182-1187.
7. Castillo O, Melin P (1999) "Automated Mathematical Modelling for Financial Time Series Prediction Combining Fuzzy Logic and Fractal Theory," Edited Book "Soft Computing for Financial Engineering," Springer-Verlag, Heidelberg, Germany, pp. 93-106.
8. Castillo O, Melin P (2001) "Soft Computing for Control of Non-Linear Dynamical Systems," Springer-Verlag, Heidelberg, Germany.
9. Chen G, Pham TT (2001) "Introduction to Fuzzy Sets, Fuzzy Logic, and Fuzzy Control Systems," CRC Press, Boca Raton, Florida, USA.
10. Du R (1998) "Engineering Monitoring and Diagnosis," Edited Book "Industrial and Manufacturing Systems," Academic Press, San Diego, California, USA, pp. 119-144.
11. Du R, Elbestawi M A, Wu SM (1993) "Automated Monitoring of Manufacturing Processes," Transactions ASME J. Eng. Ind., pp. 245-252.
12. Jang JR, Sun CT, Mizutani E (1997). Neuro-Fuzzy and Soft Computing, Prentice Hall.
13. Mandelbrot, B (1987) "The Fractal Geometry of Nature,",W.H. Freeman and Company, New York, NY, USA.
14. Melin P, Castillo O (1998) "An Adaptive Model-Based Neuro-Fuzzy-Fractal Controller for Biochemical Reactors in the Food Industry," Proceedings of IJCNN'98, IEEE Computer Society Press, Alaska, USA, Vol. 1, pp. 106-111.
15. Yager R, Filev D (1994) Generation of Fuzzy Rules by Mountain Clustering, Intelligent and Fuzzy Systems, Vol. 2, No. 3, pp. 209-219.
16. Zadeh LA (1975) "The Concept of a Linguistic Variable and its Application to Approximate Reasoning," *Information Sciences*, 8, pp. 43-80.

15 Failure Detection Using a Fuzzy Neural Network with an Automatic Input Selection Algorithm

Man Gyun Na, Young Rok Sim, and Kyung Ho Park

Department of Nuclear Engineering, Chosun University
375 Seosuk-dong, Dong-gu, Kwangju 501-759, Korea

magyna@chosun.ac.kr

Belle R. Upadhyaya, Baofu Lu, and Ke Zhao

Department of Nuclear Engineering, The University of Tennessee
Knoxville, TN 37996-2300, USA

It is well known that the performance of fuzzy neural networks applied to sensor monitoring strongly depends on the selection of inputs. In their applications to sensor monitoring, there are usually a large number of input variables related to a relevant output. As the number of input variables increases, the required training time of a fuzzy neural network increases exponentially. Thus, it is essential to reduce the number of inputs to a fuzzy neural network and moreover, to select the optimum number of mutually independent inputs that are able to clearly define the input-output mapping. In this chapter, an automatic input selection routine was described which combines the principal component analysis, correlation analysis, and genetic algorithm to select important input features. Also, whether the sensors fail or not is determined by applying the sequential probability ratio test to the residuals between the estimated signals and the measured signals. The described sensor failure detection method was verified through applications to the steam generator water level, the steam generator steam flowrate, the pressurizer water level, and the pressurizer pressure sensors in pressurized water reactors.

1 Introduction

Many neural networks and fuzzy inference methods have recently been presented to diagnose sensors. This work was started in the late 80's by Upadhyaya [1,2], was also conducted by Singer et al. [3], Hines and Uhrig, et al. [4,5], and Fantoni, et al.[6]. Through training, fuzzy-neural networks have been known to have the capability for performing complex mappings between input and output data, especially when expert diagnostic knowledge and the prior relation of fault symptom model are not clear. The direct use of transient signals in the time domain to the input of a neural network can be difficult since subtle differences may occur between different transients. Also, in recent years, the general problem of selecting a salient feature set for fuzzy neural networks has been generating a great deal of interest. It is experienced that the performance of a neural network strongly depends on the input features selected for its training. Non-salient input features to an artificial neural network can have even negative impacts on the performance.

In sensor monitoring applications, there are a large number of input signals related to an output and it would require a large amount of time to train a fuzzy neural network with all the input signals since the number of connection weights for neural networks and parameters for fuzzy inference would be extremely large. Also, by eliminating unimportant sensors and sensor parameters, the cost and time of collecting the data can be reduced. As the number of input features grows, the structure of the neuro-fuzzy system would become complex. Thus, it is essential to reduce the number of inputs to a neural network and to select optimum number of mutually independent inputs that are able to clearly define the input-output mapping. The process of selecting these important input variables is usually known as feature selection. Depending on the problem encountered, various criteria can be used to rank feature subsets. If the task is to predict the properties of the measurements like the sensor signal estimation, a criterion that evaluates the predictive ability of the selected features should be used. Many feature selection methods have been developed including principal component analysis (PCA), genetic algorithm (GA), and others [7-11]. In this chapter, an input feature selection method is described where PCA, correlation analysis, and GA are combined to select important input features and this combined feature selection method will be called a PCG method hereafter.

An important problem in sensor failure detection is to decide whether a sensor fails or not after only one abnormal observation. At every new sample of a sensor signal, a new mean and a new variance may be computed and then, these quantities may be used to check if the sensor fails or not. However, this procedure requires too many samples to obtain a meaningful mean and a meaningful variance and also, during the acquisition of the samples, a significant degradation of the process monitored may occur. Therefore, the sequential probability ratio test (SPRT) [12] is usually used. The method can detect a failure using the degree of degradation and the continuous behavior of the sensor, without having to calculate a new mean and a new variance at each sample. The signal estimated by the fuzzy neural network is compared with the measured signal, and then the sensor failure is determined by the SPRT using the residual which is the difference between the estimated signal and the measured signal.

In this chapter, the focus is on identifying important input features to a fuzzy neural network and on detecting sensor failure using fuzzy neural network and SPRT.

The failure detection algorithm will be applied to the failure detection of the steam generator water level, the steam generator steam flowrate, the pressurizer (PRZR) water level and the PRZR pressure sensors in pressurized water reactors and compared with other algorithms.

2 Input Feature Selection Methods

The number of variables has to be reduced for several reasons. But this seems to be paradoxical at first since a dimension reduction decreases the information content. A reduction of the number of variables can lead to an improved performance due to at least three reasons. First, irrelevant inputs will result in a model which is not robust. Thus, it becomes important to use only high information descriptors. Secondly, studies have shown that if colinearity is present among the variables, the prediction results can get worse. Hence, it is necessary to remove highly correlated variables. Finally, when making a model containing many input parameters, a large number of observations are required to span the complete input space. The number of required observations grows exponentially with the number of input variables, which makes a dimensional reduction necessary to get a good model. In this section, two methods using PCA and genetic algorithm are reviewed first and a new method that combines the two methods as well as correlation analysis will be described.

2.1 Principal Component Analysis

Principal component analysis attempts to reduce the dimensionality by creating new variables that are linear combinations of the original variables. Thus PCA method involves linearly transforming the input space into an orthogonal space that can be chosen to be of lower dimension with minimal loss of information. The standardized input data are projected onto the eigenvectors called the principal components of the covariance matrix of the original data. The method also makes the transformed vectors (principal components) orthogonal and uncorrelated [13,14].

Given a signal vector $\mathbf{x}$ of p dimensions, $\mathbf{x} = \begin{bmatrix} x_1 & x_2 & \cdots & x_p \end{bmatrix}^T$, its mean vector $\boldsymbol{\mu}$ and covariance matrix $\boldsymbol{\Sigma}$ are described by

$$\boldsymbol{\mu} = E\{\mathbf{x}\} = \begin{bmatrix} m_1 & m_2 & \cdots & m_p \end{bmatrix}^T, \tag{1}$$

$$\boldsymbol{\Sigma} = E\{(\mathbf{x} - \boldsymbol{\mu})(\mathbf{x} - \boldsymbol{\mu})^T\}. \tag{2}$$

Since the true mean and the true covariance matrix are seldom known, the mean and covariance matrix are replaced with the sample mean $\mathbf{m}$ and the sample covariance matrix $\mathbf{S}$. The eigenvalues $\lambda_1, \lambda_2, \cdots, \lambda_p$ of the sample covariance matrix $\mathbf{S}$ are calculated, and then ranked according to the magnitude of their associated eigenvalues:

$$\lambda_1 \geq \lambda_2 \geq \cdots \geq \lambda_p . \tag{3}$$

Their corresponding orthonormal eigenvectors $\mathbf{p}_1, \mathbf{p}_2, \cdots, \mathbf{p}_p$ are called principal components. The eigenvalues are proportional to the amount of variance (information) represented by the corresponding principal components. The transformation to the principal component space can be written as:

$$\mathbf{z} = \mathbf{x}^T \mathbf{P} , \tag{4}$$

where $\mathbf{P} = \left[\mathbf{p}_1 \; \mathbf{p}_2 \cdots \mathbf{p}_p\right]$.

The feature vector $\mathbf{z}$ can be transformed back into the original data vector $\mathbf{x}$ without loss of information as long as the number of features, m, is equal to the dimension of the original space, p. For $m < p$, some information is usually lost. The objective is to choose a small m that does not lose much information.

However, this PCA input selection method has a disadvantage of not reducing substantially the number of the input signals to a PCA block. If we use many input signals to the PCA block, a possibility that we use unreliable and faulty sensor signals will increase. Although in this method the number of inputs to the fuzzy neural network decreases, the number of the actually used signals will not decrease.

2.2 Genetic Algorithm

In optimization problems using genetic algorithms, the term *chromosome* refers to a candidate solution that minimizes a cost function, generally encoded as a bit string. Each chromosome can be thought of as a point in the search space of candidate solutions. Genetic algorithm is an optimization technique that imitates the evolutionary process of a living organism. An initial population of chromosomes is iteratively altered by mechanisms inspired by natural evolution such as selection, crossover and mutation. Thus genetic algorithms process populations of chromosomes, successively replacing one such population with another. Genetic algorithms start from many points simultaneously climbing many peaks in parallel, and hence the probability of finding a false peak is reduced compared to the methods that move from one point to another. Accordingly, genetic algorithms are less susceptible to being stuck at local optima than conventional search methods. On the other hand, genetic algorithms have a disadvantage that it requires too much computational time [8,15,16].

The genetic algorithms require a fitness function that assigns a score to each chromosome in the current population. The fitness of a chromosome (individual) depends on how well that chromosome solves the problem at hand. A fitness function that evaluates the extent to which each individual is suitable for the given objectives such as small maximum error together with small total squared error and the small number of input variables, is suggested as follows:

$$F = \exp\left(-\mu_1 E_1 - \mu_2 E_2 - \mu_3 E_3\right), \tag{5}$$

where μ_1, μ_2 and μ_3 are the weighting coefficients, and E_1, E_2 and E_3 are the overall sum of squared prediction errors, the maximum absolute prediction error and the number of input variables, respectively, defined as:

$$E_1 = \sum_{k=1}^{N} \left(y(k) - \hat{y}(k)\right)^2 , \tag{6}$$

$$E_2 = \max_{k} \left\{ \left| y(k) - \hat{y}(k) \right| \right\} , \tag{7}$$

$$E_3 = N_{input} \, . \tag{8}$$

$y(k)$ and $\hat{y}(k)$ denote the measured signal and the estimated signal, respectively.

Each chromosome is encoded as a bit string which is composed of the same bit number as the number of input variables, and one '1' in each bit string represents that the corresponding input is selected and zero '0' represents that the corresponding input is not selected.

2.3 PCG Method

To overcome the disadvantages of the above two methods, PCA, GA and a correlation between input variables and an output variable are combined to select an optimal input set for sensor signal estimation. PCA and GA are already explained in above subsections. Note that the correlation coefficient matrix of the original data set is equal to the covariance matrix of the data after the data have been standardized. This correlation matrix indicates some degree of close or distant relationship among variables (signals). The high specific (i, j) component of the correlation matrix means that the related two variables are closely related to each other. These values between the input variables and the output variable are used to initialize the chromosomes of the genetic algorithm. To run a usual genetic algorithm, each bit of the chromosomes is usually randomly assigned one or zero. However, in this chapter, it is considered that there is a high probability that the corresponding bit is assigned one in the case that a correlation between a specific input and an output is high by using the correlation matrix. On the contrary, there is a high probability that the corresponding bit is assigned zero in the case that a correlation between the specific input and the output is low. This helps reduce the computational time by not considering the inputs that are not almost related with the output from the first time. This method is not to completely exclude input signals using correlation analysis. The method reduces just probability that a signal is selected as its correlation degree with an output signal decreases.

This algorithm for input feature selection is described in Figure. 1. First, the initial chromosomes of a genetic algorithm are selected by using the correlation coefficient matrix, which reduces the computational burden of the genetic algorithm that requires much computational time. Also, the input signals substantially used for sensor signal estimation decrease in number through the genetic algorithm. A set of input signals to the PCA blocks that are closely related with estimating the output sensor signal are selected. The input signals coming into the PCA block are converted into new variables, called score vectors. For the application of the PCA method, in this chapter, the number of input features from the PCA block is selected so that above 5 percent of original information into the PCA block will not be lost. As it were, the selected features include at least 95 percent of the initial information. Also, the time-delayed signal of the first score vector which usually includes most of the information of input signals is used to describe the sequential characteristics of a signal.

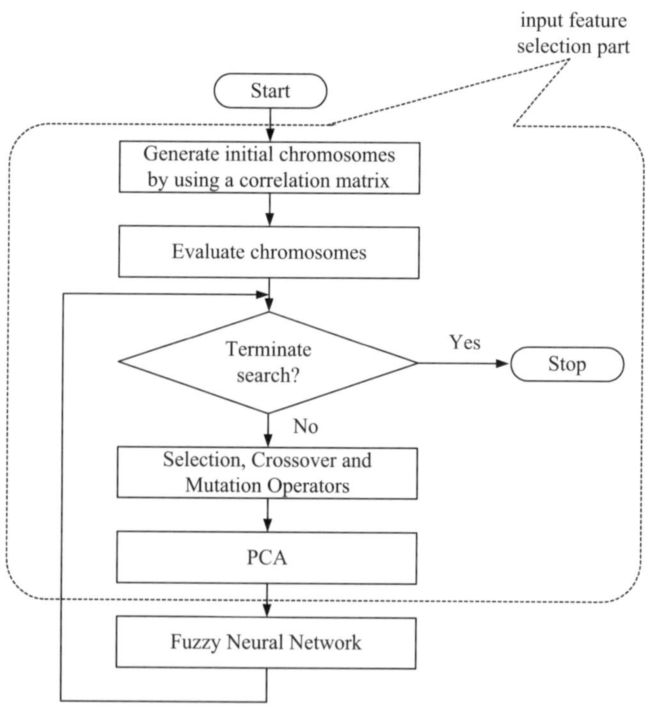

Fig. 1. Schematic diagram of the input features selection algorithm (PCG method)

3 Sensor Signal Estimation Using a Fuzzy Neural Network

3.1 Fuzzy Inference System

A system that consists of a fuzzy inference system and its neuronal training system is usually called an adaptive network-based fuzzy inference system (ANFIS) [17]. In a fuzzy inference system, the i-th rule can be described using the first-order Sugeno-Takagi type [18] as follows:

$$\text{If } x_1 \text{ is } A_{i1} \textit{ AND} \cdots \textit{AND } x_m \text{ is } A_{im}, \text{ then } \hat{y}_i \text{ is } f_i(x_1, \cdots, x_m), \tag{9}$$

where

$x_1, \cdots, x_m$ = input variables to the fuzzy neural network (m = number of input variables),

$A_{i1}, \cdots, A_{im}$ = antecedent membership function of each input variable for the i-th rule ($i = 1, 2, ..., n$),

$\hat{y}_i$ = output of the i-th rule,

$$f_i(x_1,\cdots,x_m) = \sum_{j=1}^{m} q_{ij} x_j + r_i, \tag{10}$$

q_{ij} = weighting value of the j-th input onto the i-th rule output,

r_i = bias of the i-th output,

n = number of rules.

In this chapter, the following Gaussian and sigmoid membership functions are used for each input variable:

$$A_{ij}(x_j) = \exp\left(-\frac{(x_j - c_{ij})^2}{2s_{ij}^2}\right), \tag{11}$$

$$A_{ij}(x_j) = \frac{1}{\exp\left(-\dfrac{x_j - c_{ij}}{s_{ij}}\right) + 1}, \tag{12}$$

where

c_{ij} = center position of a membership function for the i-th rule and the j-th input,

s_{ij} = sharpness of a membership function for the i-th rule and the j-th input.

The sigmoid membership function is used for the maximum and minimum center values in each input variable and the Gaussian membership function is used for other center values. The output of an arbitrary i-th rule, f_i, consists of the first-order polynomial of inputs as given in Eq. (10). The output of a fuzzy inference system with n rules is obtained by weighting the real values of consequent part for all rules with the corresponding membership grade and indicates the estimated value of the relevant sensor signal. The estimated signal is obtained as follows:

$$\hat{y} = \sum_{i=1}^{n} \overline{w}_i f_i = \mathbf{w}^T \mathbf{q}, \tag{13}$$

where

$$\overline{w}_i = \frac{w_i}{\sum_{i=1}^{n} w_i}, \tag{14}$$

$$w_i = \prod_{j=1}^{m} A_{ij}(x_j), \tag{15}$$

$$\mathbf{q} = [q_{11} \cdots q_{n1} \cdots \cdots q_{1m} \cdots q_{nm} \; r_1 \cdots r_n]^T,$$

$$\mathbf{w} = [\overline{w}_1 x_1 \cdots \overline{w}_n x_1 \cdots \cdots \overline{w}_1 x_m \cdots \overline{w}_n x_m \; \overline{w}_1 \cdots \overline{w}_n]^T.$$

3.2 Training of the Fuzzy Inference System

The back-propagation algorithm is a general method for recursively solving for parameter optimization of a fuzzy inference system. Since this conventional optimization algorithm is susceptible to getting stuck at local optima, the genetic algorithm is used in this chapter. However, the genetic algorithm requires much computational time if there are many parameters to be optimized. Therefore, the least-squares method that is a one-pass optimization method is combined for a part of the parameters. The genetic algorithm is used to optimize the antecedent parameters c_{ij} and s_{ij} (membership function parameters), and the least-squares algorithm is used to solve the consequent parameters q_{ij} and r_i (the polynomial coefficients of the consequent part). The genetic algorithm is to minimize the overall sum of squared prediction errors, the maximum absolute prediction error and the number of used sensor signals. A simple explanation on the genetic algo-

number of used sensor signals. A simple explanation on the genetic algorithm was given in the above subsection.

In this chapter, to increase the efficiency of the conventional genetic algorithm, three schemes are applied to accomplish the following good performance of the genetic algorithm: (a) initial coarse tuning and final fine tuning by changing the bit number of chromosomes versus generation; (b) prevention of an initial premature convergence without reaching optimal solutions and the acceleration of a final convergence by using two different selection methods of the crossover site that is randomly selected anywhere in a chromosome or randomly selected between only parameters in a chromosome; (c) prevention of final drifting without convergence by maintaining a certain part of chromosomes with higher fitness (refer to [19] for details).

If we fix some parameters of the fuzzy inference system by the genetic algorithm, the resulting fuzzy inference system is equivalent to a series of expansions of some basis functions. This basis function expansion is linear in its adjustable parameters. Therefore, we can use the least-squares method to determine the remaining parameters. If a total of N input-output training data are given, from Eq. (13) the consequent parameters are chosen to minimize the following cost function:

$$J = \frac{1}{2}\sum_{i=1}^{N}(y - \hat{y})^2 = \frac{1}{2}(\mathbf{y} - \hat{\mathbf{y}})^2 = \frac{1}{2}(\mathbf{y} - \mathbf{Wq})^2, \tag{16}$$

where

$$\mathbf{y} = \left[y^1\ y^2\ \cdots\ y^N\right]^T,$$

$$\mathbf{q} = \left[q_{11}\cdots q_{n1}\cdots\cdots q_{1m}\cdots q_{nm}\ r_1\cdots r_n\right]^T,$$

$$\mathbf{W} = \left[\mathbf{w}^1\ \mathbf{w}^2\ \cdots\ \mathbf{w}^N\right]^T,$$

$$\mathbf{w}^k = \left[\overline{w}_1 x_1\cdots\overline{w}_n x_1\cdots\cdots\overline{w}_1 x_m\cdots\overline{w}_n x_m\ \overline{w}_1\cdots\overline{w}_n\right]^T,\quad k = 1, 2, \cdots, N.$$

$\mathbf{y}$ is the output data vector, $\mathbf{q}$ is the parameter vector, and the matrix $\mathbf{W}$ includes the input data. An equation for minimizing the cost function is as follows:

$$\mathbf{y} = \mathbf{Wq}. \tag{17}$$

The fuzzy neural network output is represented by the $N \times (m+1)n$-dimensional matrix $\mathbf{W}$ and the $(m+1)n$-dimensional parameter vector $\mathbf{q}$. In order to solve the parameter vector $\mathbf{q}$ in Eq. (17), the matrix $\mathbf{W}$ should be invertible but is not usually a square matrix. Therefore, we solve the vector by using the pseudo-inverse as follows:

$$\mathbf{q} = \left(\mathbf{W}^T \mathbf{W}\right)^{-1} \mathbf{W}^T \mathbf{y} . \tag{18}$$

4 Sequential Probability Ratio Test (SPRT)

In sensor failure detection, at every new sample of a signal, a new mean and a new variance of the signal may be computed to check if the sensor is degraded or not. However, this procedure requires too many samples to obtain a meaningful mean and a meaningful variance. During the acquisition of the samples, a significant degradation of the monitored process may occur. So SPRT is used to detect a failure using the degree of failure and the continuous behavior of the sensor, without having to calculate a new mean and a new variance at each sample. The SPRT is a statistical model developed by Wald in 1945 [12].

The objective of sensor degradation detection is to detect the failure as soon as possible with a very small probability of making a wrong decision. In the application of sensor failure detection, the SPRT is dealing with the residual (difference between the sensor measurement and the sensor estimate). Normally the residual signals are randomly distributed, so they are nearly uncorrelated and have a Gaussian (normal) distribution $P_i(\varepsilon_k, m_i, \sigma_i)$, where ε_k is the residual signal at time k, and m_i and σ_i are the mean and the standard deviation under hypothesis i, respectively. The sensor failure can be stated in terms of a change in the mean m or a change in the variance σ^2. If a set of samples x_i, $i = 1, 2, \cdots, n$, is collected with a density function P describing each sample in the set, an overall likelihood ratio is given by

$$\gamma_n = \frac{P_1(\varepsilon_1 \mid H_1) \cdot P_1(\varepsilon_2 \mid H_1) \cdot P_1(\varepsilon_3 \mid H_1) \cdots P_1(\varepsilon_n \mid H_1)}{P_0(\varepsilon_1 \mid H_0) \cdot P_0(\varepsilon_2 \mid H_0) \cdot P_0(\varepsilon_3 \mid H_0) \cdots P_0(\varepsilon_n \mid H_0)}, \tag{19}$$

where H_1 represents a hypothesis that the sensor is degraded and H_0 represents a hypothesis that the sensor is normal. By taking the logarithm of the above equation and replacing the probability density functions in terms of residuals, means and variances, the log likelihood ratio (LLR, λ_n) can be written as the following recurrent form:

$$\lambda_n = \lambda_{n-1} + \ln\left(\frac{\sigma_0}{\sigma_1}\right) + \frac{(\varepsilon_n - m_0)^2}{2\sigma_0{}^2} - \frac{(\varepsilon_n - m_1)^2}{2\sigma_1{}^2} . \tag{20}$$

This is the form we use for deriving the sensor drift detection algorithm. For a normal sensor, the log likelihood ratio would decrease and eventually reach a specified bound A, a smaller value than zero. When the ratio reaches this bound, the decision is made that the sensor is normal, and the ratio is initialized by setting it equal to zero. For a degraded sensor, the ratio would increase and eventually reach a specified bound B, a larger value than zero. When the ratio is equal to B, the decision is made that the sensor is degraded. The decision boundaries A and B are chosen by a false alarm probability α and a missed alarm probability β; $A = \ln\left(\frac{\beta}{1-\alpha}\right)$ and $B = \ln\left(\frac{1-\beta}{\alpha}\right)$.

5 Applications

The overall sensor failure detection algorithm that combines the above-mentioned methods is described in Figure. 2. The input-output data were obtained for the load-decrease transients from the simulation of the MARS code [20] which is a unified version of the COBRA/TF and RELAP5/MOD3 codes, which consist of a total of 14 different signals. It is important to train the fuzzy inference system by using input-output data of a broad operating range since the variable selection may significantly depend on the training data. The input-output data were obtained from power decrease transients of a broad range. These data were standardized before they are presented to the fuzzy neural network. The overall sensor failure detection algorithm was applied to the steam generator water level, the steam generator steam flowrate, the pressurizer water level and the pressurizer pressure sensors. Noise was added to model the real data of the nuclear power plant. The noise is proportional to the maximum variation $\sigma_{\max}$ of each signal and is chosen from a uniform distribution on the interval ($-0.005\sigma_{\max}$, $0.005\sigma_{\max}$). Each signal consists of a total of 700 discrete time points where the sampling period is 1 sec. The fuzzy neural network was trained using one fifth of the given data in the training stage and was validated using the remaining data in the verification stage.

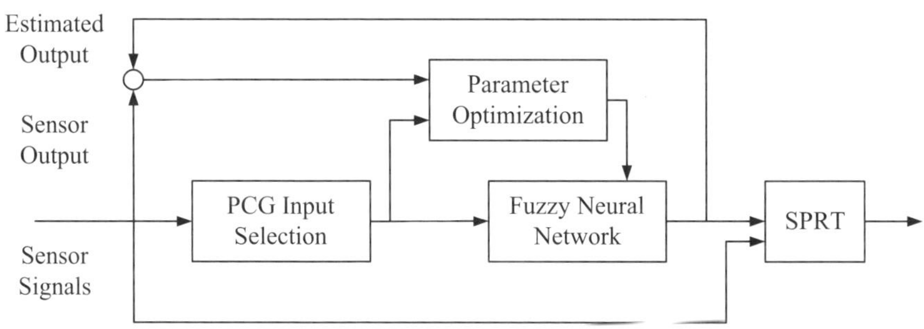

Fig. 2. Schematic diagram of the monitoring algorithm for sensors

Table 1 shows the correlation coefficient matrix that indicates the relationships among all the gathered signals. It is very important to accurately estimate the signals to determine a sensor failure. Table 2 shows the signal estimation and failure detection results of all the application cases and also, in this table the PCG input feature selection method is compared with the other two methods; genetic method and PCA method.

Table 1. Correlation coefficient matrix for gathered signals

	SF	FF	SP	ST	NL	WL	HT	CT	HF	AT	PP	PL	RP	PT
steam flowrate(SF)	1.000	0.990	-0.957	-0.957	-0.003	-0.004	0.985	-0.486	-0.320	0.954	-0.142	0.954	0.995	-0.144
feed flowrate(FF)	0.990	1.000	-0.935	-0.936	-0.092	-0.092	0.984	-0.439	-0.348	0.961	-0.090	0.960	0.988	-0.092
steam pres. (SP)	-0.957	-0.935	1.000	1.000	-0.026	-0.025	-0.908	0.717	0.044	-0.837	0.292	-0.832	-0.956	0.294
steam temp.(ST)	-0.957	-0.936	1.000	1.000	-0.026	-0.025	-0.909	0.715	0.048	-0.838	0.293	-0.833	-0.957	0.296
S/G water level(NL)	-0.003	-0.092	-0.026	-0.026	1.000	0.999	-0.037	-0.117	0.057	-0.056	-0.208	-0.050	-0.021	-0.208
S/G wide-range level(WL)	-0.004	-0.092	-0.025	-0.025	0.999	1.000	-0.037	-0.115	0.057	-0.056	-0.207	-0.050	-0.021	-0.207
hot-leg temp.(HT)	0.985	0.984	-0.908	-0.909	-0.037	-0.037	1.000	-0.366	-0.400	0.989	0.018	0.985	0.990	0.016
cold-leg temp.(CT)	-0.486	-0.439	0.717	0.715	-0.117	-0.115	-0.366	1.000	-0.618	-0.225	0.578	-0.211	-0.490	0.580
hot-leg flowrate(HF)	-0.320	-0.348	0.044	0.048	0.057	0.057	-0.400	-0.618	1.000	-0.515	-0.266	-0.545	-0.284	-0.264
RCS average temp.(AT)	0.954	0.961	-0.837	-0.838	-0.056	-0.056	0.989	-0.225	-0.515	1.000	0.113	0.998	0.959	0.111
PRZR pressure(PP)	-0.142	-0.090	0.292	0.293	-0.208	-0.207	0.018	0.578	-0.266	0.113	1.000	0.073	-0.073	1.000
PRZR water level(PL)	0.954	0.960	-0.832	-0.833	-0.050	-0.050	0.985	-0.211	-0.545	0.998	0.073	1.000	0.954	0.071
reactor power (RP)	0.995	0.988	-0.956	-0.957	-0.021	-0.021	0.990	-0.490	-0.284	0.959	-0.073	0.954	1.000	-0.075
PRZR temp. (PT)	-0.144	-0.092	0.294	0.296	-0.208	-0.207	0.016	0.580	-0.264	0.111	1.000	0.071	-0.075	1.000

Table 2. Final results for four application cases [after 30 generations training for input selection (genetic and PCG methods only) and after 50 generations training for the fuzzy neural network]

Sensors		S/G water level			S/G steam flowrate			Pressurizer water level			Pressurizer pressure		
Methods		PCA	Genetic	PCG	PCA	Genetic	PCG	PCA	Genetic	PCG	PCA	Genetic	PCG
Training data	maximum error	0.060	0.062	0.059	0.053	0.033	0.046	0.073	0.034	0.034	0.069	0.032	0.034
	sum of all squared errors	0.105	0.106	0.110	0.069	0.029	0.045	0.126	0.034	0.034	0.117	0.037	0.035
	std. dev.	0.027	0.027	0.028	0.022	0.014	0.018	0.030	0.016	0.016	0.029	0.016	0.016
	error (2 σ)	0.055	0.055	0.056	0.044	0.029	0.036	0.060	0.031	0.031	0.058	0.032	0.032
	fitness[1)]	0.638	0.774	0.811	0.653	0.834	0.800	0.612	0.831	0.849	0.620	0.852	0.882
Verification data	maximum error	0.099	0.082	0.123	0.144	0.068	0.084	0.194	0.050	0.052	0.173	0.059	0.060
	sum of all squared errors	0.527	0.571	0.498	0.421	0.188	0.223	0.608	0.176	0.187	0.853	0.213	0.211
	std. dev.	0.031	0.032	0.030	0.027	0.018	0.020	0.033	0.018	0.018	0.039	0.019	0.019
	error (2 σ)	0.061	0.064	0.060	0.055	0.036	0.039	0.066	0.035	0.037	0.078	0.039	0.039
Input number to PCA		13	N/A	1	13	N/A	4	13	N/A	3	13	N/A	1
Number of FNN inputs		5	3	2	5	4	3	5	4	3	5	3	2
Used signals		all signals	FF,WL HT	WL	all signals	FF,SP HF,PT	FF,SP PP,PL	all signals	SF,SP AT,PT	SP,HT CT	all signals	NL,HT PT	PT
Failure detection time[2)] (sec)		48	50	50	84	55	64	91	54	59	97	64	64

1) The fitness value was calculated in combination with the genetic optimization of the fuzzy neural network parameters.

2) Time after the beginning of purposely gradual degradation

5.1 Single Output Faults

In the application to steam generator narrow-range water level sensor, the maximum error and the total squared error have similar magnitudes for the three methods but PCA method is worst in the aspects of using many inputs, which is indicated by a fitness value. The steam generator narrow-range water level is almost closely related with the wide-range water level as shown in Table 1 and also as expected from the correlation coefficient matrix. The PCG input selection algorithm automatically selects the wide-range water level signal as an optimal input when it is applied to estimate the narrow-range water level. The genetic method selects other two signals (feedwater flowrate, hot-leg temperature) including the wide-range water level signal (refer to Table 2). The reason why sum of all squared errors for the verification data is much larger than that for the training data is that the number of verification data is 4 times more than that of the training data. In all three methods, the standard deviation of errors for the verification data is similar to that for the training data. Figure 3 shows the estimation of S/G water level signal using the PCG input selection algorithm. Figure 4 shows the failure detection time for three methods and a measured failed signal and estimated signals using each input selection method. The failure detection time is almost similar for three methods.

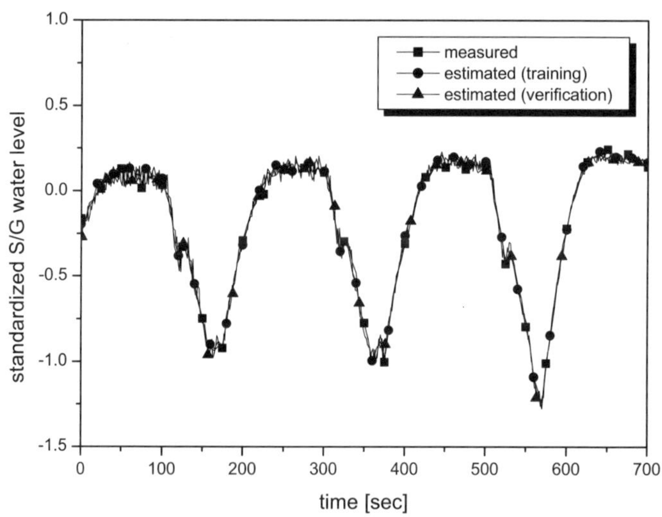

Fig. 3. Estimation of a (standardized) steam generator water level signal using the PCG method

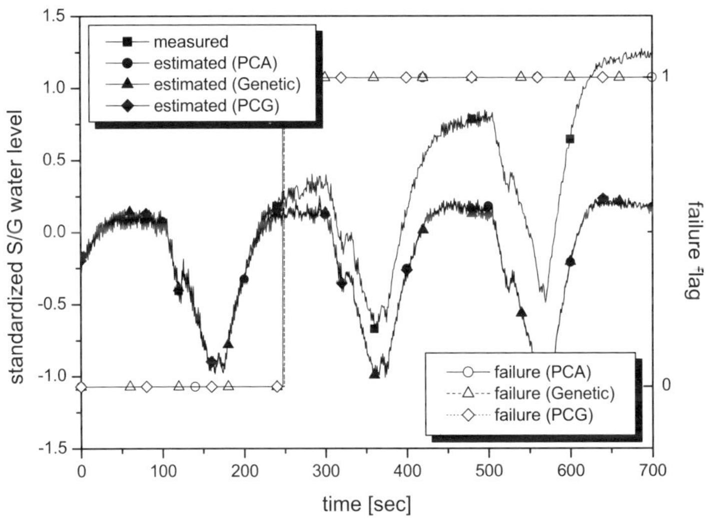

Fig. 4. Failure detection for gradual degradation of steam generator water level

In the application to steam generator steam flowrate sensor, the genetic method selects feedwater flowrate, S/G steam pressure, hot-leg flowrate, and PRZR temperature as optimal inputs, and the PCG method selects feedwater flowrate, S/G steam pressure, PRZR pressure, and PRZR water level. In all three methods, the standard deviation of errors for the verification data is similar to that for the training data. The PCG method and the genetic method have similar fitness values but PCA method has a lowest one. Figure 5 shows the estimation of the S/G steam flowrate using the PCG input selection method and Figure 6 shows the failure detection time for cases applying three methods. The detection time for the case applying the PCA method is slower than other two methods as expected from the error magnitude.

In the application to PRZR water level sensor, the genetic method selects S/G steam flowrate, S/G steam pressure, RCS average temperature, and PRZR temperature as optimal inputs, and the PCG method selects S/G steam pressure, hot-leg temperature, and cold-leg temperature. In all three methods, the standard deviation of errors for the verification data is similar to that for the training data. The PCG method and the genetic method have similar fitness values but PCA method has a lowest one. Figure 7 shows the estimation of the PRZR water level signal using the PCG input selection method and Figure 8 shows the failure detection time for cases applying three methods. The detection time for the case applying the PCA method is slower than other two methods.

In the application to PRZR pressure sensor, the genetic method selects S/G narrow-range water level, hot-leg temperature, and PRZR temperature as optimal inputs, and the PCG method selects only PRZR temperature. The PRZR pressure has a close relationship with the PRZR temperature because PRZR is in a saturated state, which is also shown in Table 1. In all three methods, the standard deviation of errors for the verification data is similar to that for the training data. The PCG method and the genetic method have similar fitness values but PCA method has a lowest one. Figure 9 shows the estimation of the PRZR pressure signal using the PCG input selection method and Figure 10 shows the failure detection time for cases applying three methods. The detection time for the case applying the PCA method is slower than other two methods.

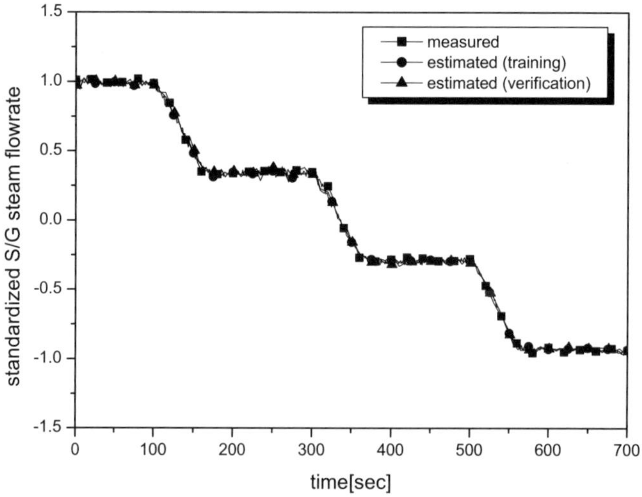

Fig. 5. Estimation of a (standardized) steam generator steam flowrate signal using the PCG method

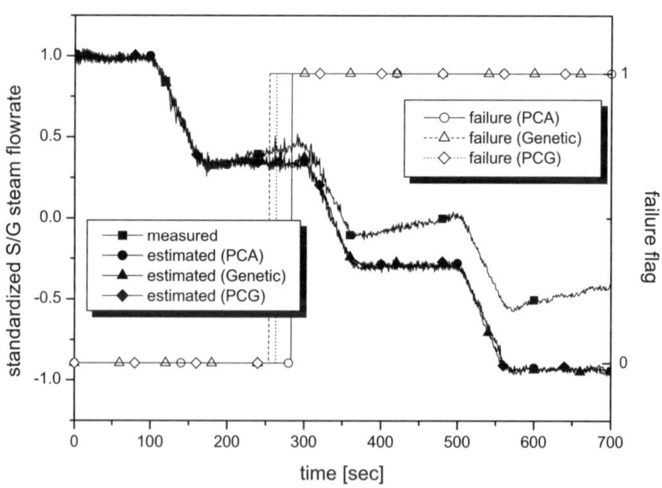

Fig. 6. Failure detection for gradual degradation of steam generator steam flowrate

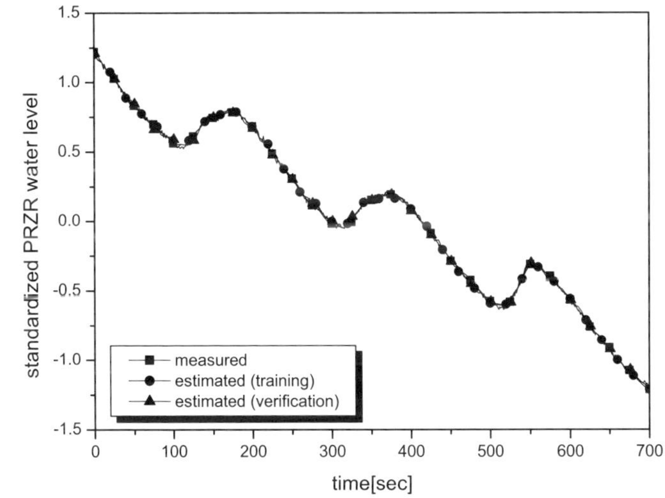

Fig. 7. Estimation of a (standardized) pressurizer water level signal using the PCG method

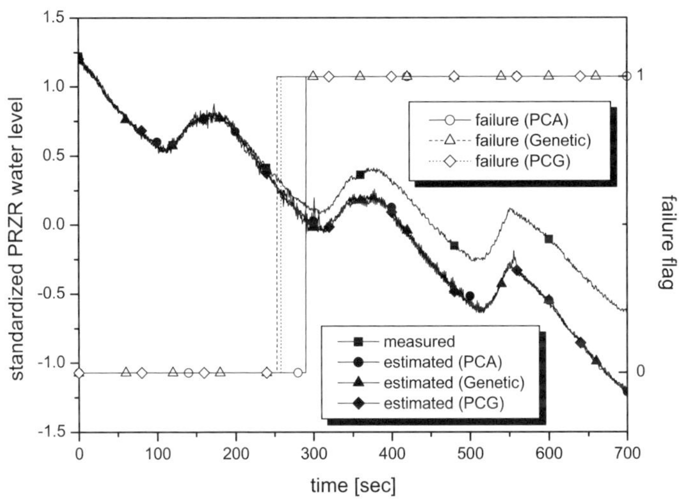

Fig. 8. Failure detection for gradual degradation of pressurizer water level

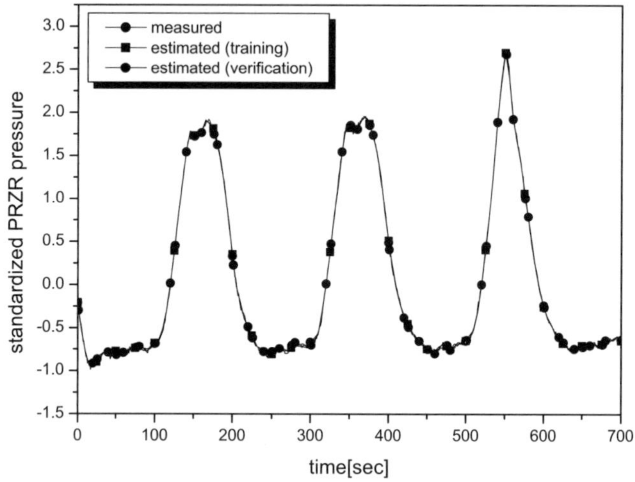

Fig. 9. Estimation of a (standardized) pressurizer pressure signal using the PCG method

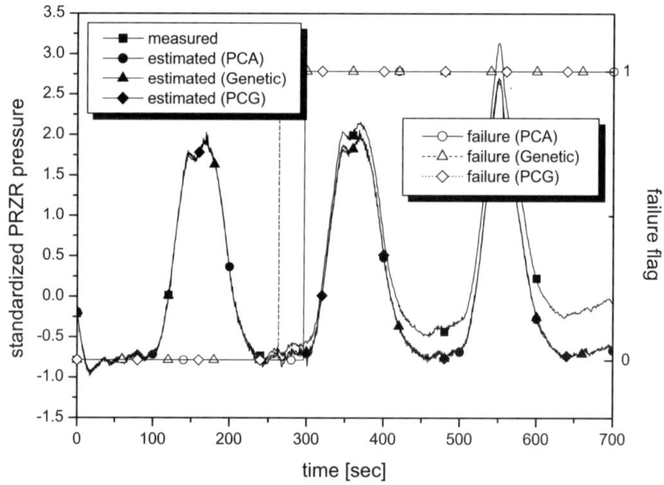

Fig. 10. Failure detection for gradual degradation of pressurizer pressure

5.2 Multiple Faults

In this subsection, two output fault cases (S/G steam flowrate and PRZR water level) were selected to check simultaneous input and output faults. The output signal and one of selected input signals were purposely degraded in a degree of 3.0×10^{-4} of the measured values each time step from 200 sec to verify the robustness of the failure detection algorithm.

Figure 11 shows the results for a case the feedwater flowrate used as a common input of three input selection methods to estimate the output signal is degraded simultaneously with the output signal, S/G steam flowrate. The failure detection for the genetic method is delayed very much due to the degradation of one of input signals and the failure detection for the PCA and PCG methods is delayed a little. As expected, the PCA method that uses all the gathered signals as input signals is affected less than other two methods.

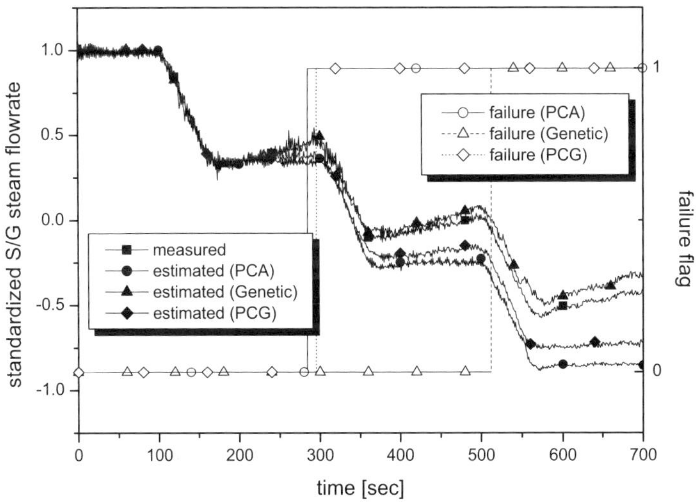

Fig. 11. Failure detection for simultaneous gradual degradation of an output (S/G steam flowrate) and one of input signals (feedwater flowrate)

Figure 12 shows the results for a case the S/G steam pressure that is used as a common input of three input selection methods to estimate the output signal is degraded simultaneously with the output signal, PRZR water level. The failure detection for the PCA, PCG, and genetic methods is delayed a little.

From the above two results and as expected also, it is known that in the case that an input which has relatively small relationship with an output is faulted, the input fault affects the failure detection of the output fault less than other inputs which have relatively large relationship with the output. And if an input, which has relatively small relationship with an output is faulted, it is known that the output fault can be detected although its detection time is delayed a little.

In the summary, from Tables 2 and 3, it is shown that although PCA method uses the largest number of input signals, PCA method has the worst result of the three methods. But the PCA method is the fastest. It is determined that the genetic and PCG methods show similar performance as the input selection methods of fuzzy neural networks with application to sensor signal monitoring. However, the training time for the genetic method is about three times slower than that for the PCG method even though calculation time depends on the test cases and the relationship between input signals and output signal. From above simulations, it is shown that a fuzzy neural network with the PCG input feature selection method actually estimates the relevant sensor signal using other sensor signals and SPRT failure detection algorithm detects the gradual degradation of sensors.

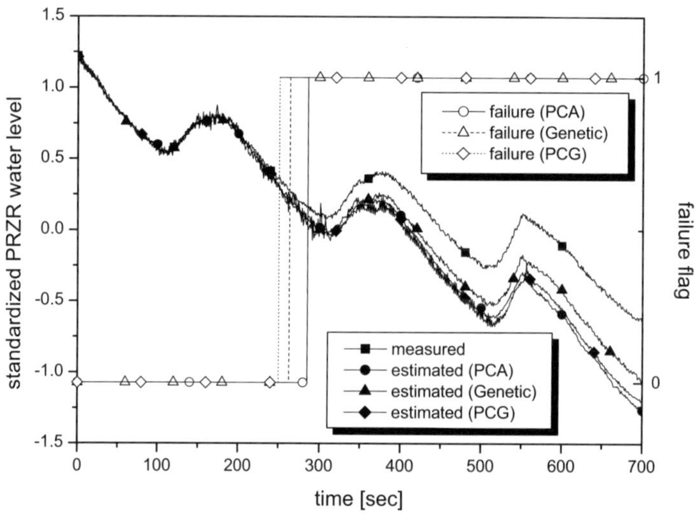

Fig. 12. Failure detection for simultaneous gradual degradation of an output (pressurizer water level) and one of input signals (S/G steam pressure)

Table 3. Failure detection time for simultaneous faults of an output and one of input signals to the fuzzy neural network

Output Sensors	S/G steam flowrate			Pressurizer water level		
Failed input	Feedwater flowrate (FF)			S/G steam pressure (SP)		
Methods	PCA	Genetic	PCG	PCA	Genetic	PCG
Failure detection time	85	313	96	86	64	51

6 Conclusions

In this chapter, a fuzzy neural network with an input features selection algorithm was described for sensor failure detection. The described PCG input selection method combines the PCA, the genetic algorithm and the correlation coefficient matrix. The real reduction of number of input signals is accomplished by the genetic algorithm, which requires the substantial computational burden. Thus, the computational burden is reduced by using the correlation coefficient matrix that provides information on the relationship between input signals and an output signal. The usage of the correlation coefficient matrix helps reduce the computation time by excluding from the first time the input signals coming into a fuzzy neural network that are almost irrelevant with the output signal. Also, by combining the PCA method, which can reduce the number of signals to the fuzzy neural network, the computation time for optimization of a fuzzy neural network can be shortened.

The developed failure detection algorithm was applied to the steam generator water level and steam pressure and the pressurizer (PRZR) water level and pressure sensors. The neuro-fuzzy inference system actually estimates the relevant output signal using other input signals. The SPRT decides fast whether a sensor is degraded or not by using the residuals between the measured signal and the estimated signal. The fuzzy neural network with the PCG input selection algorithm is superior to the fuzzy neural network with the other two input selection algorithms (PCA method, genetic method) and the final selected variables are consistent with our knowledge. Also, it is determined that the developed failure detection algorithm can decide the failure of the output even though one of selected input signals is degraded simultaneously with the output.

References

[1] B. R. Upadhyaya, et al., Development and Testing of an Integrated Signal Validation System for Nuclear Power Plants, DOE/NE/37959-35, Sept. 1989.

[2] A. Erbay and B. R. Upadhyaya, "A personal computer-based on-line signal validation system for nuclear power plants," *Nucl. Technology*, vol. 119, no. 1, pp. 63-75, July 1997.

[3] R. M. Singer, K. C. Gross, J. P. Herzog, R. W. King, and S. W. Wegerich, "Model-based nuclear plant monitoring and fault detection: theoretical foundations," Proc. 9th Intl. Conf. on Intelligence Systems Applications to Power Systems, Seoul, Korea, 1997.

[4] J. W. Hines, D. J. Wrest, and R. E. Uhrig, "Signal validation using an adaptive neural fuzzy inference system," *Nucl. Technology*, vol. 119, no. 2, pp. 181-193, 1997.

[5] J. W. Heins, Andrei V. Gribok, Ibrahim Attieh, and Robert E. Uhrig, "Regularization method for inferential sensing in nuclear power plants," in Da Ruan (ed) *Fuzzy Systems and Soft Computing in Nuclear Engineering*, Springer-Verlag, Berlin Heidelberg New York, pp.285-314, 1999.

[6] P. Fantoni, S. Figedy, and A. Racz, "A neuro-fuzzy model applied full range signal validation of PWR nuclear power plant data," FLINS-98, Antwerpen, Belgium, 1998.

[7] Tomas Eklov, Per Martensson, and Ingemar Lundstorm, "Selection of variables for interpreting multivariate gas sensor data," *Analytica Chimica Acta*, vol. 381, pp. 221-232, 1999.

[8] B. K. Lavine, A. Moores, and L. K. Helfend, "A genetic algorithm for pattern recognition analysis of pyrolysis gas chromatographic data," *J. Anal. Appl. Pyrolysis*, vol. 50, pp. 47-62, 1999.

[9] K. W. Bauer Jr., S. G. Alsing, and K. A. Greene, "Feature screening using signal-to-noise ratios," *Neurocomputing*, vol. 31, pp. 29-44, 2000.

[10] P. van de Laar and T. Heskes, "Input selection based on an ensemble," *Neurocomputing*, vol. 34, pp. 227-238, 2000.

[11] N. R. Pal, "Soft computing for feature analysis," *Fuzzy Sets and Systems*, vol. 103, pp. 201-221, 1999.

[12] A. Wald, *Sequential Analysis*, John Wiley & Sons, New York, 1947.

[13] X. Z. Wang and R. F. Li, "Combining conceptual clustering and principal component analysis for state space based process montoring," *Ind. Eng. Chem. Res.*, vol. 38, no. 11, pp. 4345-4358, 1999.

[14] J. Chen and J. Liu, "Mixture principal component analysis models for process monitoring," *Ind. Eng. Chem. Res.*, vol. 38, no. 4, pp. 1478-1488, 1999.

[15] D. E. Goldberg, *Genetic Algorithms in Search, Optimization, and Machine Learning*, Addison Wesley, Reading, Massachusetts, 1989.

[16] M. Mitchell, *An Introduction to Genetic Algorithms*, The MIT Press, Cambridge, Massachusetts, 1996.

[17] J.-S. Roger Jang, "ANFIS: adaptive-network-based fuzzy inference systems," *IEEE Transactions on Systems, Man, and Cybernetics*, vol. 23, no. 3, pp 665-685, May 1993.

[18] T. Takagi and M. Sugeno, "Fuzzy identification of systems and its applications to modeling and control," *IEEE Trans. System, Man, Cybern.*, vol. 1, pp. 116-132, 1985.

[19] M. G. Na, Neuro-fuzzy control applications in pressurized water reactors: in Da Ruan (ed) *Fuzzy Systems and Soft Computing in Nuclear Engineering*, Springer-Verlag, Berlin Heidelberg New York, pp.172-207, 1999.

[20] W. J. Lee, B.-D. Chung, J.-J. Jeong, K.-S. Ha, and M.-K. Hwang, Improved Features of MARS 1.4 and Verification, Korea Atomic Energy Research Institute, KAERI/TR-1386-99, 1999.

16 Artificial Neural Networks Modeling as a Diagnostic and Decision Making Tool

Zvi Boger[1, 2]

[1]Licensing and Safety Division
Israeli Atomic Energy Commission, P.O. Box 7061, Tel-Aviv 61070, Israel

[2]OPTIMAL – Industrial Neural Systems Ltd.
P.O. Box 9201, Be'er Sheva 84191, Israel

zboger@bgumail.bgu.ac.il

Fault diagnosis of a complex system, such as nuclear power plant (NPP), is an important part of the operator task. Artificial neural networks (ANN) model learns from the plant past behavior and is easy to develop if enough data is available for training, be it from actual plant history or from full-scale simulator fault situations. The operator is reluctant to depend on a "black-box" advice, and any reasonable explanation is welcome. Two examples of using ANN modeling for fault diagnosis are presented. The first is based on NPP full-scale simulator time history data of three instrument or component faults. The second example is the diagnosis of a seemingly erratic excessive steam turbine bearing vibration. The ANN models were able to correctly classify the causes of the faults. In the first case a knowledge extraction technique from trained ANN models, Causal Index, was used. It was able to provide plausible explanations for the ANN faults classification. In the second case, Genetic Algorithm was used as an optimization technique, with the ANN model providing the target function. It was able to identify the cause of the excessive vibration in a power plant turbine bearing.

1 Introduction

The suggested use of artificial neural networks (ANN) modeling techniques in nuclear power plants (NPP), in which the model is learned from data of the plant behavior, arouses strong emotions. "No complicated equations! No man-years of development effort!" cheer the proponents. "No detailed equations? No reliability!" counter the opponents. And the Licensing overseers add "Don't even think of using it in any safety related action." The International Atomic Energy Agency has reviewed periodically the feasibility of "intelligent" software in NPP safety systems in Expert Group meetings since 1988, and in most meetings papers were presented on possible ANN-based systems. Two reviews on the use of ANN models in NPP diagnostics were published (Reifman 1997, Boger 1998). Since then more than a hundred references are to be found when searching the International Nuclear Information System references database. However, reviewing the recent publications in this area shows that the number of actual implementations in NPP is quite small. Most of the published papers are still from the academic community, universities and national research institutes, with only occasional co-operation with electrical utility personnel, the ultimate users. But the use of ANN modeling in industrial plants is spreading, and the nuclear power plant industry has the unique advantage of having access to full-scale plant simulators used for on-going training of NPP operators. As no other modeling method fully meets the requirements of fault diagnosis, the ANN modeling is gaining support.

One of the often-cited opposition to the use of ANN modeling in NPP diagnostics is the lack of "explanation" facility, the ability of the operator to understand the basis of the ANN recommendations. In this paper two power plant diagnostic examples are presented to show that the "black-box" image of ANN model is misleading, and the trained ANN model can be analyzed to correctly explain the causes of the observed fault symptoms.

The structure of the paper is as follows: Reviews of the principal component analysis conjugate gradient (PCA-CG) algorithm for large-scale ANN modeling, and the Causal Index (CI) method of analyzing trained ANN. A novel use of genetic algorithm (GA) - ANN combination for fault analysis is proposed. Two examples of power plant diagnostics using these techniques are given: An example CI classification of NPP simulator-derived response to three fault situations. An example of actual power plant data, where the cause of an unexplained turbine fault can be found using GA manipulation analysis of a trained ANN model of this fault behavior.

2 Large-Scale ANN Modeling and Analyzing Issues

2.1 Efficient ANN Modeling and Identification of the Relevant Inputs

One of the difficulties of training large-scale ANN models can be overcome by using the PCA-CG algorithm that starts the training from non-random initial weights. Several papers (Boger 1997, Boger and Guterman 1997) describe the experience in modeling large-scale systems and extracting some knowledge from the trained ANN models using relatively simple techniques. These are:

1. Training large ANN, beginning from non-random initial connection weights;
2. Identification of the relevant inputs of a system;
3. Optimizing the number of the neurons in the hidden layer;

4. Calculation of causal indices (CI), for learning some rules of the relations between each input and output.
5. Using the outputs of the hidden layer neurons for fault detection and classification.

The first three techniques were developed in 1990 (Boger 1992), but the algorithms details were published later (Guterman 1994, Boger and Guterman 1997). With these algorithms ANN models of industrial and other large systems with hundreds of inputs and outputs have been developed on PC computers in a fraction of the time and the number of training epochs usually reported. The last technique was used to correctly classify signals even when corrupted by noise (Boger et al. 1997).

2.2 The Causal Index Method of Analyzing ANN Models

The causal index method is an easy, somewhat qualitative, method for rule extraction (Baba et al. 1990). The CI is calculated as the sum of the product of all "pathways" between each input to each output,

$$CI = \sum_{j=1}^{h} W_{ij} * W_{jk} \tag{1}$$

where there are h hidden neurons, W_{ij} are the connection weights between input i to hidden neuron j, W_{jk} are the connection weights from hidden neuron j to output k.

The CI numerical results are relative; that is, they indicate the magnitude and sign of a change in an output when a particular input is changed. Their advantage is that they do not depend on a particular input vector, but on the connection weight set that represents all the training input vectors. This is also one of their limitation, as a local situation may be overlooked in the global representation.

The PCA-CG algorithm recommends ANN architecture with a small number of neurons in the (single) hidden layer. It leads to efficient clustering, based on the hidden layer's neuron outputs. This was demonstrated in classifying research reactor neutron noise spectra (Boger 1998). It may also be used for training auto-associative ANN that will monitor deviations form normal plant behavior based of the difference between the actual inputs and their predicted values, as already done in NPP with the major inputs (Turkcan et al. 1998). A single deviation may indicate instrument fault; multiple deviations may indicate the presence of an unusual plant state and will flag it for the operator attention.

2.3 The Use of Genetic Algorithm in ANN Fault Analysis

Genetic Algorithms are part of a growing field of Evolutionary Algorithms. These algorithms try to optimize a system by adapting its parameters, generation after generation, to the environment it exists in, similar to the Darwinian evolution theory of the "Survival of the Fittest." Its use in training ANN models has been described in many papers. A list of such papers can be found on the Internet (Friedrich 2000).

Two main directions of its use are suggested. The first is finding the optimal set of connection weights during ANN model training. The second is finding the optimal ANN architecture, the number of hidden neurons and the relevant set of inputs, with the actual training done by conventional methods. The second approach was recently compared to the PCA-CG algorithm in modeling large TV program rating database (Boger and Weber 2000). The PCA-CG was found to be slightly less accurate, but less time consuming (by orders of magnitude) and more robust.

Another use of the GA-ANN combination is possible, optimizing the external inputs to a system already modeled by ANN. This is done by using one of the ANN model outputs as the "target function" used by the GA to determine the probability of this input variable set to survive into the next generation.

In the case of an ANN model of an NPP fault, the magnitude of the faulty reading can be set as the target function to maximize. When converged, the input values that leads to this maximum is divided by the mean of their "normal" values, and the significant ratio outliers then examined for possible causes of the observed faults.

3 NPP Fault Diagnosis Explanation by Casual Indices of the Trained ANN

A CI application example is the identification of an NPP component fault by analysis of the resulting transient data. Dynamic response data of 20 instruments to three simulated faults, 10, 25, 40 seconds after the fault occurrence was derived from full-scale NPP simulator (Reifman and Vitela 1997). They trained a 20-20-3 ANN, by conjugate-gradient error minimization. The same data was made available to me, and a 20-4-3 ANN was trained from the 40 seconds data. The verification was made using the earlier data, 10, 25 seconds after the fault. In 71 cases out of the 72 verification cases the faults were classified correctly. The one wrong classification was declared to be doubtful, as the hidden layer neurons' output pattern was unique, different from the other example patterns. The ANN was analyzed to find the more relevant inputs. Two progressive retraining with the reduced input list gave the minimal list of nine instruments necessary for the correct classification. The causal indices were calculated for the full and reduced ANNs (Table 1).

Table 1. Causal Indices relating the instruments readings to the faults

ANN	Full			Reduced		
Component Fault	**Feedwater pump trip**	**Feed flow tx fail hi**	**Turbine CV fail close**	**Feedwater Pump trip**	**Feed flow tx fail hi**	**Turbine CV fail close**
Plant Parameter						
Quench tank pressure	1	-1	2			
Hot &cold leg A dtemp	0	1	0			
Hot &cold leg B dtemp	2	3	-5	15	-5	-10
Pressurizer pressure	-7	8	-1			
Pressurizer water level	1	-5	5			
Pressurizer level	0	6	-7			
Reactor power %	3	5	-8			
Steam gen. A water level	-4	-4	9	5	-12	7
Steam gen. B water level	0	-10	11	4	-12	8
Steam gen. A exit pres.	-1	-5	5	3	-11	8
Steam gen. B exit pres.	-1	-5	5			
HP turbine throttle. pres.	0	-6	5	2	-10	8
HP turbine exhaust pres.	3	-1	-1			
Feedwater loopA CVdP	-22	25	-4	-25	29	-4
Feedwater loop B CVdP	-30	32	-2	-32	36	-3
Feedwater loop A inletP	-18	13	4	-17	7	10
Feedwater loop B inletP	-18	11	6	-17	7	10
Letdown water flow	3	-4	1			
Makeup water flow	-7	11	-4			

It may be seen that the CI values in the reduced ANN are much more distinct than those in the full ANN. This effect may be used in verification of an initial classification by the full ANN, and can be used in an explanation facility. It may be seen that the thermo-hydraulic causal relationships are correct. For example, when the feed water flow transmitter fails high, the control system decrease the water flowrate to "correct" the flow, which would then decrease the water level in the steam generators. Conversely, a turbine CV closure will increase the water level in the steam generators, as no steam leaves them.

4 Turbine Bearing Fault Diagnosis Explanation by Genetic Algorithm Manipulation of a Trained ANN

4.1 Description of the Problem

An erratic steam turbine bearing fault occurred in a Hungarian power plant. Here are some quotations from a paper describing the problem and its solution by ANN modeling (Por and Vasko 1997). "Several units of conventional power plants have a combined turbine-generator purchased from different companies. To connect the shaft of the two important components ... an additional bearing (t5) was planned and added (see Fig.1). This works without major problems in most of the normal conditions. But it was observed several years ago, that in some cases that bearing (near to coupling) went to excess vibration approaching allowed limits. Typically such excess vibration appeared several times during unloading the given unit to about half of the power. Several attempts were made to solve this problem. Vibration measurements, calculations and finite element analysis showed as a possible source of the excess vibration the rotational frequency and the shaft eigenfrequency resonance. Therefore additional masses were added to the base of the bearing holder to avoid that resonance. Some of the observed resonance peaks were eliminated but exactly that one which appeared during unloading remained. One of the curiosities of this vibration was that it showed up not always during unloading."

4.2 ANN Modeling

To diagnose the fault an ANN with two hidden layers, each with a large number of hidden neurons, was trained by Por and Vasco. They selected set of 30 plant measurements as inputs, and the vibration amplitude of the bearing as the output. The model predicted accurately the excessive vibration when the turbine power was decreased to 50%. No obvious cause was found, until each input was gradually changed over its range, keeping all other inputs constant. Then it was observed that the excessive bearing vibration occurred at a particular water level in the condenser tank.

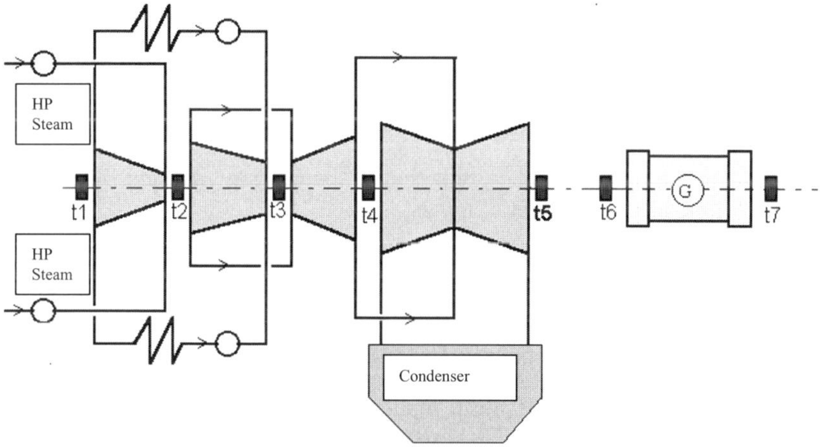

Fig.1. Schematic representation of turbine-generator ensemble. Main components and bearings are marked. The bearing marked t5 had excess vibration

4.3 A Physical Explanation to the ANN Solution

"The condensator is attached to the whole concrete construction of the turbine supporting construction from below. That also means that changing the water level one changes also the overall mass of the construction. Therefore resonance vanishes, but we call the attention to the fact that before we got this results using artificial intelligence modeling technique for prediction how different parameters can affect on vibration of the given bearing, nobody suspected that the condensator water level can be a cause of the excess vibration" (Por and Vasko 1997).

4.4 Alternative ANN Modeling and Analysis

A set of 2552 time history vectors of the whole 135 instrument readings was provided to me for the training. A134–5–1 ANN was trained from this database, with the turbine t5 bearing vibration as an output. The ANN modeling accuracy was very good, but the input reduction technique, or the CI method, did not identify the condenser water level as the causing input.

However, simple GA procedure did succeed:

1. Pre processing the original data by subtracting the column mean and dividing the result by the column standard deviation;
2. Definition of a set of number (in this case, 100) randomly selected input vectors;

3. Multiplying each input vector by a vector coefficients with random normal distribution, with a zero mean and (one, in this case) standard deviation;
4. Clipping the resulting vectors to the [-3, 3] limit (to avoid getting out of the original input value range). Penalizing the evaluation function in proportion to the number of the clipped values;
5. Evaluating the ANN model prediction of the bearing vibration by presenting each of the all resulting set of inputs;
6. Selecting candidate input set for the next generation, with a probability proportional to the ratio of the predicted vibration to the mean vibration value in data set;
7. Repeating for predetermined number of generation (in this case, 500);
8. Calculate the ratio of the mean inputs values of the "best" input set to the mean value of the original database inputs;
9. The parameters with the outstanding ratios are the probable causes of the faults.

The original 100 "seeds" of the GA were chosen from a reduced 279 input vector set closest to the 50% power, less than 130 MW (out of the range of 120 – 200 MW). The GA optimization converged easily, and the identified outliers' ratios are given in Table 2.

Table 2. The GA-found outlier maximal-vibration to mean input value ratios

Outliers input values names		**units and ranges**			**Ratio**
4.BL.	Feed Water Reducing Valve DP	bar	0.00	50.00	99.2
4NE22T002	UH2 Front Steam Temperature Left Side	^{o}C	50.00	600.00	2.58
4NA71T001	TH3 Back Steam Temperature Right Side	^{o}C	50.00	600.00	2.09
4RA02T007	Fresh Steam Temp. Turbine E Right Side	^{o}C	50.00	600.00	4.10
4RB03P001	Steam Pressure RII Reducer Back Right Side	bar	-1.00	3.00	2.47
4SJ51P002	Turbine Cover Oil Pressure	bar	0.00	10.00	2.12
4SB03V001	Turbine 3 Bearing Horizontal Vibration	mm/s	0.00	15.00	3.90
4SD01L001	Condenser Water Level	mm	2000	3500	8.86

It may be seen that the other major outliers values are related to the reduced power operating conditions, and the last one, the condensate water level, is the cause of the excessive t5 bearing vibration.

5 Discussion and Conclusions

The examples given in this paper show that the ability to easily train large-scale ANN models can be used for fault classification in power plant, provided that enough training data are available. The ANN model may be analyzed by the CI technique, or by the GA-based optimization, to provide a plausible explanation of its recommendation. Thus a hybrid fault diagnosis module may be developed – the ANN model pointing out suspected fault causes, and an expert rule system that will check the validity of these causes by checking the existence of pre-defined conditions for each suspected cause.

The first example given in this paper is of a relatively small scale, 25 instrument reading with only three faults. However, it was able to diagnose the faults 10 seconds after their occurrence. The second example shows the feasibility the GA-based optimization, using the ANN model to provide the "utility" function, for fault detection.

The key for a more realistic demonstration is the availability of NPP operating data, and more important, the availability of the accumulated "fault" dynamic data from the full-scale NPP simulators. Using such data, the ANN modeling techniques described in this paper may provide the sought-after reliable NPP fault detection capabilities.

Acknowledgements

Thanks are due to Dr. Jaques Reifman, Reactor Analysis Division, Argonne National Laboratory, for providing the NPP simulator data, and to Dr. Gabor Por, Institute of Nuclear Techniques, Technical University of Budapest, Hungary, for providing the turbine bearing vibration data, the text and the figure in his published paper.

References

Baba K, Enbutu I, Yoda M (1990) Explicit representation of knowledge acquired from plant historical data using neural network, Proceedings of the International Joint Conference on Neural Networks. San Diego, 3: 155-160.

Boger Z (1992) Application of neural networks to water and wastewater treatment plant operation. Instrument Society of America Transactions 31(1):25-33.

Boger Z (1997) Experience in industrial plant model development using large-scale artificial neural networks. Information Sciences - Applications 101(3:4):203-212.

Boger Z, Guterman, H (1997) Knowledge extraction from artificial neural networks models. *Proceedings of the IEEE International Conference on Systems Man and Cybernetics, SMC'97*, Orlando, Florida, pp. 3030-3035.

Boger Z, Ratton L, Kunt TA, Mc Avoy TJ, Cavicchi RE, Semancik S (1997) Robust classification of "artificial Nose" sensor data by artificial neural networks. Proceedings of the IFAC ADCHEM '97 Conference, Banff, Canada.

Boger Z (1998) Implementing artificial neural networks in nuclear power plants diagnostic systems: Issues and challenges. Proceedings of the IAEA Technical Committee Meeting on Diagnostic Systems in Nuclear Power Plants, Istanbul, Turkey, IAEA-IWG-NPPCI--98/2, pp 149-160.

Boger Z, Weber R (2000) Finding an optimal artificial neural network topology in real-life modeling - two approaches. Proceedings of the International Conference on Neural Computation 2000, NC2000', Berlin, Germany.

Guterman H (1994) Application of principal component analysis to the design of neural networks. Neural, Parallel and Scientific Computing 2:43-54.

Friedrich CM (2000) http://www.tussy.uni-wh.de/~chris/gann/chrisLit/chrisLit.html

Por G, Vasko J (1997) Application of an artificial intelligence software to model the excess vibration of a bearing in a conventional power plant. Proceedings of the ESPRIT ICIMS-NOE annual conference, ASI'97. Budapest, Hungary.

Reifman J, Vitela JE (1994) Accelerating learning of neural networks with conjugate gradients for nuclear power applications. Nuclear Technology 106:225-241.

Reifman J (1997) Survey of artificial intelligence methods for detection and identification of component faults in nuclear power plants. Nuclear Technology 119:76-97.

Turkcan E, Ciftcioglu O, van der Hagen THJJ (1998) Surveillance and fault diagnosis for power plants in the Netherlands: Operational Experience. Proceedings of the IAEA Technical Committee Meeting on Diagnostic Systems in Nuclear Power Plants, Istanbul, Turkey, IAEA-IWG-NPPCI--98/2, pp 53-70.

17 A New Approach for Transient Identification with "Don't Know" Response Using Neural Networks

Antônio C. de A. Mol [1], Aquilino S. Martinez [2] and Roberto Schirru [2]

[1] Divisão de Confiabilidade Humana - Instituto de Engenharia Nuclear – CNEN - Caixa Postal 68550, Ilha do Fundão s/n, 21945-970, Rio de Janeiro, Brasil

mol@ien.gov.br

[2] COPPE / UFRJ – Caixa
Postal 68509, lha do Fundão s/n, 21945-970, Rio de Janeiro, Brasil

{aquilino, schirru}@lmp.ufrj.br

In the last years, many different approaches based on neural network (NN) have been proposed for transient identification in nuclear power plants (NPPS). Some of them focus the dynamic identification using recurrent neural networks, however, they are not able to deal with unrecognized transients. Other kind of solution uses competitive learning in order to allow the "don't know" response. In this case, dynamic features are not well represented.

This work presents a new approach for neural network based transient identification which allows either dynamic identification or "don't know" response. Such approach uses two multilayer neural networks trained with backpropagation algorithm. The first one is responsible for the dynamic identification. This NN uses, as input, a short set (in a movable time window) of recent measurements of each variable avoiding the necessity of using starting events. The other one is used to validate the instantaneous identification (from the first net) through the validation of each variable. This net is responsible for allowing the system to provide "don't know" response.

In order to validate the method, an NPP transient identification problem comprising 15 postulated accidents, simulated for a pressurized water reactor, was proposed in the validation process it has been considered noising data in order to evaluate the method robustness. Obtained results reveal the ability of the method in dealing with both dynamic identification of transients and correct "don't know" response.

1 Introduction

Nuclear power plants contain complex systems, which are generally monitored by human operators. During an abnormal event, the operator comes across a great amount of information from the instruments, which represents a specific type of event. The simple amount of data alternating itself can confuse the operator, negatively influencing his diagnosis and therefore on his work.

The identification of these abnormal events is considered extremely important as support to the operation at a nuclear power plant [1]. As the power plant conditions change according to these events, if the transient is recognized, information as alarms can be selected and provided to the operator.

A method used for the transient identification compares data sampled from the power plant with knowledge previously prepared, as a tree of cause-effect, for example [2]. In this method, however, a comprehensive knowledge of the plant state is indispensable for the transient identification, as there is a wide range of anomalies with different progression speeds.

There are works [3-8] that explore multilayer Artificial Neural Networks (ANN) with backpropagation training, as this kind of ANNs has an excellent capacity of approximation and generalization. However, diagnoses based only on this kind of ANNs can lead to wrong classifications when encountering transients outside the training scope.

There are methods that use ANN Classifiers [9,10] on the identification of these abnormal events, taking advantage of the fact that these ANNs are able to generate a "don't know" answer to transients outside the training scope. Nevertheless, one of the great problems of these ANNs is that they are not capable to directly work with dynamic systems, where there is a direct dependence upon time, that is, they can only map a static input vector for an output value.

Other methods use recurrent ANNs [11], which are a different class of ANNs able to work with dynamic systems due to their internal architectures that use feedbacks on the connections among their neurons. However, these ANNs models are inherently difficult to train and, moreover, their responses become unpredictable when encountering inputs outside the training scope. This last limitation of the recurrent networks also makes this type of ANN not to directly meet the important requirement of any diagnosis system, that is being able to obtain the "don't know" output.

In this work, it is presented the proposal of a system for the transient classification, which explores the excellent performance of multilayer ANNs with 'backpropagation" training algorithm. This system also has a diagnosis validation procedure so as to obtain the "don't know" output for transients outside the training scope, thus preventing wrong or not reliable classifications. Another important characteristic of the system is that it is independent of the time variable, making it robust with reference to this limitation.

In Section 2, it is presented a brief review of neural networks, in particular the "Jump" network. In Section 3, it is shown the utilization of this ANN on noisy dynamic systems. Section 4 contains the description of the procedures for the transient identification and validation. In Section 5, it is presented an application of this method to an example case of transients of a PWR-type nuclear power plant. Finally, in Section 6, it is carried out the analysis of the results and the conclusion of the work.

2 Artificial Neural Networks

Artificial neural networks are computational techniques which present a mathematical model based upon the brain structure and which acquire knowledge through experience.

Figure 1 shows the architecture of a "jump"-type multilayer ANN with an input layer, two intermediate layers and an output layer. For this kind of ANN, a neuron of any layer of the network is connected to all other neurons of the further layers (there is no feedback). The signal flow through this network propagates "forward," from left to right, passing by all the layers.

Figure 2 presents a part of the "jump" network. Two types of signal are identified in this network: (1) the "stimulus signal", which is the input signal from the first layer of the network, propagating "forward" (neuron by neuron) through the network, emerging at the output of the network as output signal; (2) "error signal," which is the error originated at a network output neuron and propagated back (neuron by neuron) through the network.

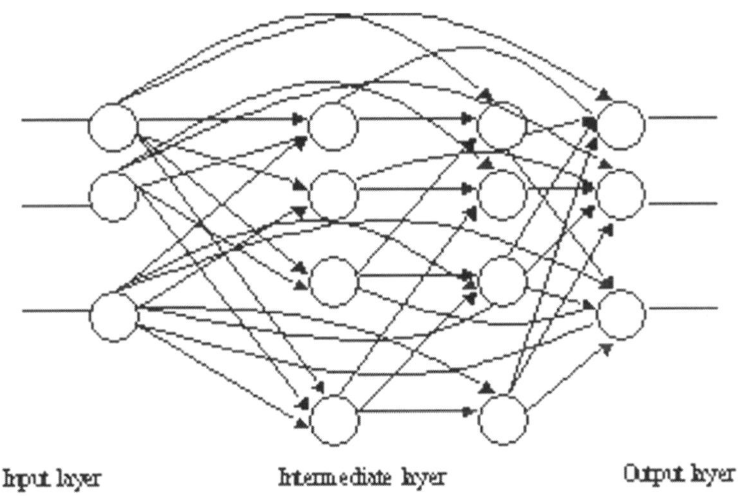

Fig. 1. Jump-ANN

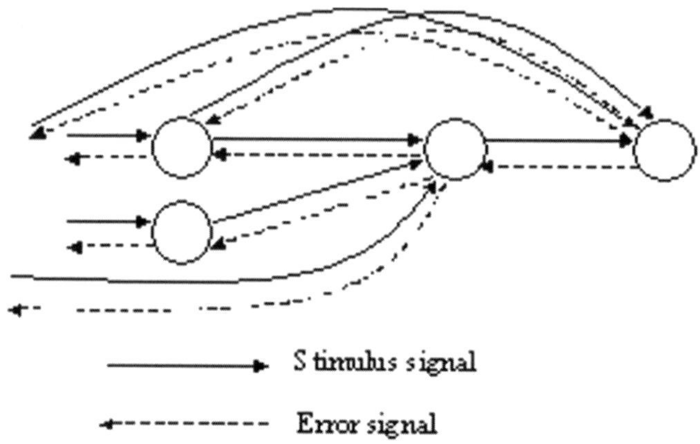

Fig. 2. Types of signal in "Jump" Neural Network

The most important property of the neural networks is the ability to learn from their environment and thus improve their performance. This is achieved by an interactive setting process applied to their synapses called *training*. It is possible to say that knowledge occurs when the neural network reaches a generalized solution for a class of problems. A knowledge method used successfully in many types of ANNs is the "backpropagation," which has as general rule of update of synapses [12] ($\Delta w_{ji}(n)$) using the following equations:

$$\Delta w_{ji}(n) = \eta * \delta_j(n) * y_i(n) \tag{1}$$

$$\delta_j(n) = \frac{dF(a_j(n))}{da_j(n)} \cdot \sum \delta_k(n) \cdot w_{kj}(n) \tag{2}$$

where: k represents any neuron of the further layers relative to the layer of the neuron j, $\delta(n)$ is the local gradient, y_i is the output of the neuron i, $F(.)$ is the activation function, $a_j(n)$ is the activation of the neuron j and w_{ki} the synapse between the neurons k and i.

The great difference between the "jump" network used in this work and a feedforward network, concerning the training process by the "backpropagation" algorithm, is due to the local gradient term $\delta_j(n)$, for the "jump" network, undergoes direct influence of all the neurons of the layers subsequent to the layer of neuron j. This direct influence makes the synapse changes of the first layers of the network more sensitive to the errors presented at the output of the network, thus making it possible for the "jump" network to work better when separating patterns in very complex spaces, which is the case of the space of the state variables during some abnormal events.

3 Artificial Neural Networks in Noisy Dynamic Systems

The ANN presented in the previous section has a good capacity of generalization. However, since this type of ANN does not have recurrence in its internal structure (where there is a direct dependence upon time), it is not able to directly work with dynamic systems. In this way, aiming at overcoming this limitation, it was adopted the system of "movable temporal window," that is, the time dependence was incorporated to the external architecture of the network, by means of the sequential presentation in time of the recent history of the state variables used in the transient identification, as is represented in Figure 3. In order to define the number of elements of this "window," it is necessary to consider that this window must be large enough to identify the dynamic behavior of the system, however without affecting the identification of fast transients.

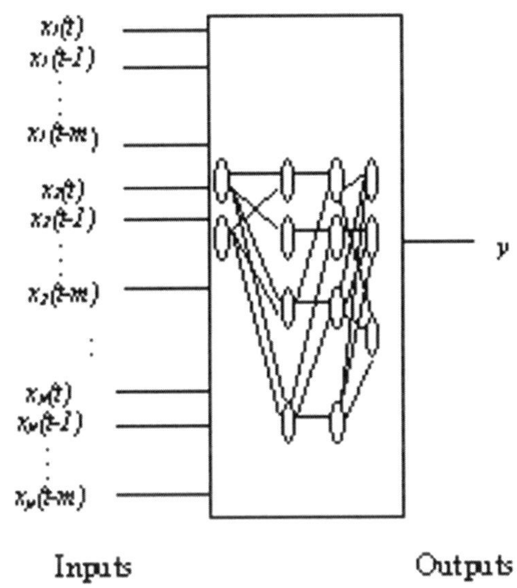

Fig. 3. ANN with movable temporal windows

So as to increase the robustness to noise of this network, it is necessary that a great amount of patterns be presented to the network on the training stage. This is done adding examples of patterns with overlapped noises and forcing, through training, the network to recognize these new examples as belonging to the respective original classes (without noise). So, it was added to each state variable a blank noise with normal distribution and three standard deviations equal to 5% of the maximum value of the respective variable. This value of 5% is quite reasonable if compared to the actual values presented in a PWR-type nuclear power plant, which had values lower than 1%.

4 Criteria for the Transient Identification and Validation

The first criterion of identification and validation is based on a type of ANN architecture represented in Figure 3, using as input the recent history of the process variables of the power plant. As this type of architecture allows that the presented value at its output can assume continuous values, and on the other hand, the transient classification system adopted represents every transient by a discrete integer value (that is, transient A is represented by the value 1, transient B is represented by the value 2, and so on successively for all anticipated transients), the values presented at the network output, for each transient, oscillated around the anticipated discrete value. In this way, the radius of the range zone (R_{ev}) of each transient was determined as being the greatest deviation between the expected value and the value presented at the network output for the respective transient.

$$R_{ev} = \text{Maximum} \mid y_{ev} - \hat{y}_{ev} \mid \tag{3}$$

where: *ev* indicates the transient, y_{ev} is the expected value, $\hat{y}_{ev}$ is the value obtained by the ANN.

Once the ANN response becomes unpredictable when encountering inputs representative of transients outside the training scope, making this type of network not able to obtain a "don't know" output for these transients, then it becomes indispensable to define a validation procedure for the response provided by the network. In such case, two criteria were defined for the validation procedure.

The first criterion is defined as the criterion of identification and validation of the network (CIVN). According to this criterion, the identified transient (y) will be the one closer to the value obtained at the output of the ANN, and this transient will be considered valid if, and only if, the absolute deviation between these two values $\mid y_{ev} - \hat{y} \mid$ is equal or lower than the R_{ev} of the transient.

The second validation criterion, denoted as the criterion for the validation of variables (CVV), is complementary to the first one and its purpose is to punctually validate all the state variables used as network inputs. For this purpose, an auto associative neural simulator is used to estimate the value of every state variable. This simulator estimates the value of the variables using as inputs the measured values of the own process variables and the type of transient identified by the CIVN, as shown in Figure 4.

Likewise the ANN of the CIVN, the simulator network was also trained with a great amount of patterns. During this stage, the largest Δx_{max} deviation, between the expected value of the variable and the value presented at the network output, is determined for each state variable. In this way, during the operation, a variable is considered valid if, and only if, the absolute value of the difference between the variable value $X(t)$ and the reference value for this same variable $X_{ref}(t)$, generated by the neural simulator, is equal or lower than Δx_{max} of the respective variable. Finally, according to the CVV, the diagnosis is considered valid if all variables are also valid. Figure 5 presents the block diagram of the system of transient identification and validation.

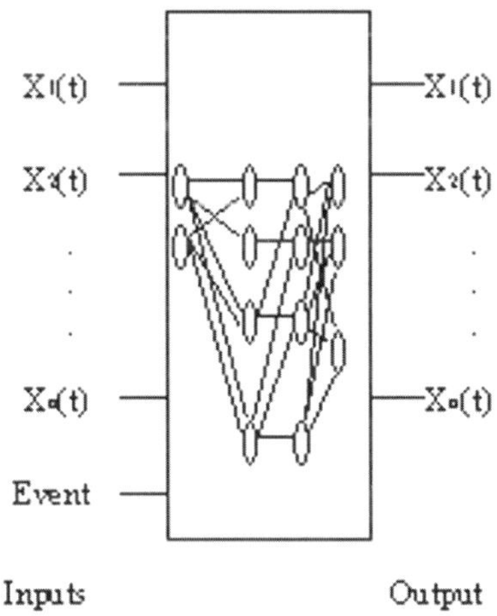

Fig. 4. Auto associative neural simulator

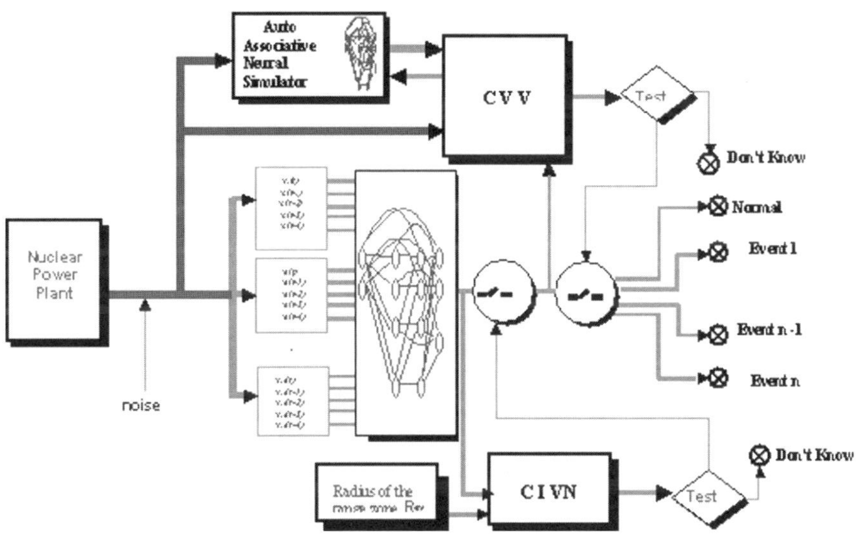

Fig. 5. Architecture of system

After specifying the CIVN and CVV criteria, it is possible to describe the procedure for the transient identification and validation using the following algorithm presented as a flowchart, shown in Figure 6.

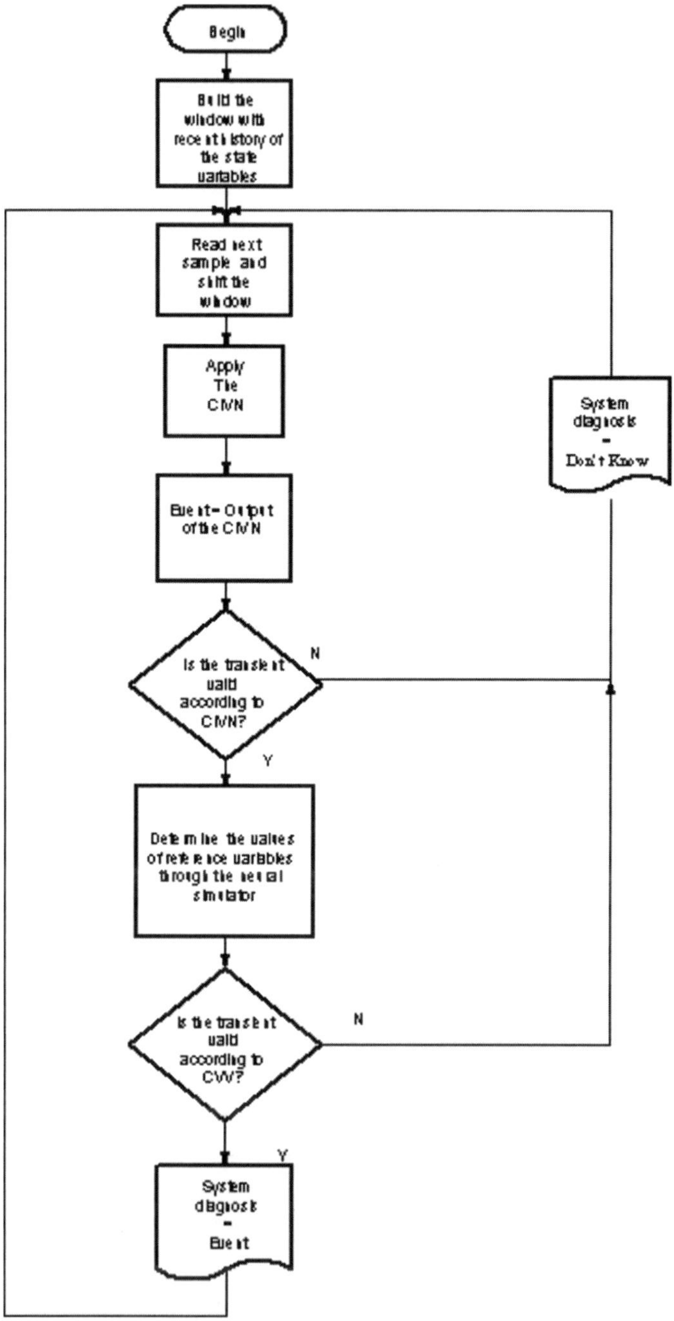

Fig. 6. Flowchart the procedure for the identification and validation of transients

5 Application of the Method to the Transient Identification

In this section, the method previously described is applied to the transient identification, which belongs to the set of accidents postulated for a PWR-type nuclear power plant, generated by computational simulation, operating at 100% power. So, besides the normality condition, fifteen transients were generated: (a) complete loss-of-electrical power (BLACKOUT); (b) loss of coolant of thc primary system (LOCA); (c) steam generator tube rupture (SGTR); (d) main feed water break (MFWBR); (e) shutdown of the reactor and turbine (TRIP); (f) main and auxiliary feed water isolation (MEFWISO); (g) main feed water isolation (MFWISO); (h) main steam line isolation (MSTMISO); (i) steam line break (STMLIBR); (j) loss-of-electrical power without reactor shutdown (BLACKSEM); (k) main feed water break without reactor shutdown (MFWBRSEM); (l) main steam line isolation without reactor shutdown (MSTMISEM); (m) main and auxiliary feed water isolation without reactor shutdown (MEFWISEM); (n) main feed water isolation without reactor shutdown (MFWISEM); (o) shutdown of the turbine without reactor shutdown (TRIPTUR). Each one of these transients was described by values of 17 state variables, sampled at intervals of 1 second, making a total of 60 seconds of simulation. The 17 state variables are presented in Table 1.

Table 1. State variables used in the test scenarios

Variable Number	State Variables	Unit
1	Core flow rate	%
2	Hot-leg temperature	°C
3	Cold-leg temperature	°C
4	Core flow rate	Kg/s
5	Steam generator level (wide range)	%
6	Steam generator level (narrow range)	%
7	Steam generator level	Mpa
8	Feedwater flow	Kg/s
9	Steam flow	Kg/s
10	Steam generator break flow	Kg/s
11	Primary flow	Kg/s
12	Pressurizer pressure	Mpa
13	Thermal power	%
14	Nuclear power	%
15	Subcooling Margin	°C
16	Pressurizer level	%
17	Average temperature	°C

The simulated transients were divided into two groups in relation to the occurrence or not of the reactor shutdown signal. The first group contains the accidents where the reactor shutdown has occurred, that is: BLACKOUT, LOCA, SGTR, MFWBR, TRIP, MEFWISO, MFWISO, MSTMISO and STMLIBR. Consequently, the other accidents, together with the normality condition, have formed the second group, that is: BLACKSEM, MFWBRSEM, MSTMISEM, MEFWISEM, MFWISEM and TRIPTURSEM. This division has made it possible for the diagnosis system to be divided in two independent subsystems, specifying each of them to one of the previous groups.

After defining the input variables and the two groups of accidents, the training set for the ANN of the CIVN was formed for each group. The first training set contained 480 patterns without noise and 2400 patterns with noise, representing all the accidents of the first group, except the MFWBR accident which will serve as an unknown transient to check the "don't know" classifications. The second training set contained 360 patterns without noise and 1800 patterns with noise, representing the entire first group, except the MFWBRSEM accident which will serve as an unknown transient to check the "don't know" classifications.

In order to define all the ANN parameters, it was necessary to carry out several simulations, once there is not a well-defined general criterion for choosing these parameters. After these tests, the ANN, which presented the best result, according to each group, has acquired the following configuration:

First group:

- input layer containing 85 neurons with functions of linear activation (17 "movable temporal windows" with five elements each);
- 2 intermediate layers with 34 neurons with activation functions of the logistic type;
- output layer with 1 neuron with activation function of the logistic type;

Second group:

- input layer containing 85 neurons with functions of linear activation (17 "movable temporal windows" with five elements each);
- 2 intermediate layers with 18 neurons with activation functions of the logistic type;
- output layer with 1 neuron with activation function of the logistic type;

After the ANNs training, the radii of the range zone (R_{ev}) were obtained for the accidents of each one of the groups, as shown in Tables 2 and 3.

Likewise two training groups were formed for the ANNs of the CIVN, two training groups were also formed for the ANNs of the simulator, containing the same number of patterns and for the same accidents. So, after some simulations, the following configurations were obtained for the ANNs:

Table 2. Range radius of each accident of the first group

Accident	ANN Code	R_{ev}
BLACKOUT	2	0,10
LOCA	3	0,09
SGTR	4	0,1
MEFWISO	5	0,1
MFWISO	6	0,14
MSTMISO	7	0,14
STLIBR	8	0,16
TRIPTUR	9	0,05

Table 3. Range radius of each accident of second group

Accident	ANN Code	R_{ev}
NORMAL	1	0,035
BLACKSEM	2	0,069
MEFWISEM	3	0,15
MFWISEM	4	0,11
MSTMISEM	5	0,08
TRIPTURSEM	6	0,17

First group:

- input layer containing 18 neurons with functions of linear activation (17 state variables and the ANN output of the CIVN);
- 2 intermediate layers with 20 neurons with activation functions of the logistic type;
- output layer with 17 neurons with activation function of the logistic type;

Second group:

- input layer containing 18 neurons with functions of linear activation (17 state variables and the ANN output of the CIVN);
- 2 intermediate layers with 15 neurons with activation functions of the logistic type;
- output layer with 17 neurons with activation function of the logistic type;

After the training of the ANNs of the simulators, the maximum deviations of each (Δx_{max}) variable were obtained for the accidents of each one of the groups. In Tables 4 and 5 the maximum deviations are shown for each accident.

Table 4. Value of the maximum desviations Δx_{max} for the first group

Variable Accident	1	2	3	4	5	6	7	8	9
BLACKOUT	1.8	1.1	1.2	132	1.3	0.5	0.1	5.0	10
LOCA	0.3	0.6	0.8	217	0.2	0.2	0.1	4.0	2.7
SGTR	0.5	1.5	1.3	639	0.4	0.3	0.1	2.9	6.0
MEFWISO	0.7	2.3	0.8	461	0.7	0.7	0.1	22	14
MFWISO	0.8	1.5	0.7	316	0.6	0.8	0.5	6.1	11
MSTMISO	0.9	2.1	1.0	274	1.3	0.5	0.1	11	25
STLIBR	0.3	0.6	1.2	99	0.5	0.2	0.1	7.6	3.1
TRIPTUR	0.4	0.5	0.2	186	0.7	0.4	0.1	11	8.3

Variable Accident	10	11	12	13	14	15	16	17
BLACKOUT	1.3	2.0	0.1	0.5	0.6	0.4	0.6	0.5
LOCA	3.5	0.4	0.2	0.3	0.3	0.3	1.3	0.5
SGTR	2.8	0.3	0.2	0.4	0.3	0.2	0.9	0.7
MEFWISO	1.5	0.7	0.1	1.0	0.9	0.3	1.0	1.2
MFWISO	1.8	0.5	0.1	0.5	0.5	0.3	0.4	0.9
MSTMISO	1.1	0.9	0.1	0.8	0.5	0.4	0.4	1.2
STLIBR	0.9	0.3	0.1	0.4	0.4	0.1	0.5	0.6
TRIPTUR	1.2	0.3	0.1	0.4	0.2	0.4	0.3	0.3

Table 5. Values of the maximum desviation Δx_{max} for the second group

Variable Accident	1	2	3	4	5	6	7	8	9
NORMAL	0.2	0.5	0.9	289	0.3	0.2	0.1	2.6	5.1
BLACKSEM	1.9	1.1	1.5	334	1.2	0.6	0.1	12	39
MEFWISEM	0.5	0.6	1.5	434	1.0	0.9	0.1	19	11
MFWISEM	0.4	0.7	1.0	394	1.0	0.5	0.1	6.1	9.4
MSTMISEM	0.5	0.8	2.5	448	0.9	0.8	0.1	8.4	18
TRIPTURSEM	0.4	0.6	1.0	382	0.9	0.3	0.1	6.1	45

Variable Accident	10	11	12	13	14	15	16	17
NORMAL	0.1	0.3	0.1	0.2	0.2	0.1	0.3	0.8
BLACKSEM	0.1	1.4	0.1	2.3	2.4	0.7	0.8	1.5
MEFWISEM	0.1	0.5	0.2	0.4	0.3	0.6	0.5	0.8
MFWISEM	0.1	0.4	0.1	0.3	0.3	0.3	0.5	0.3
MSTMISEM	0.1	0.6	0.1	0.6	0.5	0.4	0.7	1.3
TRIPTURSEM	0.1	0.5	0.1	0.4	0.4	0.2	0.6	0.7

The end test of the system was carried out in two steps, where each one aimed at evaluating the response of the system for each group. In this way, a total of 840 new patterns representative of the transients included in the training stage (adding, for each accident, new noises to the state variables) and 120 patterns representative of the MFWBR and MFWBRSEM transients, which were not included in the training, were generated for both groups.

For each transient of the first group, a set of 60 patterns was used, where all patterns represent the condition of the accident. Figure 7 shows the graphs for the responses of the ANN of the CIVN for this group and Table 6 presents the response of the system.

For each transient of the second group, a set of 60 patterns was also used, however the first ten patterns of this set represent the normality condition and the last 50 patterns represent the accident condition. Figure 8 shows the graphs for the responses of the ANN of the CIVN for this group and Table 7 presents the response of the system.

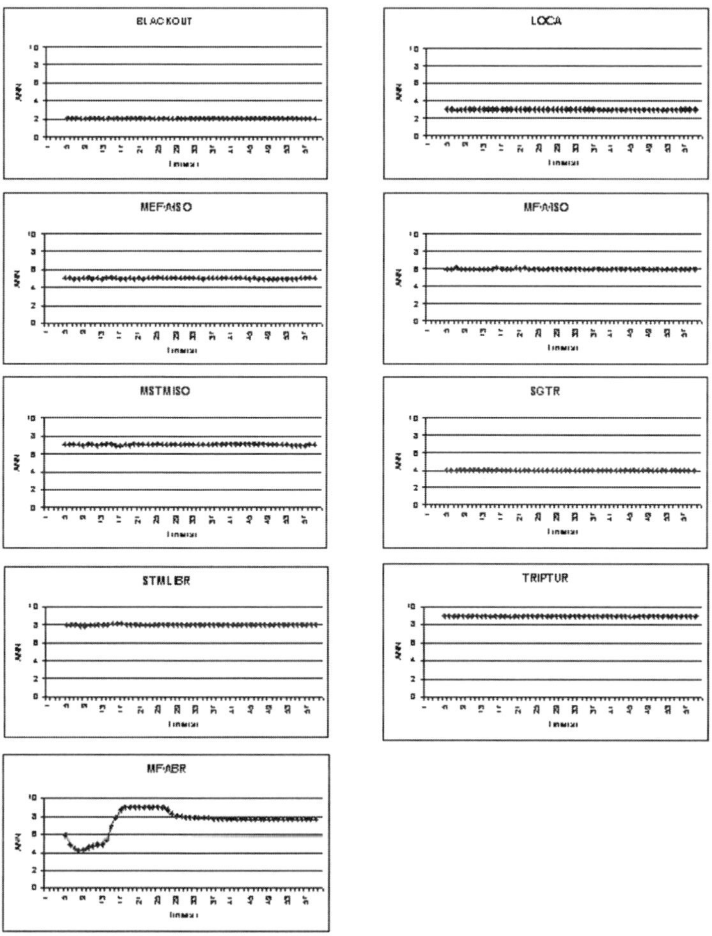

Fig. 7. Responses of the ANN of the CIVN for the first group

Table 6. Response of the system for the first group

Accident Time	2	3	4	5	6	7	8	9	MFWBR
1	-	-	-	-	-	-	-	-	-
2	-	-	-	-	-	-	-	-	-
3	-	-	-	-	-	-	-	-	-
4	-	-	-	-	-	-	-	-	-
5	2	3	4	5	6	7	8	9	Don't Know (CVV)
6	2	3	4	5	6	7	8	9	Don't Know (CIVR)
7	2	3	4	5	6	7	8	9	Don't Know (CIVR)
8	2	3	4	5	6	7	8	9	Don't Know (CIVR)
9	2	3	4	5	6	7	8	9	Don't Know (CIVR)
10	2	3	4	5	6	7	8	9	Don't Know (CIVR)
11	2	3	4	5	6	7	8	9	Don't Know (CIVR)
12	2	3	4	5	6	7	8	9	Don't Know (CIVR)
13	2	3	4	5	6	7	8	9	Don't Know (CIVR)
14	2	3	4	5	6	7	8	9	Don't Know (CIVR)
15	2	3	4	5	6	7	8	9	Don't Know (CVV)
16	2	3	4	5	6	7	8	9	Don't Know (CVV)
17	2	3	4	5	6	7	8	9	Don't Know (CIVR)
18	2	3	4	5	6	7	8	9	Don't Know (CVV)
19	2	3	4	5	6	7	8	9	Don't Know (CVV)
20	2	3	4	5	6	7	8	9	Don't Know (CVV)
21	2	3	4	5	6	7	8	9	Don't Know (CVV)
22	2	3	4	5	6	7	8	9	Don't Know (CVV)
23	2	3	4	5	6	7	8	9	Don't Know (CVV)
24	2	3	4	5	6	7	8	9	Don't Know (CVV)
25	2	3	4	5	6	7	8	9	Don't Know (CVV)
26	2	3	4	5	6	7	8	9	Don't Know (CVV)
27	2	3	4	5	6	7	8	9	Don't Know (CIVR)
28	2	3	4	5	6	7	8	9	Don't Know (CIVR)
29	2	3	4	5	6	7	8	9	Don't Know (CVV)
30	2	3	4	5	6	7	8	9	Don't Know (CVV)
31	2	3	4	5	6	7	8	9	Don't Know (CVV)

Accident Time	2	3	4	5	6	7	8	9	MFWBR
32	2	3	4	5	6	7	8	9	Don't Know (CVV)
33	2	3	4	5	6	7	8	9	Don't Know (CVV)
34	2	3	4	5	6	7	8	9	Don't Know (CVV)
35	2	3	4	5	6	7	8	9	Don't Know (CVV)
36	2	3	4	5	6	7	8	9	Don't Know (CIVR)
37	2	3	4	5	6	7	8	9	Don't Know (CIVR)
38	2	3	4	5	6	7	8	9	Don't Know (CIVR)
39	2	3	4	5	6	7	8	9	Don't Know (CIVR)
40	2	3	4	5	6	7	8	9	Don't Know (CIVR)
41	2	3	4	5	6	7	8	9	Don't Know (CIVR)
42	2	3	4	5	6	7	8	9	Don't Know (CIVR)
43	2	3	4	5	6	7	8	9	Don't Know (CIVR)
44	2	3	4	5	6	7	8	9	Don't Know (CIVR)
45	2	3	4	5	6	7	8	9	Don't Know (CIVR)
46	2	3	4	5	6	7	8	9	Don't Know (CIVR)
47	2	3	4	5	6	7	8	9	Don't Know (CIVR)
48	2	3	4	5	6	7	8	9	Don't Know (CIVR)
49	2	3	4	5	6	7	8	9	Don't Know (CIVR)
50	2	3	4	5	6	7	8	9	Don't Know (CIVR)
51	2	3	4	5	6	7	8	9	Don't Know (CIVR)
52	2	3	4	5	6	7	8	9	Don't Know (CIVR)
53	2	3	4	5	6	7	8	9	Don't Know (CIVR)
54	2	3	4	5	6	7	8	9	Don't Know (CIVR)
55	2	3	4	5	6	7	8	9	Don't Know (CIVR)
56	2	3	4	5	6	7	8	9	Don't Know (CIVR)
57	2	3	4	5	6	7	8	9	Don't Know (CIVR)
58	2	3	4	5	6	7	8	9	Don't Know (CIVR)
59	2	3	4	5	6	7	8	9	Don't Know (CIVR)
60	2	3	4	5	6	7	8	9	Don't Know (CIVR)

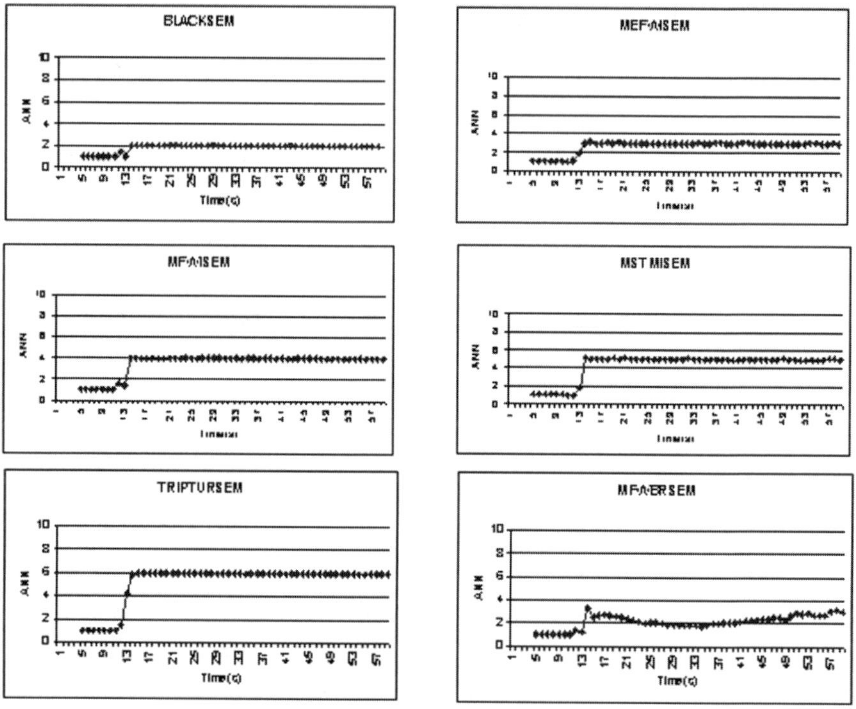

Fig. 8. Responses of the ANN of the CIVN for the second group

6 Conclusions

In this work, a system for the identification of transients was presented, based on ANNs, aiming at helping the operator to make a decision relative to the procedure to be followed in situations of accidents/transients at nuclear power plants.

Using the results presented in Tables 6 and 7, it was verified that the "jump"-type ANN, with “backpropagation” training, was able to fast diagnose all the accidents postulated for a PWR nuclear reactor, even including noise. Particularly, Table 6 and the graphs shown in Figure 7 demonstrated that the ANN of the CIVN was able to identify all the accidents of the first group, soon after the formation of the first "temporal window" (first five seconds of the accident).

Table 7. Response of the system for the second group

Accident Time	2	3	4	5	6	MFWBRSEM
1	-	-	-	-	-	-
2	-	-	-	-	-	-
3	-	-	-	-	-	-
4	-	-	-	-	-	-
5	1	1	1	1	1	1
6	1	1	1	1	1	1
7	1	1	1	1	1	1
8	1	1	1	1	1	1
9	1	1	1	1	1	1
10	1	1	1	1	1	1
11	Don't Know (CVV)	1	1	1	Don't Know (CVV)	1
12	Don't Know (CIVR)	Don't Know (CIVR)	Don't Know (CIVR)	Don't Know (CVV)	Don't Know (CIVR)	Don't Know (CIVR)
13	Don't Know (CVV)	Don't Know (CVV)	Don't Know (CIVR)	Don't Know (CIVR)	Don't Know (CIVR)	Don't Know (CIVR)
14	2	3	4	5	Don't Know (CIVR)	Don't Know (CIVR)
15	2	3	4	5	6	Don't Know (CIVR)
16	2	3	4	5	6	Don't Know (CIVR)
17	2	3	4	5	6	Don't Know (CIVR)
18	2	3	4	5	6	Don't Know (CIVR)
19	2	3	4	5	6	Don't Know (CIVR)
20	2	3	4	5	6	Don't Know (CIVR)
21	2	3	4	5	6	Don't Know (CIVR)
22	2	3	4	5	6	Don't Know (CIVR)
23	2	3	4	5	6	Don't Know (CIVR)
24	2	3	4	5	6	Don't Know (CIVR)
25	2	3	4	5	6	Don't Know (CIVR)
26	2	3	4	5	6	Don't Know (CIVR)

Accident Time	2	3	4	5	6	MFWBRSEM
27	2	3	4	5	6	Don't Know (CVV)
28	2	3	4	5	6	Don't Know (CIVR)
29	2	3	4	5	6	Don't Know (CIVR)
30	2	3	4	5	6	Don't Know (CIVR)
31	2	3	4	5	6	Don't Know (CIVR)
32	2	3	4	5	6	Don't Know (CIVR)
33	2	3	4	5	6	Don't Know (CIVR)
34	2	3	4	5	6	Don't Know (CIVR)
35	2	3	4	5	6	Don't Know (CIVR)
36	2	3	4	5	6	Don't Know (CIVR)
37	2	3	4	5	6	Don't Know (CVV)
38	2	3	4	5	6	Don't Know (CVV)
39	2	3	4	5	6	Don't Know (CVV)
40	2	3	4	5	6	Don't Know (CVV)
41	2	3	4	5	6	Don't Know (CIVR)
42	2	3	4	5	6	Don't Know (CIVR)
43	2	3	4	5	6	Don't Know (CIVR)
44	2	3	4	5	6	Don't Know (CIVR)
45	2	3	4	5	6	Don't Know (CIVR)
46	2	3	4	5	6	Don't Know (CIVR)
47	2	3	4	5	6	Don't Know (CIVR)
48	2	3	4	5	6	Don't Know (CIVR)
49	2	3	4	5	6	Don't Know (CIVR)
50	2	3	4	5	6	Don't Know (CIVR)
51	2	3	4	5	6	Don't Know

Accident Time	2	3	4	5	6	MFWBRSEM
						(CVV)
52	2	3	4	5	6	Don't Know (CIVR)
53	2	3	4	5	6	Don't Know (CVV)
54	2	3	4	5	6	Don't Know (CIVR)
55	2	3	4	5	6	Don't Know (CIVR)
56	2	3	4	5	6	Don't Know (CIVR)
57	2	3	4	5	6	Don't Know (CVV)
58	2	3	4	5	6	Don't Know (CVV)
59	2	3	4	5	6	Don't Know (CVV)
60	2	3	4	5	6	Don't Know (CVV)

For the accidents of the second group, Table 7 and the graphs presented in Figure 8 showed that the ANN of the CIVN was also able to successfully treat the transition from the normality condition to the accident condition, identifying all the accidents of this group in the maximum time of 4 seconds after their beginning.

The results obtained have also demonstrated that the proposed system has successfully overcome, using the criterion of validation of the variables (CVV), the great limitation of this type of network that is the incapacity of supplying a reliable "don't know" answer for transients outside the training scope.

Another important point studied in this work is that the system has shown to be independent of a signal which indicated the beginning of the transient, thus making it robust in relation to this limitation.

References

1. JEONG E, FURUTA K, KONDO S, "Identification of Transient in Nuclear Power Plant Using Adaptive Template Mtching with Neural Network," *Proceedings of the International Topical Meeting on Nuclear Plant Instrumentation, Control and Human_Machine Interface Technologies*, *243-250 (1996)*

2. LONG B, "Technical Assessment of Disturbance Analysis Systems," *Nuc. Saf. , 21 ,1, 38 (1980).*

3. BARTLETT EB, UHRIG RE, "Nuclear Power Plant Status Diagnostcs Using an Artificial Neural Network," *Nuclear Technology, Vol. 97, 272 (1992).*

4. CHEON SW, CHANG SH, "Application of Neural Networks to a Connectionist Expert System for Transient Identification in Nuclear Power Plants," *Nucl. Techol., 102, 177 (1993)*

5. OHGA Y, SEKI H, "Abnormal transient Identification in Nuclear Power Plants Using Neural Network and Knowledge Processing," *Nucl. Technol., 101, 109 (1993)*

6. DHANWADA CV, BARTLETT EB, "A New Method for Nuclear Plant Diagnostics Using Neural Networks," *Trans. Am. Nuc. Soc., 66, 116 (1992)*

7. KIM K, ALJUNDI TL, BARTLETT EB, "Confirmation of Artificial Neural Networks: Nuclear Power Plant Fault Diagnostics," *Trans. Am. Nucl. Soc., 66, 112 (1992)*

8. PARLOS AG et al, "Nonlinear Identification of Process Dynamics Using Neural Networks," *Nucl. Technol.,97, 79 (1993)*

9. FURUKAWA H, UEDA T, KITAMURA M, "Use of Self-Organizing Neural Networks for Rational Definition of Plant Diagnostic Symptoms," *Proceedings of the International Topical Meeting on Computer-Based Human Support Systems, 441-448 (1995)*

10. BARTAL Y, LIN J, UHRIG RE, "Nuclear Power Plant Transient Diagnostics Using Artificial Neural Networks that Allow "Don't Know" Classifications," *Nuclear Technology, Vol. 110, (June), 346-449. (1995)*

11. ROVERSO D, "A Neural Model for Transient Identification in Dynamic Processes." *In: Report HWR-516, OECD HALDEN REACTOR PROJECT, Institutt for Energiteknikk, Norway. (1998)*

12. HAYKIN S, "Neural Networks a Comprehensive Foundation," *2 ed. Prentice Hall (1999)*

18 Planning Surveillance Test Policies Through Genetic Algorithms

Celso M. F. Lapa[1,3], Cláudio M. N. A. Pereira[1,2,3] and P. F. Frutuoso e Melo[2]

[1] Comissão Nacional da Engenharia Nuclear, IEN
Ilha do Fundão s/n, 21945-970, P.O. Box 68550, Rio de Janeiro, Brasil

{lapa, cmnap}@ien.gov.br

[2] COPPE/UFRJ – Prog. Eng. Nuclear –
Caixa Postal 68509, 21945-970, Rio de Janeiro, Brasil

frutuoso@lmn.con.ufrj.br

[3] Universidade Iguaçu, FACET –
Mathematical and Computer Science Department
Av. Abílio Augusto Távora, 2134, Nova Iguaçu, RJ, Brazil

Nuclear power plant systems are comprised of both online and cold-standby components. Cold-standby components differ from the online ones, as they may be unavailable due to unrevealed failures. The usual procedure employed to reveal failures is to submit the components to surveillance tests. The surveillance tests policy may deal with two conflicting scenarios: the test frequency must be sufficiently high in order to minimize the occurrence of failures but, on the other hand, it must be low enough due to its influence on the component unavailability. Obtaining an optimum surveillance test policy at the system level, which involves consideration of many interdependent components, is a very difficult optimization problem. Hence, in this work we propose the use of genetic algorithms for searching the optimum surveillance tests policy, due to its robustness and efficiency in complex problem solving. Here, the probabilistic model considers: a) wear-out effects on cold-standby components when they undergo surveillance tests; b) that when a failure is revealed during a surveillance test, corrective maintenance is performed and, c) that components are distinct (that is, each has distinct test parameters, such as: outage time, wear-out factors, etc.); d) that tests are not necessarily periodic.

1 Introduction

As opposed to the case of active electromechanical components, aging of standby components is not so intense. However, the operational performance of these latter is affected by the occurrence of unrevealed failures. Whenever the occurrence rate of these events is high, the affected components will have a reduced capacity of satisfactorily responding to real demands.

In order to solve this problem, it is necessary to establish a periodic test policy so that the component operational state may be known in order to minimize the chance of not responding to a real demand and of performing corrective maintenance actions, when applicable.

However, industrial facilities hardly have isolated components. The most common case is that of complex systems with structurally and physically interdependent components. In this sense, the determination of a global intervention policy, concerning the system or even the facility level, is clearly a nonlinear optimization problem with many degrees of freedom, which generates a multi-modal search space and which admits a great number of technically possible solutions.

2 Optimization Methodology

Genetic algorithms (HOLLAND, 1975; GOLDBERG, 1989) have been used as an optimization tool for approaching the problem under discussion due to two reasons:

- They have been employed quite successfully for solving similar problems (MUÑOZ, 1997, LAPA *et al.*, 1999 and LAPA *et al.*, 2000a); and
- Standard optimization techniques by linear programming, based on the gradient search method are not applicable to multi-modal nonlinear problems that may exhibit numerical and combinatorial features at the same time. Even deterministic methods for global optimization, such as the branch and bound method, the interval method (HAMSEN, 1992), or integer programming (NEMHAUSER, 1988) are not applicable when no previous knowledge of the search space characteristics is available.

3 Background

Some recently published papers have addressed preventive maintenance optimization of nuclear power plant systems, focusing on system availability (HILSMEIER and ALDEMIR, 1995; VAURIO, 1997), or comparing employed models and tests among some standard policies (MARTOREL *et al.*, 1996, DUTHIE *et al.*, 1998), or applying expert knowledge to determine good maintenance policies (VAN NOORTWIJK *et al.*, 1992).

In order to avoid the optimization difficulties inherent to huge search spaces, many applications have considered systems with a few components (HARUZZAMAN and ALDEMIR, 1996; BILLINTON AND PAN, 1998). RAJE *et al.* (2000) have followed the same trend by considering standby systems subjected to periodic testing.

PARK *et al.* (2000) contributed to the solution of the class of problem under discussion by including components with very small degradation degrees.

CHIANG and YUAN (2001) approached maintenance optimization problems aiming at obtaining system availabilities by Markovian methods.

DIJKHUIZEN and HEIJEN (1999) optimized the distribution of availability intervals instead of optimizing the preventive maintenance policy.

All the cited references have a common feature: they all considered systems composed by a few components and even so they faced difficulties from the point of view of optimization.

Typically, safety-related nuclear systems have many redundancies and components with a great number of combination and alignment alternatives among them.

As already pointed out, the establishment of a maintenance policy that involves a system with the characteristics just mentioned, is a very difficult task. A powerful and appropriate optimization technique is then necessary to allow the determination of an optimal combination of all system parameters.

MUÑOZ *et al.* (1997) were the first to propose the use of genetic algorithms as an optimization tool for maintenance scheduling activities.

LAPA *et al.* (1999, 2000a, 2000b) applied genetic algorithms in order to optimize inspection and maintenance intervals with a new approach: instead of searching for an optimal intervention frequency which means equally spaced interventions, they employed the optimization tool to search for time points in which preventive maintenance interventions should be performed. In this sense, it is understood that equally spaced intervention actions do not necessarily furnish the optimal policy.

4 Purpose

The purpose of this work is to present a methodology for optimizing inspection policies of operational tests that takes into account arbitrary times between successive inspections, much in the same way that has been previously applied to preventive maintenance policies. The following assumptions were considered:

- A component in cold standby is not significantly affected by aging effects, so that its standby failure rate is considered constant (LEWIS, 1996) and if it undergoes exhaustive testing (i. e., small times between successive tests) it will undergo aging;
- After a test, the component would return to an as good as new condition if only the constant standby failure rate model were considered. However, as the test itself results in component aging, then the component returns to an as bad as old condition after a test;
- As long as the component undergoes a failure during a test, it would become unavailable, not only during the test itself but also for the necessary time for performing its repair;
- Tests do not necessarily last the same time;
- When a given component is scheduled for testing, all components that are operationally related to it become unavailable;
- Tests are not necessarily periodic.

5 Reliability Modeling

Let $A_t(t)$ be the availability of a repairable component that is in a cold standby redundancy and subject to a periodic testing policy. We assume also that the component has not undergone any test until time t, where t is the time at which the component is in a standby mode. Taking into account that the cold standby failure rate $\lambda^C(t)$ is assumed to be constant and also that the component is subject to N tests of period T_t, its availability may be written as (LEWIS, 1996 and VAURIO, 1997):

$$A_t(t) = R[t-(N \cdot T_t)] \tag{1}$$

where $R(t)$ is the component reliability.

It is seen that the component is unavailable during the test and *"R(t) = A(t)"* (if $A(t)$ is the non-repairable component availability), then Eq. (1) becomes:

$$A_t\ (t,\Delta_t) = \begin{cases} A\,[t-(k\cdot T_t)] \quad or \\ 0, \quad if \quad kT_t \le t \le kT_t + \Delta_t \end{cases} \tag{2}$$
$$where\ k = 1 \to N$$

Given that a periodic test can be scheduled to any time $T_t(k)$ and also letting $\Delta_r(k)$ be the time to perform a repair when necessary after the k-th test, Eq. (2) may be recast into:

$$A_t\ [t,T_t(k)] = \begin{cases} A\,[t-T_t(k)] \ or \\ 0, \quad if \quad T_t(k) \le t \le T_t(k) + \{\Delta_t(k) + [1-A_t(t-T_t(k)-1)]\Delta_r(k)\} \end{cases} \tag{3}$$
$$where\ k = 1,.., N$$

$1 - A_t(t - T_t(k) - 1)$ is the probability that the component is down, in a standby mode, until the scheduled test at $T_t(k)$ is performed. In this sense, $[1 - A_t(t - T_t(k) - 1)]\Delta_r(k)$ is the mean downtime of the component due to failures at scheduled periodic tests.

Integrating $[1 - A_t(t - T_t(k) - 1)]\Delta_r(k)$ over the mission time, one obtains the contribution to the downtime due to failures revealed during the tests. The component unavailability, taking into account Eq. (3) is given by:

$$\overline{A}_t\ [t,T_t(k)] = \begin{cases} 1 - A\,[t-T_t(k)] \quad or \\ 1, \quad if \quad T_t(k) \le t \le T_t(k) + \{\Delta_t(k) + [1-A_t(t-T_t(k)-1)]\Delta_r(k)\} \end{cases} \tag{4}$$
$$where\ k = 1,.., N]$$

At each periodic test, the component undergoes a demand that lasts the test duration time, so that its failure probability is increased due to this contribution. Letting $\lambda^R(t)$ be the component online failure rate, one may evaluate the component unavailability contribution due to operation period that includes the *k* tests until *t*:

$$\overline{A}_r[k,\Delta_t(k)] = \overline{A}_r\ [\sum_{i=1}^{k} \Delta_t(k)] \tag{5}$$

In the case where the test duration is the same (a reasonable approximation in most cases) Eq. (5) becomes:

$$\overline{A}_r[k,\Delta_t] = \overline{A}_r\ (k\ \Delta_t) \tag{6}$$

Taking into account the unavailability contribution due to the test-induced wear-out, the overall unavailability is given by [from Eq.(4)]:

$$\overline{A}_t\ [t,T_t(k)] = \begin{cases} 1 - A[t-T_t(k)] + \overline{A}_r[k\ \Delta_t(k)]A[t-T_t(k)] \quad or \\ 1, \quad if \quad T_t(k) \le t \le T_t(k) + \{\Delta_t(k) + [1-A_t(t-T_t(k)-1)]\Delta_r(k)\} \end{cases} \tag{7}$$
$$where\ k = 1,..., N$$

Eq. (7) indicates that at each periodic test the cold-standby component unavailability is set equal to zero, $\{1\text{-}A_t[t\text{-}T_t(k)]\}$. However, at each test, a contribution of $\bar{A}_r[\Delta_t(k)]A[t\text{-}T_t(k)]$ should be added, which results in an increasing saw-toothed curve. This model is displayed in Figure 1.

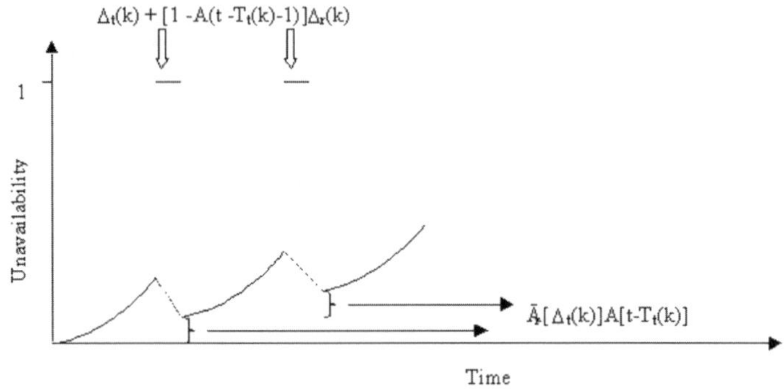

Fig. 1. Unavailability behavior of a component subject to a periodic test policy

The unavailability model just presented considers the behavior of a single component, given a test policy.

However, the main purpose of this investigation is to treat multi-component systems. Moreover, the unavailability evaluation should be dynamic, as long as safety systems can be aligned in different ways when component tests are to be performed. In this sense, different system configurations are possible.

The approach followed in this work is one that considers a single component but may be extended to consider multi-component systems (LEVITIN and LISNIANSK 1999.)

The chosen approach is based upon the assumption that any system failure can be represented by a block diagram of component failures connected in series and/or in parallel.

With a few assumptions concerning basic events independence and by using known properties of probability spaces, this method can evaluate the failure probability of a given system.

The system availability may be represented as a function of the failure probabilities of its components [Eq. (7)] and these latter, as already discussed, depend directly on the components test and repair policies:

$$A_{sis} = f\{A^1_t[t, T_t, (k)], A^2_t[t, T_t, (k)], \ldots, A^x_t[t, T_t, (k)]\} \tag{8}$$

Other techniques, like fault trees or the device of stages (NUNES *et al.*, 2000) could be employed. Of course, different assumptions would have to be employed. Fault trees would replace the block diagram approach and could also consider features not treated by this latter. The device of stages has been employed to model systems under aging with different functional restraints.

6 Genetic Modeling

The genotype must encode all possible scheduling combinations, for all system components. In order explain the codification, a system, comprised of C components, will be considered. Consider also that this system must operate for a period of 480 days. Assuming that 15 days is a good time step for this problem, each component has, then, 30 (480/15 – 2) possible base dates to undergo testing.

Following the classic paradigm of genetic algorithms, a fixed length binary string was used, so that, the genotype which encodes this problem will have *[(30+3).C]* bits. For each component, each of the 30 possible dates is represented by one bit. "1" means that the component is on line and ready to start. "0" means that it was selected to undergo testing at the corresponding base date (multiple of 15). Three additional bits are considered as a time delay (offset) that must be added to the base date, in order to obtain the real date of the test. The delay allow components with same base date to undergo tests on different days. Figure 2 illustrates the genotype and its decoding for each component, represented by a vector whose elements are the times at which the tests occurs.

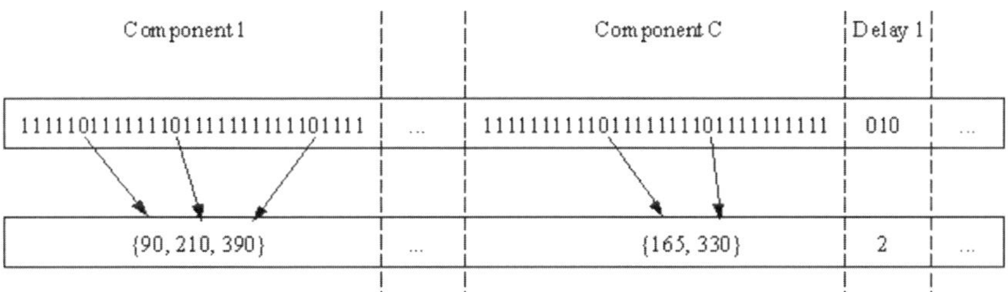

Fig. 2. Example of the genotype which encodes the scheduling

The objective function used to evaluate the fitness of each individual (genotype) is the average system unavailability considering the surveillance test scheduling policy encoded by the genotype and calculated according to the probabilistic model previously described:

$$Fit = \frac{1}{T_{mis}} \int_{t=0}^{Tmis} [\overline{A}(t)_{sis}]\, dt \tag{9}$$

where T_{mis} is the mission time.

6 Results

The aim of this work was to investigate the advantages and shortcomings of a search problem approached by traditional methods and also by genetic algorithms.

A classical standby electromechanical system has been chosen for the analysis. This is due to the fact that this kind of system plays an important role in nuclear plant safety. The system considered in the analysis is made up by two electrical motors, two pumps, and three valves, as displayed in Figure 3. These components are arranged in two redundant trains and due to safety reasons these two trains are also connected so that there are two more discharge possibilities.

The operational data employed, such as test duration times, repair times, standby failure rates and online failure rates are fictitious. However, they were based on typical figures published in the literature (HARUNUZZAMAN and ALDEMIR 1996).

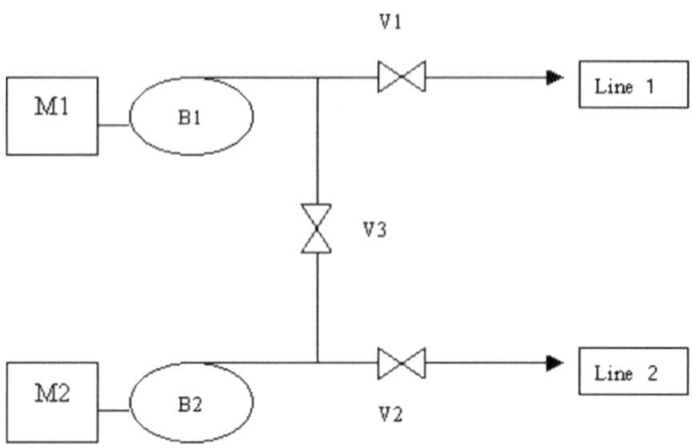

Fig. 3. Block diagram of electromechanical system employed for comparing the optimization methodologies

This work was split in two parts: initially, the GA searched for a constant interval policy. The time between tests was set equal to 15 days and a mission time of 480 days was considered. A delay not greater than 7 days was considered also, in order to take into account that test policies with the same test frequency could consider staggered testing. In a second stage, the GA searches for the optimal test scheduling policy for which the times between tests are not necessarily constant. Using the conditions displayed in Table 1, the steps discussed were repeated three times, each using a different randomization seed.

Table 1. Operational conditions of the case studies

	Standby Failure Rate (day^{-1})	Online Failure Rate (hr^{-1})	Test Duration Time (hr)	Repair Time (hr)
V1	0.000395	0.000095	2	8
V2	0.000395	0.000095	2	8
V3	0.000395	0.000095	2	8
B1	0.00053	0.000142	4	72
B2	0.00053	0.000142	4	72
M1	0.00053	0.000675	4	72
M2	0.00053	0.000675	4	72

The three simulations suggested virtually identical policies, specially for the policies obtained by means of the constant frequency methodology. For the flexible interval methodology, the results were close and qualitatively consistent, thus confirming the GA robustness for this kind of problem.

Table 2 presents the policies proposed by the GA for both methodologies obtained in one of the simulations. Surprisingly, in all simulations for both approaches, the GA indicated a system configuration that considers pumps and valves in a crossed alignment. This option clearly indicates that testing components of different trains and using the train linking line is a good alternative to the procedure of testing each train at a time, when the train linking line is available.

Table 2. Periodic test policies suggested by the GA

Arbitrary Intervals - $1-\bar{A}_{sis}$ = 3,8 E-03
V1 : 30 60 90 120 150 165 195 225 255 285 315 345 375 405 420 435 450 465
V2 : 32 62 92 122 152 182 212 242 272 287 302 332 362 392 407 422 437 452 467
V3 : 30 60 90 120 150 165 195 225 255 285 315 345 375 405 420 435 450 465
B1 : 32 77 122 152 182 212 242 287 302 332 362 392 422 452 467
B2 : 37 67 97 142 187 217 262 292 322 352 382 412 427 457
M1: 77 122 182 242 287 332 362 392 422 452 467
M2: 67 142 217 262 322 352 382 412 427 457
Constant Interval - $1-\bar{A}_{sis}$ = 4,1 E-03
V1 : 36 66 96 126 156 186 216 246 276 306 336 366 396 426 456
V2 : 30 60 90 120 150 180 210 240 270 300 330 360 390 420 450
V3 : 34 64 94 124 154 184 214 244 274 304 334 364 394 424 454
B1 : 30 60 90 120 150 180 210 240 270 300 330 360 390 420 450
B2 : 36 66 96 126 156 186 216 246 276 306 336 366 396 426 456
M1: 60 120 180 240 300 360 420
M2: 66 126 186 246 306 366 426

It can be seen from the convergence plots displayed in Figure 4 that the traditional methodology that considers constant test intervals (full curve) converges to the final search figure before the thousandth generation, while the simulation based on the proposed methodology reaches the final search figure after the 35,000-th generation (dashed curve).

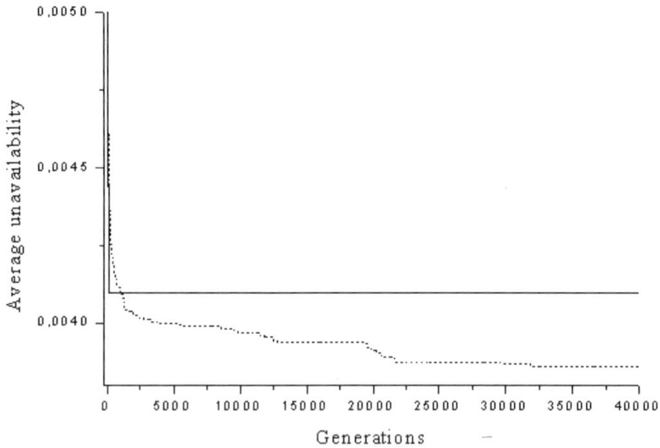

Fig. 4. Comparison between the convergence patterns of the searches performed by the two methodologies

The reason for the difference in the number of generations involved is that the search performed according to the traditional approach employs much less possible candidate solutions than those of the flexible interval approach.

It is reasonable to admit that the GA has come quite close to a global optimum by means of the traditional search methodology and that the search by using the new methodology still has relevant knowledge to be unraveled. This knowledge could help to establish better policies by means of more powerful computers with more extensive simulations and considering greater populations.

However, the initial aim of the work was to investigate the potential of the new methodology and, to this end, more extensive simulations are not necessary. The results obtained so far allow for some useful conclusions.

The results presented in Table 2 and specially those displayed in Figure 4 clearly indicate that the GA can find test policies by taking into account flexible test intervals, thus giving better results, even though much more computational effort is demanded.

7 Conclusions

In the context of a global analysis of the results, one can clearly see that the GA approach is powerful. This is confirmed by the convergence pattern displayed in Figure 4. The numerical differences between the unavailabilities obtained are not so important, as long as they are case dependent.

For more complex applications, it is reasonable to expect that a greater number of components and also longer mission times and seasonal restraints impose further restrictions on periodic tests with constant times between tests, so that even more flexible test intervals be found as optimal solutions.

It is worth noting that, no matter the adopted time discretization of the search space, when constant test frequencies are searched, all possible solutions are contained in the candidate solutions space generated by the probabilistic methodology of flexible intervals.

In this sense, if for a given specific system the best policy is that of a constant test frequency, the proposed methodology will be able to find this solution.

However, if the problem at hand has to do with components that need to undergo some flexible test interval policy, due to reasons concerning their failure rate evolution in time, it is readily seen that the traditional methodology does not give satisfactory solutions.

This feature will be even more relevant when it is intended to design test policies for longer time periods and also when more complex systems are involved.

In this case, the involved components will pass through different periods of their operational life and will be more intensively affected by aging even due to the tests themselves.

In this sense, not only policies that take into account each operational period are needed, but also they should incorporate the effects of their very application in an interactive manner. The optimal policy depends on the components availabilities and these latter are influenced by the test itself in a permanent and auto-corrective way.

Another significant advantage of the proposed methodology is its ability to schedule test policies that fit to possible seasonal features or to specific time restraints.

Let us imagine, for example, an industrial facility that has production lines whose unavailability in the months of November and December will cause serious problems due to Christmas-related demands.

The proposed methodology can design test policies that takes the aforementioned restraint into account, by just imposing penalties to the test scheduling assigned to both months. A policy obtained by the GA under these circumstances will globally fit all the remaining tests so that tests and or maintenance actions are not necessary in these months.

If these restraints were put into a problem of searching test frequencies, they would significantly impart the global solution, as long as a point restraint would influence all the remaining test and/or maintenance interventions. This is due to the fact that the optimal test interval would have to be such that interventions were performed neither in November nor in December and, furthermore, it would have to be greater than 61 days.

Other operational restraints, like the number of repair crews, safety-related restrictions, or due to operational procedures can be easily incorporated to the optimization problem.

References

BILLINTON R, PAN J (1998) Optimal Maintenance Scheduling in a Two Identical Component Parallel Redundant System, *Reliability Engineering & System Safety*, v.59, pp. 309-316.

CHIANG JH, YANG J (2001) Optimal Maintenance Policy for a Markovian System under Periodic Inspection . *Reliability Engineering and System Safety,* v.71, pp.165-172.

DIJKHUIZEN G, HEIJDEN M (1999) Preventive Maintenance and the Interval Availability Distribution of an Unreliable Production System, *Reliability Engineering and System Safety,* v.66, pp.13-27.

DUTHIE JC, ROBERTSON MI, CLAYTON AM, LIDBURY DPG (1998) Risk-based Approaches to Ageing and Maintenance Management, *Nuclear Engineering and Design* , v.184, pp.27-38.

GOLDBERG DE (1989) *Genetic Algorithms in Search, Optimization and Machine Learning*, New York, Addison-Wesley.

HAMSEN E (1992) *Global Optimization Using Interval Analysis*, New York, Marcel Dekker.

HARUNUZZAMAN M, ALDEMIR T (1996) Optimization of Standby Safety System Maintenance Scheduling in Nuclear Power Plants, *Nuclear Technology*, v.113, pp. 354-367.

HILSMEIER TA., ALDEMIR T, VESELY WE (1995) Time-dependent Unavailability of Aging Standby Components Based on Nuclear Plant Data, *Reliability Engineering and System Safety*, v.47, pp.199-205.

HOLLAND JH (1975) *Adaptation in Natural and Artificial Systems*, Ann Arbor, University of Michigan.

LAPA CMF, PEREIRA CMNA, MOL ACA (1999) Application of Genetic Algorithms to the Optimization of Reliability-Centered Preventive Maintenance Politices of Nuclear Systems (in Portuguese), *Proceedings of the 15th Brazilian Congress of Mechanical Engineering,* Associação Brasileira de Ciências Mecânicas, Rio de Janeiro, RJ.

LAPA CMF, PEREIRA CMNA, MOL ACA (2000a) Maximization of a Nuclear System Availability through Maintenance Scheduling Optimization Using Genetic Algorithm, *Nuclear Engineering & Design,* v.196, pp. 95-107.

LAPA, CMF, PEREIRA,CMNA, FRUTUOSO PFM (2000b) Coupled Emergency Diesel Generators-Component Coolant Water System Maintenance Scheduling Optimization Using Genetic Algorithms", Proceedings of the Fourth FLINS (Fuzzy Logic and Intelligent Techniques in Nuclear Science) Conference on Intelligent Techniques and Soft Computing in Nuclear Science and Engineering, pp. 519-526, Bruges, Belgium.

LEWIS EE. (1996) *Introduction to Reliability Engineering*, New York, John Wiley & Sons.

LEVITIN G, LISNIANSKI A (1999) Joint Redundancy and Maintenance Optimization for Multistate Series-parallel Systems, *Reliability Engineering and System Safety*, v.64, pp.33-42.

MARTORELL S, MUNÕZ A, SERRADELL V (1996) Age-Dependent Models for Evaluating Risks & Cost of Surveillance & Maintenance of Components, *IEEE Transactions on Reliability*, v.45, n.3, pp. 433-441.

MUNÕZ A, MARTORELL S, SERRADELL V (1997) Genetic Algorithms in Optimizing Surveillance and Maintenance of Components, *Reliability Engineering & System Safety*, v.57, n.2, pp. 107-120.

NEMHAUSER GL, WOLSEY LA (1988) *Integer and Combinatorial Optimization*, New York, John Wiley and Sons.

NUNES MEC, PEREIRA CMNA, FRUTUOSO E MELOPF (2000) Optimization of the Device of Stages Through Genetic Algorithms for Non-Markovian Systems Reliability Evaluation: an Application to Nuclear Safety Systems, *Proceedings of the Fourth FLINS (Fuzzy Logic and Intelligent Techniques in Nuclear Science) Conference on Intelligent Techniques and Soft Computing in Nuclear Science and Engineering*, pp. 527-534, Bruges, Belgium.

PARK DH, JUNG GM, YUM JK (2000) Cost Minimization for Periodic Maintenance Policy of a System Subject to Slow Degradation, *Reliability Engineering & System Safety*, v.68, pp.105-112.

PEREIRA CMNA, LAPA CMF (2000) Evolutionary Computation at the Institute of Nuclear Engineering (in Portuguese), Proceedings of the XII National Meeting on Reactor Physics and Thermo-hydraulics, Associação Brasileira de Energia Nuclear, Rio de Janeiro, RJ.

VAN NOORTWIJK J, DEKKER R, MAZZUCHI T (1992) Expert Judgment in Maintenance Optimization, *IEEE Trans. Reliability*, v.41, pp. 427-432.

VAURIO JK (1997) On Time-dependent Availability and Maintenance Optimization of Standby Units under Various Maintenance Policies. *Reliability Engineering and System Safety*, v.56, pp.79-89.

19 A Possibilistic Approach for Transient Identification with "Don't Know" Response Capability Optimized by Genetic Algorithm

José Carlos S. de Almeida [1], Roberto Schirru [2] and Cláudio M. N. A. Pereira [1,2]

[1] Comissão Nacional de Energia Nuclear, IEN –
Ilha do Fundão s/n, 21945-970, P.O.Box 68550, Rio de Janeiro, Brasil

{Jcsa, cmnap}@ien.gov.br

[2] Universidade Federal do Rio de Janeiro, PEN/COPPE –
Ilha do Fundão s/n, 21945-970, P.O.Box 68509, Rio de Janeiro, Brasil.

{schirru, claudio}@lmp.ufrj.br.

This work describes a possibilistic approach for transient identification based on the minimum centroids set method, proposed in previous work, optimized by genetic algorithm. The idea behind this method is to split the complex classification problem into small and simple ones, so that the performance in the classification can be increased. In order to accomplish that, a genetic algorithm is used to learn, from realistic simulated data, the optimized time partitions, which the robustness and correctness in the classification are maximized. The use of a possibilistic classification approach propitiates natural and consistent classification rules, leading naturally to a good heuristic to handle the "don't know" response, in case of unrecognized transient, which is fairly desirable in transient classification systems where safety is critical. Application of the proposed approach to a nuclear transient identification problem reveals good capability of the genetic algorithm in learning optimized possibilistic classification rules for efficient diagnosis including "don't know" response. Obtained results are shown and commented.

1 Introduction

Transient identification in industrial plants is often a very hard task and may involve a great amount of human cognition. Observing and analyzing the behavior of state variables (such as pressure, temperature, flux, etc.), the expert (commonly the operator) can diagnose malfunctions in the system. In some situations, however, the time needed for the expert to conclude its diagnosis may be considerably large. Aiming to support the operator in his decision, enhancing the transient diagnosis effectiveness and quickness, computer based systems have been widely used. Besides, intelligent soft computing technologies, such as expert systems (ES) (Schirru et al. 1999b) and artificial neural networks (ANN) (Bartal et al. 1995; Bartlett and Uhrig, 1992) has been proposed.

This work extends previous research on genetic algorithms applied to transient identification systems design. In the first work (Pereira et al. 1998; Schirru et al. 1999a) it was introduced the minimum centroids set (MCS) method for partitioning a complex identification problem into smaller and easier ones, in order to enhance the identification response. In that work the classification was based on the Euclidean distance between samples and centroids (classes prototypes). The work proposed here has important additional features. The use of a possibilistic approach in the classification leads to more natural cognition, providing information of the cluster dimension in the classification formula. Such fact could naturally point to a simple heuristic that comprises the "don´t know" response in case of unknown transient.

2 Possibilistic Approach

In 1965, L. A. Zadeh (1978) introduced the fuzzy sets theory, to which a more flexible definition for membership function over the domain of discourse was conceived. The membership value of a sample to a class is indicated by a real number between 0 and 1. The membership function, $\mu_A(\vec{x})$, of the sample $\vec{x}$ to the set A maps each possible element in the [0,1] interval. Several types of membership functions are possible. In 1985, Zimmermann and Zysno (1985) has shown, through empirical studies, that a good model for such membership is expressed by Eq. 1.

$$\mu(\vec{x}) = \frac{1}{1 + d(\vec{x}, \vec{x}_0)} \tag{1}$$

where $\vec{x}$ is a sample point and $\vec{x}_0$ is a centroid of a class.

By this approach, a sample is considered to belong to the class that contains the centroid to which the sample membership is the highest. Note that, according to Eq. 1, if a sample $\vec{x}$ is equidistant from centroids of two different classes β_1 and β_2 ($d_1 = d_2$), the membership values are the same even if the cluster have different size and shape. Figure 1 illustrates this fact.

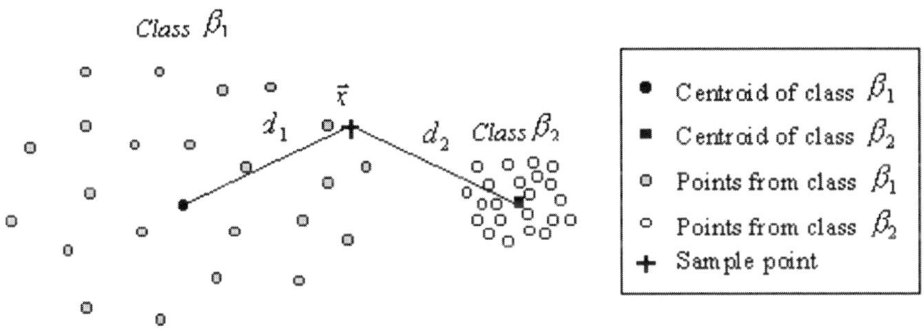

Fig. 1. A sample with same membership to two different classes according to Eq. 1

Observing Figure 1, the human cognition may lead to the conclusion that sample $\vec{x}$ belongs to class β_1, however, by Eq. 1 such fact is not contemplated. In order to solve this kind of problem, a possibilistic approach (Krishnapuram and Keller 1993) is proposed. Based on the method proposed by Krishnapuram and Keller, it was included in Eq. 1 an extra parameter η, which carries the information of the cluster dimension, the membership of a sample $\vec{x}$ to a class represented by the centroid $\vec{x}_0$ can be written as in Eq. 2:

$$\mu(\vec{x}) = \frac{1}{1 + \dfrac{d(\vec{x}, \vec{x}_0)}{\eta_0}} \tag{2}$$

where η is a positive number which depends on the cluster characteristics. Assigning to η the value of the distance from the class centroid to the most distant point belonging to that class, then, its membership value, according to Eq. 2, is 0.5, whichever is the class. By this formulation, $\vec{x}$, in Figure 1, will be considered to belong to class β_1, as desired.

By using such formulation, a frontier that depends on the cluster dimension can naturally be used to carry out the "don't know "classification.

3 Centroids Optimization via Genetic Algorithms

3.1 The Minimum Centroids Set Method

The Minimum Centroids Set Method (Pereira et al. 1998; Schirru et al. 1999a) for transient identification consists on partitioning the time axis into the minimum number of sub-regions (that define sub-classes of patterns) which prototypes (centroids) lead to the maximum performance in the classification task.

Consider for example the two hypothetical transients represented by one-variable (v) time excursion each one. Figure 2 the variable excursion sampled in time, from t_1, to t_2, for both transient.

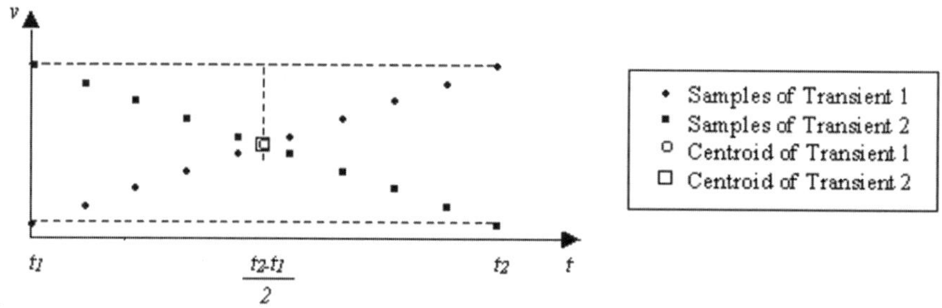

Fig. 2. Two hypothetical transients represented by one-variable time excursion

Note that, because of the symmetry, the centroids of the classes are overwritten. In this case, the transients would not be discriminated. The idea behind the MCS method is to answer the question: " What is the minimum set of time partitions that lead to the maximum efficiency in the classification?". In this sample case, the MCS method should find a partition at $\frac{t_2 - t_1}{2}$, as shown in Figure 2.

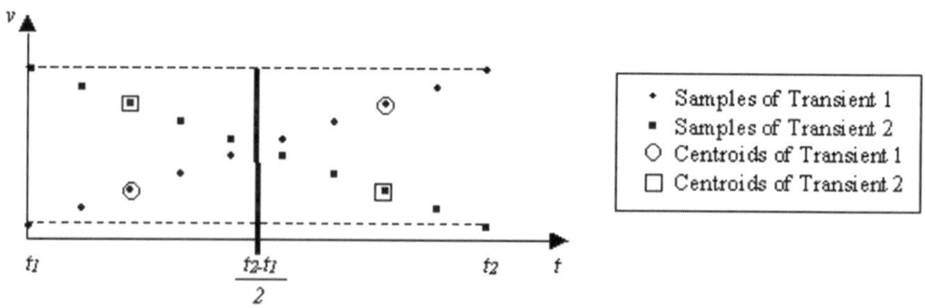

Fig. 3. Optimum time partition for the two hypothetical transients

Note that the centroids set of proposed partition lead to a perfect classification (no misclassification) for the whole samples. In other words, such centroids are the Voronoi vectors (Pereira and Schirru 2000), which perfectly separate the sub-classes.

To find the optimum partition, however, is not a simple task when the number of transients, as well as the number of variables considered increase. Hence, the use of a powerful optimization technique is requires. Due to its ability in the dealing with such kind of problem, the genetic algorithm (GA) was used as the optimization tool.

3.2 Genetic Modeling

Considering the classical GA (Holland 1975; Goldberg 1989) paradigm, a fixed-length binary string was used to encode the problem solution candidates. The genotype represents a discrete time axis where each set of contiguous '1' or '0' is considered to be a partition. Figure 4 exemplifies the genotype and respective time partition.

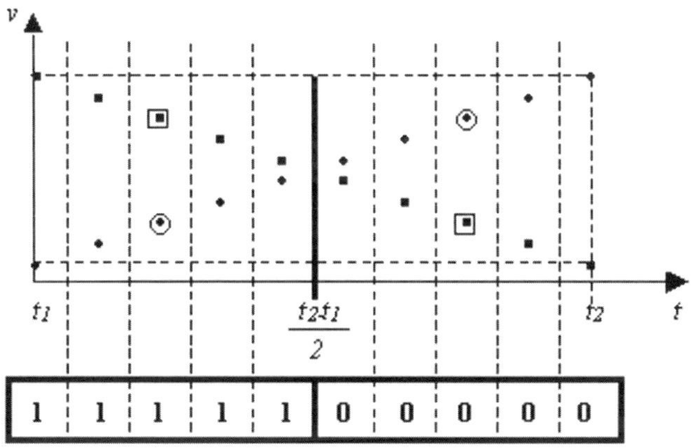

Fig. 4. Genotype and its decoding into the time partitions

The time position of the centroid for each subclass is the average time of all samples in the partition.

The objective function, which evaluates the fitness of a solution candidate, considers both the effectiveness in classification (number of correct classifications for a given training/test set) as the main objective and the number of partitions which may penalize the fitness as it increases. A simple way to do so is given by Eq. (3).

$$f = k_E \cdot E - k_P \cdot P \tag{3}$$

where E is the effectiveness of the classification (number of correct classifications) for the training/test set, obtained by the application of the possibilistic classification rules, P, the number of partitions, k_E and k_P positive constants which assign importance of E and P respectively. As the idea is to give relative importance, k_E was set to unity (1.0) and the f expression becomes:

$$f = E - k_P \cdot P \tag{4}$$

As the constant k_P is defined according to the relative importance of E and P, it is problem-dependent and may be defined by the an expert. In next section it can be seen an example of k_P definition for a nuclear power plant transient diagnosis system.

4 Method Application

Here is presented some validation tests of the proposed method as well as its application to a nuclear power plant transient identification problem.

4.1 Definition of k_P

Due to the nature of the application (where safety is critical), increasing the effectiveness in the classification is too much more important than decreasing the number of time partitions. Hence, the constant k_P was chosen so that:

$$E_{\max} - K_P \cdot C_{\max} > (E_{\max} - 1) - K_P \cdot 1 \tag{5}$$

where E_{max} is the maximum number of correct classifications (100%) and P_{max} the maximum number of clusters. In other words if the E_{max} can be reached even if it is required to use the maximum number of clusters, it is considered better than any other solution where $E < E_{max}$.

In the validation tests as well as in the nuclear application, the time window considered was 1 minute, and the sample rate was 1 second. Consequently, there are 61 sample points for each variable. So P_{max} =61 and:

$$E_{\max} - K_p \cdot 61 > (E_{\max} - 1) - K_p \cdot 1 \Rightarrow 60 \cdot K_p < 1 \tag{6}$$

$$K_P < 0.01666... \tag{7}$$

4.2 Verification Tests

In order to verify the proposed method, it was proposed 2 hypothetical scenarios, were the optimum solution is well known. Each scenario comprises 2 different transients, represented by one-variable time series each one, as shown in Figure 5.

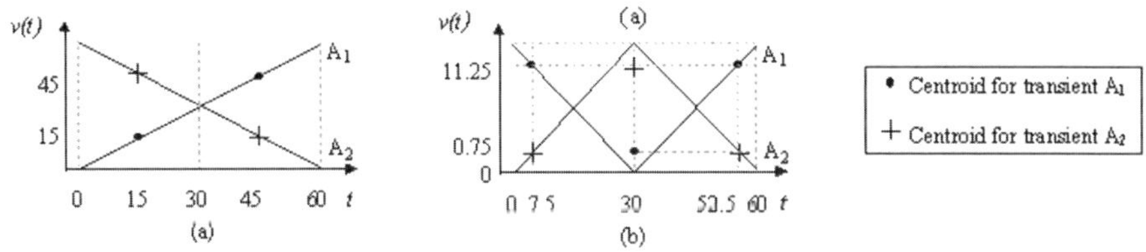

Fig. 5. Test scenarios and their respective centroids. (a) Scenario 1; (b) Scenario 2

More than 10 experiments with different genetic parameters were made for each problem and in more than 90% of then lead to the optimum result, demonstrating efficacy and good robustness.

4.3 Application to a Nuclear Transient Identification Problem

Problem Description

Three nuclear transients were considered: the Black-out (BLKOUT), a Lost Of Coolant Accident (LOCA) and the Steam generator tube rupture (SGTR). Each transients is represented by 61 points per variable, generated by precise simulation with time-step of one second (by hypothesis the transients can be characterized in one minute). The variables are: the primary flux, nuclear power, thermal power, cold-leg temperature, hot-leg temperature, average primary temperature, sub-cooling margin, primary pressure, primary flow, steam generator wide range, steam generator narrow range, steam pressure, feed-water flow, break flow, pressurizer level and steam flow.

Instantaneous Classification

After optimization, the proposed identification system found 6 partitions that has led to 98% of correct classification for the training set. In order to verify the generalization and robustness of the instantaneous classification a 1% gaussian noise were generated using the training patterns. The optimized system was applied to the new samples and the result was 80% of correct classification. Note that the system was not trained using noising data, however it could satisfactorily generalize the classification. The idea is that minimizing the number of partitions lead to more representative (with more patterns) clusters, which centroids are able to accommodate variations in the sample data to be classified.

In order to test the system's ability for "don't know" response, it was submitted to two other transients (not used for training). The reactor trip (TRIPREA) and the feedwater isolation (without trip) (MEFWISEM) were then simulated by the same tool. Applying the system to such transients resulted in 95% of don't know classification for both. Table 1 resumes the results.

Table 1. Instantaneous classification results

Transient	Noise (%)	Classification	Correct classifications (%)
LOCA	0	LOCA	96
SGTR	0	SGTR	100
BLKOUT	0	BLKOUT	98
LOCA	1	LOCA	61
SGTR	1	SGTR	90
BLKOUT	1	BLKOUT	89
MEFWISEM	0	DON'T KNOW	95
TRIPREA	0	DON'T KNOW	95

Evidence Accumulation

As it could be seen, instantaneous classification is reasonably susceptible to noising data. One of the techniques that can be used to enhance reliability on the identification task is the evidence accumulation. The idea is to compute and process sequence of instantaneous classification during a period of time. For example, the evidence of being occurring a given transient, β_k, is increased every time the instantaneous classification response is "β_k."

Defining a decision function d_{β_k} at time instant *i* as:

$$d_{\beta_k}(i) = \begin{cases} 1 & \text{, if the sample belongs to } \beta_k \\ 0 & \text{, otherwise} \end{cases} \tag{8}$$

The evidence accumulation function, E_{β_k}, at time instant *i* can be given by:

$$E_{\beta_k}(i) = \frac{1}{(i+1)} \sum_{j=0}^{j=i} w_i(j)\, d_{\beta_k}(j) \tag{9}$$

where $w_i(j)$ is a weighting factor of instantaneous classification. It may depend on the time of instantaneous classification (for example, reinforcing last classification events), or maybe increasing when consecutive classification events points to the same transient. Such choice will probably depend on the problem.

As the objective here is to illustrate a real identification task, a simple evidence accumulation approach was used. Then, it was used a constant $w_i(j)$ set to 1.0.

Figure 6 shows the graphs of evidence accumulation for the tests described above.

It can be observed that, by the use of the evidence accumulation, the identification becomes more clear. Note that for the training patterns (without noise), after some seconds, the evidence has pointed (confirmed) the correct transient. The same have been occurring with the "unknown" transients (TRIPREA and MEFWISEM) where the evidence has pointed to "don't know".

When considering the 1% noise data, the classification is a little bit confused in the earlier seconds for the BLKOUT and the SGTR, but it could as well reinforce the correct transient evidence. The LOCA, however, although it tends to point to the correct classification, is more confuse and the value of E_{β_k} is not so outstanding. But note that, fortunately, the second great amount of evidence is for the "don't know" classification. Such fact demonstrates that, in this case, the system response seems to be conservative.

5 Conclusions

The use of a possibilistic classification approach coupled with the MCS method, optimized by GA has been demonstrating to be a promising method for transient identification. Its robustness and accuracy was demonstrated in the hypothetical tests where it could find the global optimum solutions.

The effectiveness obtained in the nuclear application for the training patterns was quite good. Robustness has also been demonstrated by the fact that noising data could as well be handled (remember that the training patterns did not include noise). The "don't know" response was properly provided when the transient was out of the scope of the training set. Furthermore, systems response seems to be conservative, in the sense that confusing data (for example, due to excessive noise as in the LOCA with 1%) tend to respond "don't know."

The future investigations are related to the use of slim membership functions in order to better adjust the clusters shape. Multi-objective and parallel genetic algorithms are also intended to be used in order to enhance the quality of the solutions, as well as the capability of the method to solve more complex and bigger problems.

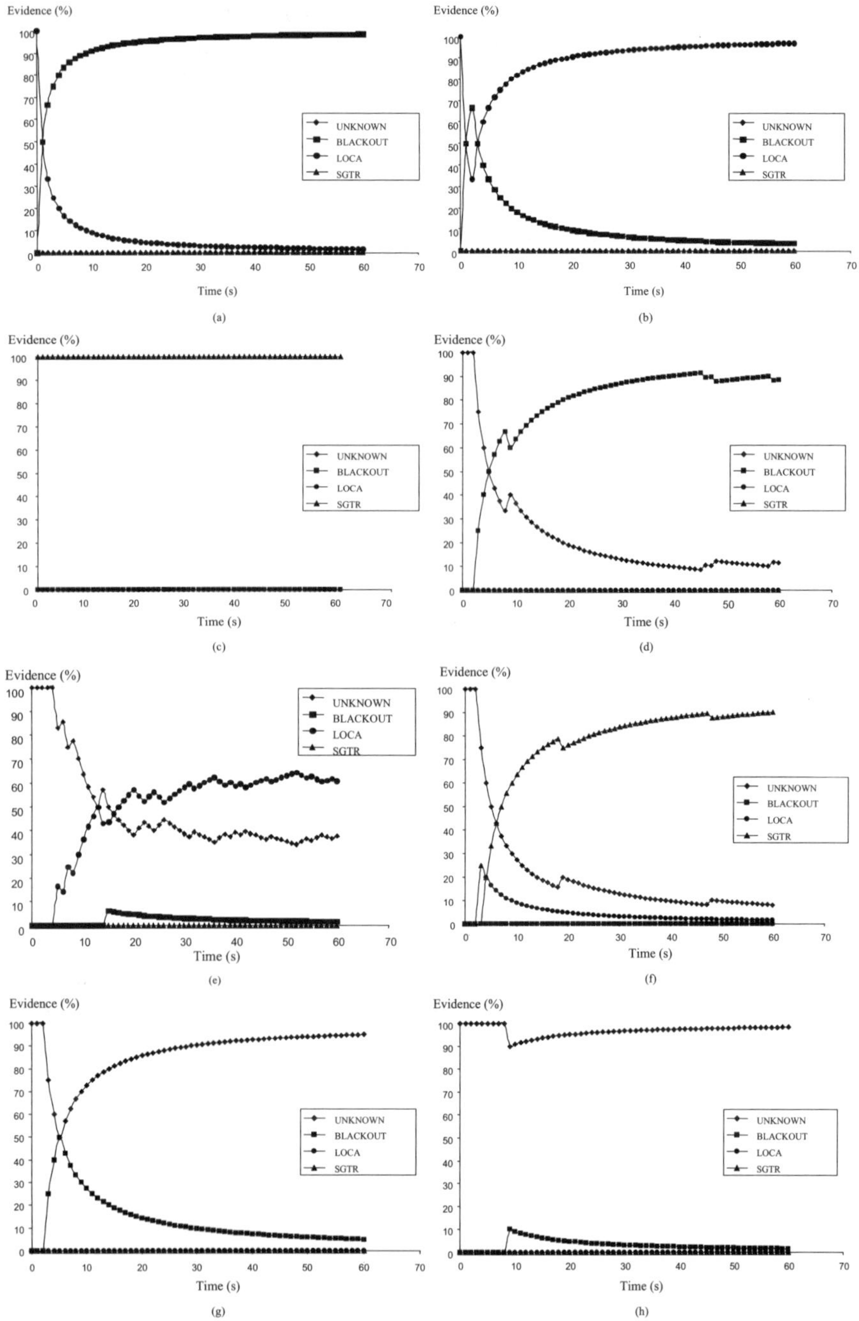

Fig. 6. Evidence accumulation. (a) BLKOUT (without noise); (b) LOCA (without noise); (c) SGTR (without noise); (d) BLKOUT (1% noise); (e) LOCA (1% noise); (f) SGTR (1% noise); (g) MEFWISEN (without noise); (h) TRIPREA (without noise)

References

Bartal Y, Lin J, Uhrig RE (1995) Nuclear Power Plant Transient Diagnostics Using Artificial Neural Networks that Allow "Don't Know" Classifications, *Nuclear Technology*, Vol. 110, (June), pp. 346-449.

Bartlett EB, Uhrig RE. (1992) Nuclear Power Plant Status Diagnostics Using an Artificial Neural Network, *Nuclear Technology*, Vol. 97, pp. 272-281.

Goldberg DE (1989) *Genetic Algorithms in Search Optimization and Machine Learning*, Addison Wesley.

Holland JH (1975) *Adaptation in Natural and Artificial Systems, An Arbor*, University of Michigan.

Krishnapuram R, Keller JM (1993) A Possibilistic Approach to Clustering, *IEEE Transactions on Fuzzy Systems*, Vol. 1, No. 2.

Pereira CMNA, Schirru R (2000) Designing Optimized Pattern Recognition Systems by Learning Voronoi Vectors Using Genetic Algorithms. In: Ebecken, N. F. F (ed), *Data Mining II*, Computational Mechanics Publications, WIT Press, England.

Pereira CMNA, Schirru R, Martinez AS (1998) Learning an Optimized Classification System from a Data Base of Time Series Patterns Using Genetic Algorithms, In: Ebecken, N. F. F (ed), *Data Mining, Computational Mechanics Publications*, WIT Press, England, pp. 20-34.

Schirru R, Pereira CMNA, Martinez AS (1999a) Genetic Algorithms Applied to Nuclear Power Plant Operation, In: Da Ruan (ed), *Fuzzy Systems and Soft Computing in Nuclear Engineering*, Springer-Verlag, pp. 335-350.

Schirru R, Martinez AS, Pereira CMNA., Domingos,RP, Machado MD, Machado L (1999b) Intelligent Soft Computing in Nuclear Engineering in Brazil, *Progress in Nuclear Energy*, 35 (3-4), pp. 367-391.

Zadeh, LA (1978) Fuzzy Sets as a Basis for Theory of Possibility. *Fuzzy Sets and Systems*, Vol. 1, pp.3-28.

Zimmermann HJ, Zysno P (1985) Quantifying Vagueness in Decusion Models. *European Journal of Operational Research*, Vol. 22, pp. 148-158.

20 Regularization of Ill-Posed Surveillance and Diagnostic Measurements

Andrei V. Gribok, J. Wesley Hines, Aleksey Urmanov, and Robert E. Uhrig

The University of Tennessee
Nuclear Engineering Department
Knoxville, TN 37996-2300, USA

rhurig@utk.edu

Most data-based predictive modeling techniques have an inherent weakness in that they may give unstable or inconsistent results when the predictor data is highly correlated. Predictive modeling problems of this design are usually under constrained and are termed ill-posed. This paper presents several examples of ill-posed diagnostic problems and regularization methods necessary for getting accurate and consistent prediction results. The examples include plant-wide sensor calibration monitoring and the inferential sensing of nuclear power plant feedwater flow using neural networks, and non-linear partial least squares techniques, and linear regularization techniques implementing ridge regression and informational complexity measures.

1 Introduction

The accurate knowledge of the state of nuclear power plants is of vital importance for plant safety and efficiency optimization. For the past fifteen years, The University of Tennessee's Nuclear Engineering (UTNE) Department has been applying both conventional and artificial intelligence based predictive techniques to assist in gathering and presenting this information to operators and engineers (Upadhyaya 1985, 1989, 1992, Uhrig 1991). Over the past five years, the focus has been on optimizing the predictive models to assure their accuracy and repeatability using several categories of regularization techniques (Gribok, 2000, 2001, Hines, 1999, 2000, Rasmussen 2000, Urmanov 2001). These techniques range from principal component regression and training with noise for autoassociative neural networks, to ridge regression and complexity regularization for linear models. This paper will present a historical overview of the use of regularization methods in surveillance and diagnostics at UTNE. The methods range from a heuristic approach used with non-linear models such as neural networks, to systematic approaches for linear models, and finally to informational approaches applicable to both linear and non-linear modeling techniques.

2 Ill-Posed Data-Based Prediction

Traditional approaches used to validate plant sensors are operating correctly involve the use of redundant sensors coupled with periodic instrument calibrations. Since few of the sensors are actually out of calibration when calibrated, the end result is that many instruments are unnecessary maintained. An alternative condition-based technique is desirable. When implementing these methods, the instruments are calibrated only when they are determined to be out of calibration. On-line, real-time sensor calibration monitoring identifies faulty sensors which permits reduced maintenance efforts and increases component reliability.

Inferential sensing is the prediction of a sensor value through the use of correlated plant variables. Most calibration monitoring systems produce an inferred value and compare it to the sensor value to determine the sensor status. There are a number of techniques, which were proposed for on-line inferential sensing during recent years. Most notable are neural nets (NN) (Hines, 1997,1998) and multivariate state estimation techniques (MSET) (Gross, 1997, Zavaljevski, 1999). Both of these methods use related sensors as inputs to estimate a model (sets of weights), which is subsequently used to infer the sensor's value based on the input values. A peculiar feature of any on-line sensor monitoring system is that the system should not only accurately infer the sensor's value but it should also be robust to moderate changes in input values. This means that the sensor validation system should resolve a subtle compromise between accuracy and robustness. Recently, the role of regularization in this process was realized (Hines, 1999, 2000, Gribok, 2001, Zavaljevski, 1999).

In data-based surveillance and diagnostic systems, the relationships found in past plant data is modeled for use in making present or future predictions. For example, the inferential measurement of feedwater flow is based on its relationship to other correlated plants parameters. The problem with using other highly correlated parameters as predictors is that they are not only highly correlated with feedwater flow, but they are also correlated with each other. If this degree of correlation is high, the data matrix becomes ill-conditioned and the problem of drift detection becomes ill-posed in the sense of Hadamard (1923). Hadamard defined a well-posed problem as a problem that satisfies the three following conditions:

1. The solution for the problem exists.
2. The solution is unique.
3. The solution is stable or smooth under small perturbations of the data; i.e., small perturbations in the data should produce small perturbations in the solution.

If any of these conditions are not met, the problem is termed ill-posed and special considerations must be taken to ensure consistent reliable solutions. To understand the essence of ill-posed problems, let us consider the linear least squares model whose objective is to find a linear combination of predictor variables **x** that accurately models the response variable **y**.

The ordinary least squares (OLS) solution is given by:

$$w=(X^T*X)^{-1}*X^T*y \tag{1}$$

which is the minimizer of the ordinary least squares functional:

$$E(w) = (y-Xw)^T(y-Xw) \tag{2}$$

We can see that this solution involves the inversion of what is commonly called the Hessian matrix ($X^{T*}X$). If the predictors (X) are highly correlated, this matrix is ill-conditioned and slight changes in the predictor data cause significant changes in the solution weights. A measure of condition is the ratio of the largest eigenvalue to the smallest eigenvalue. The numerical inversion of ill-conditioned matrices causes unstable solutions means that the problem is ill-posed. The difficulty of constructing prediction models with correlated data is not constrained to linear regression models but also occurs, with even greater instabilities, in non-linear techniques such as neural networks. The following sections will present several ill-posed surveillance and diagnostic problems, the difficulties that accompany them, and methods used to produce reliable models.

3 Heuristic Regularization Approaches

Some of our earliest dealings with ill-posed problems occurred during our work with developing a neural network based sensor monitoring system using Florida Power Corporation's Crystal River Nuclear Power Plant data. This system used an autoassociative neural network: one in which individual outputs are trained to emulate the corresponding inputs over an appropriate dynamic range. The signals from several (20-30) correlated plant sensors constitute the inputs and are used to create best estimates for each sensor. The scheme is to have each sensor estimate predicted by the values of other sensors that are related (correlated) to it. As a result, any specific network output should show virtually no change when a corresponding input has been distorted by noise, faulty data, or missing data. The difficulty is to force the network model to rely equally on all of the pertinent data rather than just on a few inputs or the corresponding input alone. Such a network is said to be robust and its robustness can be visualized through a simple sensitivity analysis plot.

Sensitivity is defined as the change of the prediction over the corresponding change in a predictor input. The sensitivity of the network is shown by cumulative bar plots that show the change in each of the networks outputs due to a 5% perturbation applied sequentially to each of its inputs. The upper plot of Figure 1 shows the sum of all output changes for each signal perturbation. For the unregularized solution, a perturbation of 5% in each input of the network results in a cumulative change in the outputs of between 10% and 900% with the contributions to the overall change in each output varying greatly for different inputs. The desired solution would have each prediction depending equally on the inputs that are correlated with it.

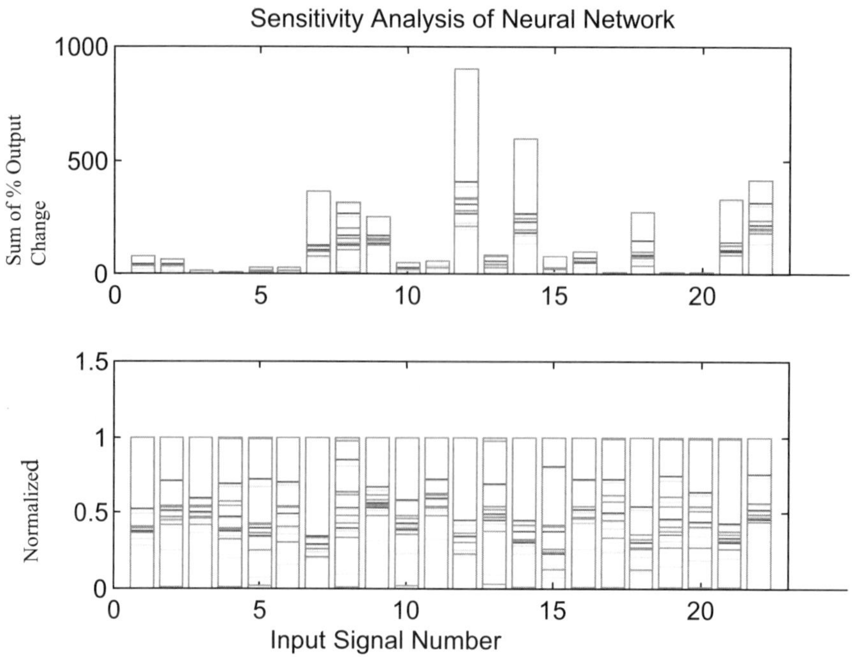

Fig. 1. Sensitivity analysis without regularization

3.1 Robust Training

To obtain what we term a robust solution, one that is equally dependent on all inputs and thus equally resistant to a faulty input, we used common sense and intuition to create a training data set with each input having artificially inserted faults of equal magnitude. We trained the network so the output would be accurate even though the input was distorted. This caused the network to rely on all inputs and not be overly dependent on a single input. We later discovered that this is a regularization technique known as "training with jitter." A valuable discussion of this technique can be found in Chapter 7 of *Neural Smithing* (Reed 1999).

The sensitivity results for a network trained with jitter (see Figure 2) showed considerable improvements over the non-robustly trained network. Each of the 22 sensors predicted by the network respond almost equally to each 5% input perturbation with the output cumulative change varying from 2% to 5%. When normalized, the sensitivities due to each input are approximately the same. The robustly trained network is stable to perturbations in the inputs; hence the robust network produces accurate predictions when the inputs contain faulty sensor signals.

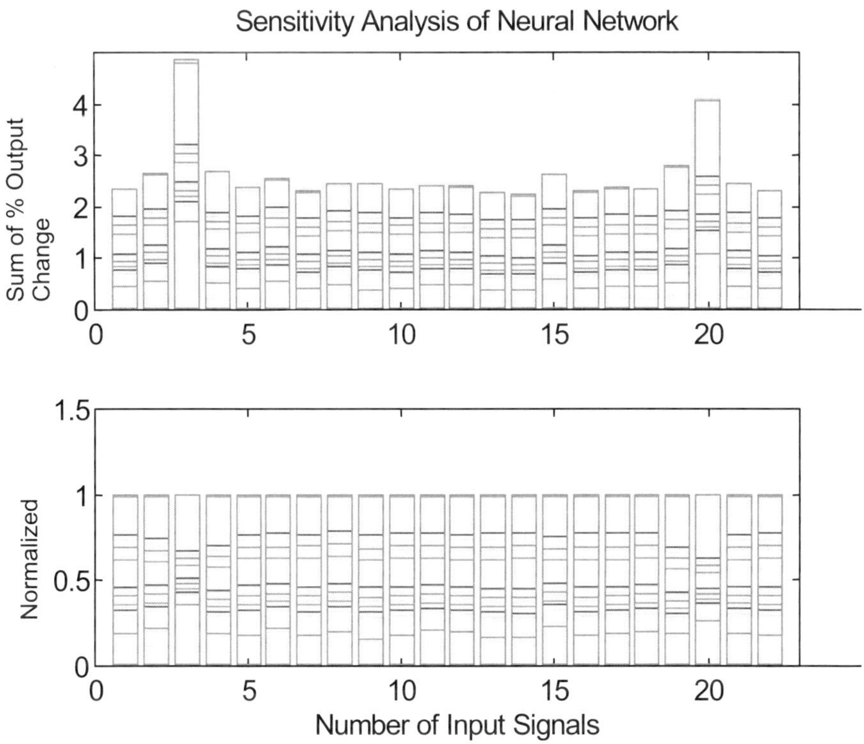

Fig 2. Sensitivity analysis with regularization

3.2 Network Architecture with Inherent Regularization Properties

The AANN architecture is a five-layer feedforward network proposed by Kramer (1991) and shown in Figure 3. It consists of an input layer, 3 non-linear hidden layers, and a linear output layer. The first of the hidden layers is the *mapping layer* with dimension greater than the number of input/outputs. The second hidden layer is called the *bottleneck layer* and serves to break up the input signals so that none of them directly "pass through." The dimension of this layer is required to be the smallest in the network. The third hidden layer is called the *demapping* layer and has the same dimension as the mapping layer. Kramer points out that five layers are necessary for such nets in order to model non-linear processes. This architecture is sometime called a non-linear principal component analysis (NLPCA) network because the outputs of the bottleneck layer may be thought of as the nonlinear principal components of the data when the architecture is properly defined. This architecture has regularization properties similar to principal component regression and tends to make the outputs use the input information equally.

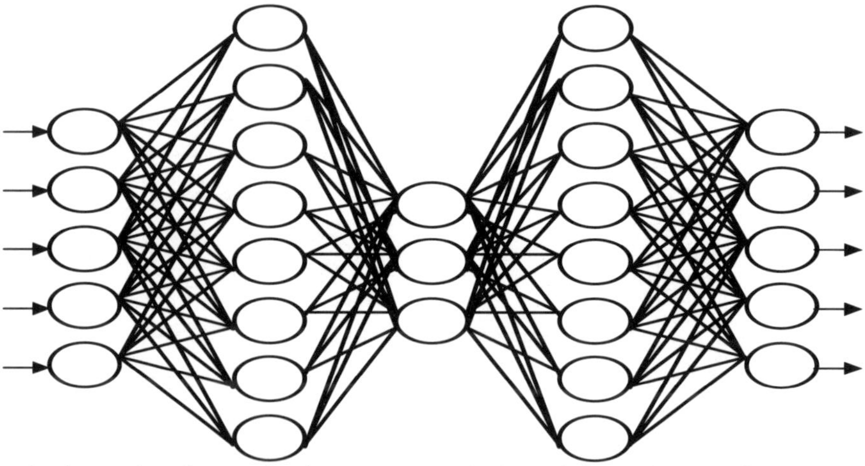

Fig. 3. Five layer auto associative neural network architecture

The last regularization methods used in the AANN based system is called principal component regression or truncated singular value decomposition. Because the activation functions of the neurons of the output layer are linear, the training of their weights can be speeded up by performing an ordinary least squares (OLS) solution. Without this step, this neural network training is extremely slow due to the high number of hidden layers that must be trained. Performing an OLS solution increased training speed by a factor of ten; however, it resulted in very large output weights and poor generalization performance. This is a common result of solving ill-posed problems with OLS and will be discussed in more detail in the next section. To regularize the solution, we performed a truncated singular value decomposition (TSVD) solution, which resulted in a robust network. More information on this method can be found in Hansen (1989).

3.3 Conclusions to Heuristic Approaches for AANN Development

Several methods were used to obtain consistent, accurate, and robust results from the AANN design; although no single method, or any combination of methods, provides theoretically defendable results (Hines 2000). At the time of this research (Hines 1997a, 1997b), the methods were not identified as regularization methods and were not applied in a systematic regularization framework. Subsequent research identified the true nature of this ill-posed problem and formal regularization techniques were used for its solution (Hines 1999).

An example result of using an AANN for plant-wide sensor calibration modeling is given in Figure 4. The sensor with an artificially induced drift and its best estimate are plotted in the upper portion. It should be noted that the actual signal is fairly noisy and the ANN estimate is fairly noise free. This occurs due to the canceling of the noise contained in each of the 22 predictor variables. The middle section shows a plot of the residual between the sensor and its estimate, and the lower section show a plot of the Sequential Probability Ratio Test (SPRT) decision logic output. When its value goes high, the residual has been determined to be non-zero and the signal is determined to be faulted. This fault was identified by the time it reached 0.2%.

3.4 Non-Linear Partial Least Squares

After we realized the necessity for solving ill-posed problems using regularization procedures, we investigated non-linear techniques that have inherent regularization properties. One of these techniques, non-linear partial least squares (NLPLS), was already being utilized in the Chemical Engineering field for inferential sensing (Qin 1992). Altering it for autoassociative sensing was straightforward [Rasmussen 2000a] and we compared its performance with the AANN (Rasmussen 2000b, Hines 2001). It was determined that the NLPLS method is superior to AANN when the relationships are mostly linear. However, highly non-linear systems should still use the AANN design. The NLPLS method also does not require grouping the variables into sets of size 20-30. When monitoring small sets of sensors (<30), this is not a concern for AANN. Lastly, the NLPLS model architecture development is completely automated and without local minima, whereas the AANN design is very heuristic, requires oversight, and local minima in the training error surfaces can cause poor results.

4 Systematic Approach to Regularization

Whereas well-structured approaches for regularizing neural networks do not exist, systematic approaches for regularizing linear systems are available. This section describes a systematic approach to regularizing linear models and these approaches can also be used for non-linear kernal regression models such as MSET and radial basis function networks.

The problems, which arise in inferential sensing, are twofold. First, the use of correlated sensors as inputs to an inferential system leads to a significantly non-uniform eigenvalue spectrum of the data matrix. This causes some weights to be unreasonably large and hence decreases the system's robustness to minor input perturbations. The second problem is that these techniques, being statistically based, can significantly overfit the training data set, again providing poor performance on the test data. Regularization techniques can remedy both of these problems and produce a stable and accurate estimator. Currently, neural nets, as well as MSET, can be regularized and include regularization parameters either in an explicit form (weight decay parameter for NN) or in both explicit and implicit forms (regularization parameter and kernel width for MSET).

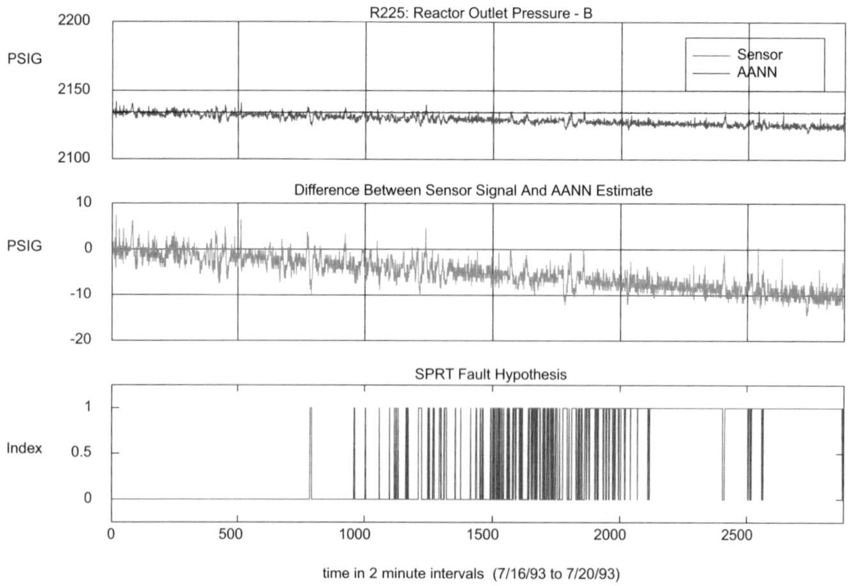

Fig. 4. Demonstration of detection of simulated calibration drift of 0.1% per day of the instrument's maximum scale deflection in reactor outlet pressure

To illustrate the problems with the ordinary least squares (OLS) solution for sensor validation, let us consider a simple linear multiple regression model fit to a set of plant data. The linear approach is taken for the sake of simplicity, but the problem also exists when using nonlinear models. Eighty-two variables recorded at a fossil power plant and arranged in a data matrix X (1000 samples x 82 inputs), are used as predictor variables to infer the value of a response variable Y, which is the sensor under surveillance. The amount of correlation in the data matrix can be quantified through the condition number (ratio of the maximum eigenvalue to the minimum eigenvalue) of the correlation matrix (Belsley, 1980). In this case the condition number is 550,000, which is an unacceptable value; acceptable values should be significantly less than 100. The OLS solution is shown in Figure 5.

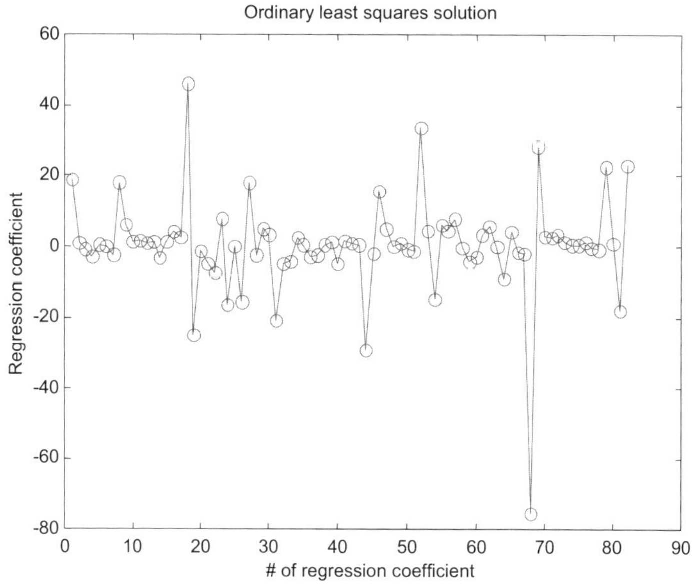

Fig. 5. Regression coefficients produced by the OLS solution

We see that the solution has a number of large magnitude weights and inference would rely heavily on these weights to make predictions on future unseen data. As a result of the highly correlated data, the confidence intervals on the regression coefficients are very wide and the values are not statistically significant. Using the OLS model to make predictions on new unseen data (see Figure 6), we can see that after some period of decent performance the prediction becomes horrible. The reason for this is that one predictor sensor: #53, changed its behavior drastically as shown in Figure 7, thus changing the prediction.

This example shows that although the OLS solution may fit the training data quite well, it may not perform well in the future because of the lack of robustness and stability. Dropping signal #53 from the list of predictor variables does not help because other sensors having non-stationary behavior would also adversely affect the predictions. As has been pointed out, the main reason for unstable, non-robust prediction is the statistical inconsistency of the regression coefficients obtained from highly correlated data sets.

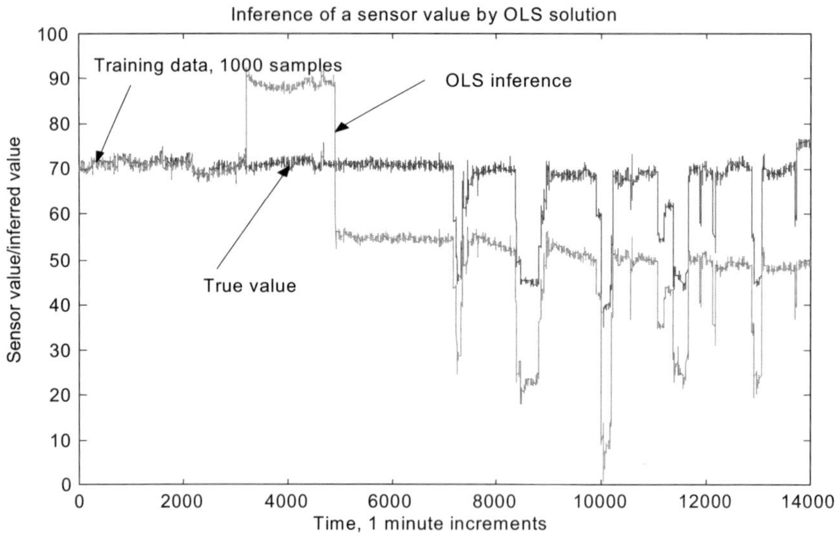

Fig. 6. Inference of a sensor value using OLS solution

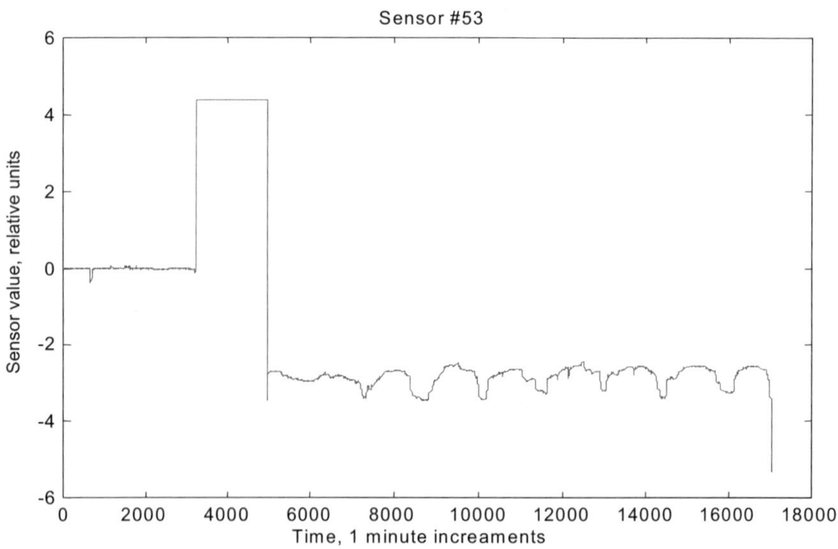

Fig. 7. Plot of sensor #53

The simplest form of regularization, known as ridge regression (RR), can be used to stabilize the solution. The ridge solution is calculated as (Hoerl, 1970)

$$w = (X^T * X + \lambda I)^{-1} * X^{T*} Y \tag{3}$$

which is the minimizer of the following regularized functional:

$$E(w) = (Y - Xw)^T (Y - Xw) + \lambda w^T w \tag{4}$$

Ridge regression is a shrinkage estimator: it shrinks large regression coefficients towards zero simultaneously reducing their variance, thus making them statistically significant. The parameter λ is called the ridge or regularization parameter and its optimal value has to be estimated from the data. The prediction on the previous set of data using ridge regression is shown in Figure 8.

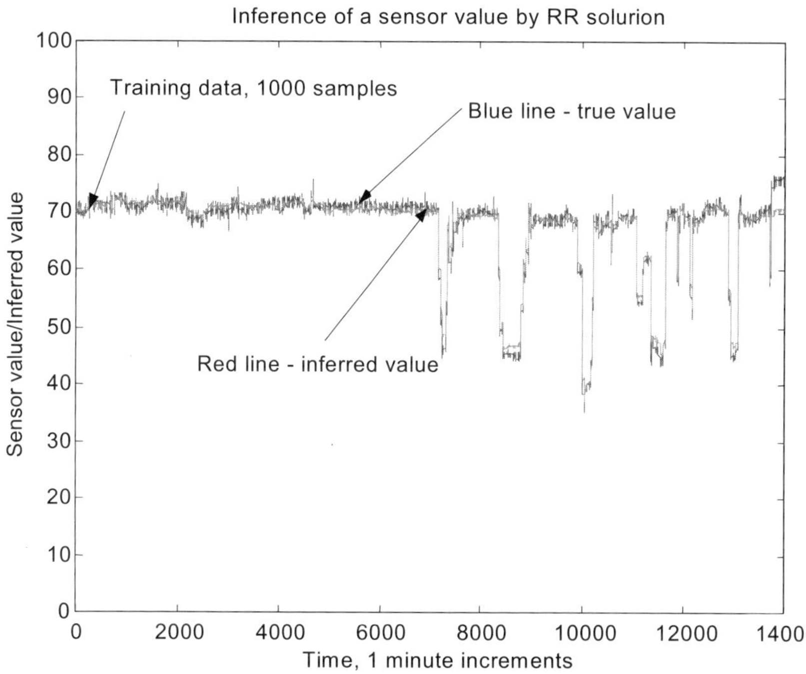

Fig. 8. Inference by regularized solution

As we can see the prediction is very stable and was not affected by changes in sensor # 53 (compare to Fig. 5). This simple but instructive example shows the benefits which regularization provides for sensor validation systems. However, ridge regression is a linear technique and can be used effectively only when the system under consideration is essentially a linear one. These techniques can be extended to work for kernal regression techniques such as MSET and Radial Basis Function networks.

5 Informational Approach

In the previous section we assumed that the solution of the problem is unique and we made it stable using regularization techniques. Now we are going to relax that assumption and consider the uniqueness of the solution. To demonstrate the informational approach to uniqueness, we use the venturi meter drift detection problem.

In the United States, the operating limit of a nuclear power generating plant is tied to the plant's thermal power production. The majority of pressurized water reactors (PWR) utilizes venturi meters to measure feedwater flow rate, which is used to estimate reactor thermal power. However, venturi meters are susceptible to measurement drift due to corrosion products building up near the meter's orifice. This fouling increases the measured pressure drop across the meter, which in turn results in an over-estimation of the flow rate. Consequently, reactor thermal power is also overestimated. To stay within regulatory limits, reactor operators have to derate their plants or justify a compensating process. On average, the amount of derating varies from insignificant to 3% of full power. For example a derating of 2% in an 800 MWe unit will cost the utility ~$20,000 per day. To overcome this problem, an inferential sensing system has been developed at the University of Tennessee to infer the "true" feedwater flow rate (Hines, 1999; Gribok, 2001). Twenty four plant variables have been selected as predictor variables, based on engineering judgment and their high correlation with feedwater flow rate and a linear regression model has been chosen as a predictive tool. The predicted value of feedwater flow rate is not affected by fouling because the linear regression model is built on the data corresponding to the initial operation time period before fouling starts. The difference between the predicted value and the actual measurement defines a drift due to fouling. Using regularization methods, a stable feedwater flow rate estimation was achieved, Gribok (2001).

One of the simplest examples of non-unique solution is variable selection. In regression we have a number of potential input variables (regressors) and have to decide which ones should be used in a model to predict the output or response variable. It is not unusual to have from 10 to hundreds of regressor signals with some being redundant sensors, and others highly correlated. Each particular combination of regressors yields some solution. Different combinations produce different solutions. This describes the non-uniqueness of solutions. In Figure 9, we demonstrate how different variable subsets can provide different predictions of the drift value.

When we understand the physical data generating process and have an analytical expression for the dependence of the response on the input variables, we can choose a physically meaningful combination. However, in practice, it is common for the analytical expression to be either unknown or unavailable, and the problem of variable selection becomes challenging. For example, to choose the best combination of variables among 24 variables for prediction of the drift, we have to try all 2^{24} = 16,777,216 possible variable subsets. This number is prohibitively large. On the other hand, we do not want to use all available input variables because in most applications prediction accuracy is not improved by simply using all available predictors, more often the opposite effect is achieved (Linhart, 1986). In other words, the prediction accuracy of the full model is worse (or at least not better) than those of the subset models because the variance of the regressor coefficients increases monotonically with the number of variables used in prediction (Miller, 1990)

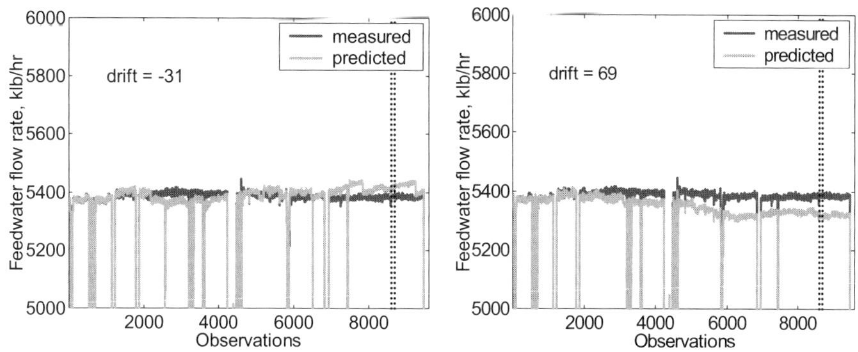

Fig. 9. Predicted values of the venturi meter drift can vary from negative (left) to positive (right), depending on the variable subset chosen for prediction. This demonstrates non-uniqueness of the solution

In addition, we have to find some criterion for assessing these combinations or models. The simplest criterion is the sum of squared errors, which is also termed the explanation error. However, it has been recognized that this criterion produces unacceptable results and is overoptimistic. It favors the model including all possible input variables, including irrelevant and redundant ones. This happens because any additional variable can have its corresponding coefficient set to zero with the explanation error staying the same. In practice, with finite data sets, the explanation error will always be minimal for the full combination. This degrades the prediction error (the error in predicting future observations) because the model fits or learns not only the informative part of the data but also noise and accidental dependencies between the response and the irrelevant input variables included in the model.

Technically speaking, the explanation error is a biased estimate of the prediction error. Since in many applications the ultimate goal is to get good predictive abilities on unseen data, we need to use the prediction error as the criterion in choosing the unique solution. Unfortunately, the prediction error cannot be calculated because of the unknown data-generating process and/or unavailable future observations at the moment of model building. However, it can be successfully estimated by using the explanation error and correcting for its bias. One method to correct for the bias is termed complexity penalization, which penalizes the excessive complexity (number of variables) of the model.

Complexity is a relatively simple and intuitive framework for model selection. A heuristic justification of using complexity is that in real life simple events are much more probable to happen than more complex ones. Consequently, when collecting data generated by some unknown process, it is reasonable to assume that among possible underlying phenomena responsible for data generating, the simplest one is the most probable. This is an interpretation of the famous Occam's razor, which says that among equally good explanations the simplest one is most probable. In statistical model selection, complexity methods became useful when more powerful computers became available and researchers received an opportunity to try more complicated models. It soon became evident that highly complex models do not perform better. Due to their excessive complexity and flexibility, they not only fit the informative part of the data but also fit the noise. This drastically reduces the precision of model parameter estimation and produces very unstable results. Therefore, we must be able to intelligently control the complexity of models. Ideally, we must use models of specified complexity depending on the particular application (Vapnik, 1995), but it is extremely difficult to evaluate the needed complexity of the model to solve a given problem.

To avoid this difficulty, researchers use so-called complexity penalizing methods that tradeoff the complexity of models against the explanation ability. Such methods were developed for model selection in regression analysis, in principal component and ridge regression, for architecture selection of artificial neural network models and in other fields. Model complexity can be assessed in many ways. The two most popular methods are descriptional complexity, due to independent works of Kolmogorov, Solomonoff, and Chaitin (Li, 1997); and covariance complexity, due to Emden (1971). Descriptional complexity is not computable but can be successfully approximated. Examples of using descriptional complexity are the minimum description length (MDL) criteria by Rissanen (1989), and the minimum message length (MML) by Wallace (1987). Covariance complexity is similar to the system theory definition of complexity and states that the complexity is related to the degree of interdependency between different components of the system, such as inputs, or the parameter estimates in statistical model selection. The covariance complexity criterion is termed information complexity (ICOMP) and was developed by Bozdogan (1996).

The usual form of a complexity-based model selection criterion is:

Prediction Error (Model) = Explanation Error (Model) + Complexity (Model)

and an example of typical behavior of the terms in the criteria is shown in Figure 10. There are other criteria also in such form: AIC (Akaike, 1973), GIC (Konishi, 1996), SIC (Schwaz, 1978) to name a few. Once we decide which model selection criterion we should use, the model is chosen to minimize the criterion value. Such a model will have smaller prediction errors on future data than those models corresponding to larger values of the criterion.

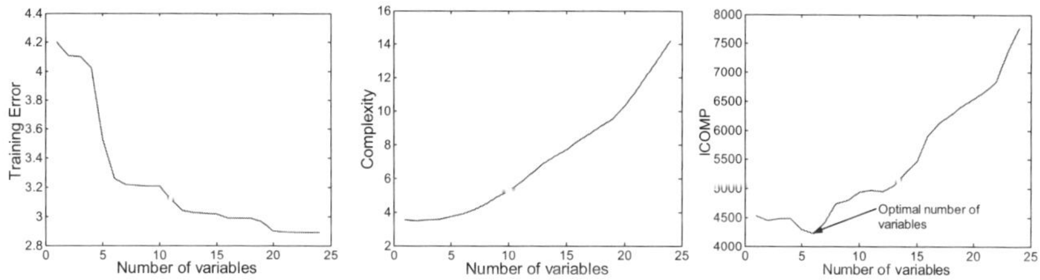

Fig. 10. As variables are added to the model, training error decreases (left), the complexity of the model increases (center), and the value of the complexity-based model selection criterion ICOMP (right)

Figure 10 (right) shows that the optimal model has a minimum ICOMP value for 6 input variables. This 6-variable model will have the smallest prediction error on unseen data although the 24 variable model has the minimum training error.

The MDL criterion chose the 7-variable model to be optimal. This model was deemed to have the best balance between model's complexity and explanation ability (Urmanov, 2000) and hence the minimal prediction error. The values of the MDL criterion for different number of variables are shown in Fig. 11.

In Figure12 (left) we show the estimated probability densities of the drift predicted by using the 7-variable subset model and the full model (all 24 variables). As mentioned above, the variance of the 7-variable model prediction is less than that of the full model prediction. This results in more accurate predictions, but obviously some bias has been introduced. Complexity-penalized model selection techniques determine when to stop trading off reduced variance against increased bias. The 7-variable subset model represents a unique solution for the drift prediction that gives the drift value of 35.7 klb/hr. Yet, the solution is still unstable and must be stabilized using regularization techniques described in the previous section. For example, a stabilized solution using ridge regression is shown in Fig. 12 (right). When using ridge regression and other stabilization techniques, we must chose the regularization parameter value which controls the degree of regularity (or stability) of the solution. Regularization parameter selection can be performed by several methods: Mallows' (1973) CL, GCV (Golub, 1979), the L-curve (Hansen, 1992), and Morozov's (1966) Discrepancy Principle. The ridge regression prediction using the 7-variable subset model is very stable; a variance reduction of 20 times is achieved. However, the stabilization introduces a bias relative to the unstable solution obtained by using this 7-variable subset model and results in a drift value of 26.9 klb/hr. Please note that because the true value of drift is not known, the bias is just the difference between the two solutions, not an error from the true solution.

It is not unusual for different model selection methods to give different optimal models. This poses another tough problem of the non-uniqueness of the solution with respect to a model selection method. In other words, we have to decide which model among those chosen using different model selection methods has the best performance on unseen data.

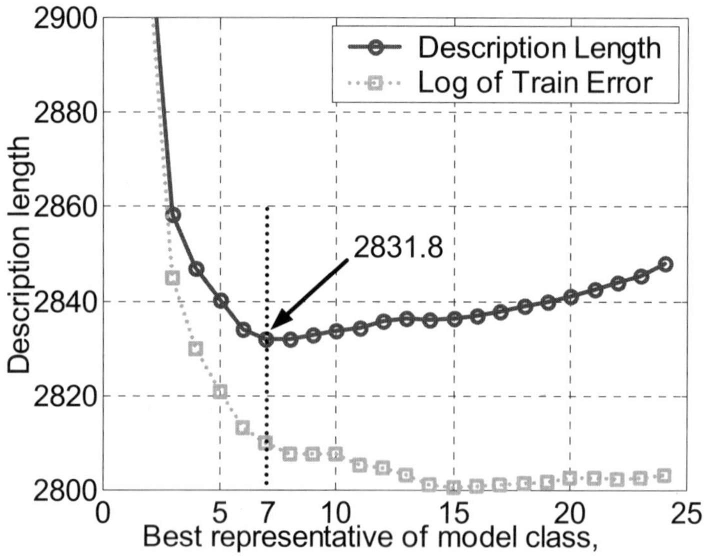

Fig. 11. Minimum description length (MDL) achieved by using models with different number of variables

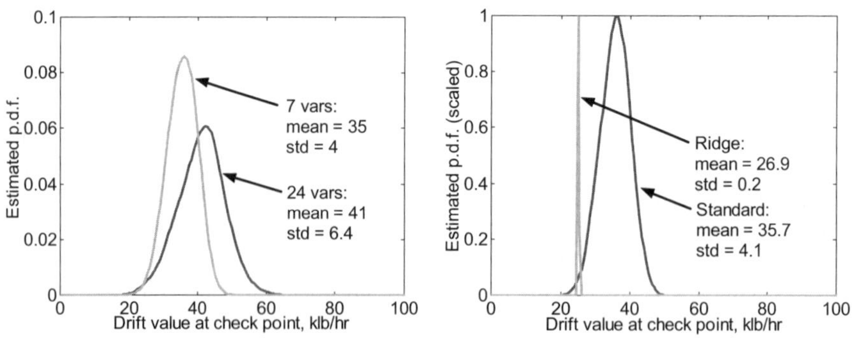

Fig. 12 Estimated predictive probability densities of the venturi meter drift for a 7-variable regression model without stabilization (left) and with stabilization (right)

Using complexity-based model selection with model regularization, we obtained a unique stable solution for prediction of the feedwater flow sensor drift. The resulting solution is biased from one given by the full model including all 24 variables. However, since the true drift value is unknown either drift value can be closer to the true one. The benefit is that the unique solution can be proven to give you better predictions on unseen data and the regularization gives you repeatable solutions.

6 Conclusions

This paper presented a historical overview of data-based prediction techniques applied to sensor monitoring at The University of Tennessee. Early work focused on heuristic techniques used to get accurate stable predictions using Autoassociative Neural Networks. These techniques were not theoretically founded and models with inherent regularization properties, such as Non-Linear Partial Least Squares, were used. Next, systematic approaches to regularization such as ridge regression were investigated and shown to provide consistent results. Finally, informational methods were used to solve the problem of uniqueness and select the optimal inputs for a model. The techniques, when used in combination, provide a viable framework for correctly solving ill-posed data based prediction problems.

References

Akaike H (1973) Information theory and an extension of the maximum likelihood principle, in 2nd International Symposium on Information Theory, Ed. B.N. Petrov and F. Csaki, pp. 267-281. Budapest: Akademiai Kiado

Belsley DA, Kuh E, Welsch RE (1980) Regression Diagnostics: Identifying Influential Data and Sources of Collinearity, John Wiley @ Sons, New York

Bozdogan H (1996) Informational complexity criteria for regression models, Information Theory and Statistics Section on Bayesian Stat. Science. ASA Annual Meeting, Chicago, IL, Aug. 4-8

Golub GH, Heath M, Wahba G (1979) "Generalized cross-validation as a method for choosing a good ridge parameter," *Technometrics*, **21**, No. 2, pp. 215-223

Gribok AV, Hines JW, Uhrig RE(2000) "Use of Kernel Based Techniques for Sensor Validation in Nuclear Power Plants," American Nuclear Society International Topical Meeting on Nuclear Plant Instrumentation, Controls, and Human-Machine Interface Technologies (NPIC&HMIT 2000), Washington, DC, November

Gribok AV, Attieh I, Hines JW, Uhrig RE (2001) "Regularization of Feedwater Flow Rate Evaluation for Venturi Meter Fouling Problems in Nuclear Power Plants," *Nuclear Technology*, **134** *pp 3-14*

Gross KC, Singer RM, Wegerich SW, Herzog JP, Van Alstine R, Bockhorst FK (1997) "Application of a Model-based Fault Detection System to Nuclear Plant Signals," Proc. of the Intl. Conf. on Intelligent System Application to Power Systems, Seoul, Korea. pp.60-65

Hansen PC (1989) *Regularization, GSVD and Truncated GSVD*, BIT 29, pp. 491-504

Hansen PC (1992) "Analysis of discrete ill-posed problems by means of the L-curve," SIAM Review, Vol. 34, No. 4, pp. 561-580

Hines JW, Uhrig RE, Black C, Xu X (1997) "An Evaluation of Instrument Calibration Monitoring Using Artificial Neural Networks," published in the proceedings of the 1997 American Nuclear Society Winter Meeting, Albuquerque, NM, November 16-20

Hines JW, Uhrig, RE, Wrest DJ (1998) "Use of Autoassociative Neural Networks for Signal Validation," *Journal of Intelligent and Robotic Systems*. 21:143-154

Hines JW, Gribok AV, Attieh I, Uhrig RE (1999) "Regularization Methods for Inferential Sensing in Nuclear Power Plants," *Fuzzy Systems and Soft Computing in Nuclear Engineering*, Ed. Da Ruan, Springer, 1999

Hines JW, Gribok AV, Attieh I, Uhrig RE (2000) "Neural Network Regularization Techniques for a Sensor Validation System," American Nuclear Society Annual Meeting, San Diego, California, June 4-8, 2000

Hines JW, Rasmussen B (2001) "On-Line Sensor Calibration Verification: A Survey," Proceedings of the 14th International Congress and Exhibition on Condition Monitoring and Diagnostic Engineering Management, Manchester, England, September, 2001

Hoerl AE, Kennard RW (1970) "Ridge regression: biased estimation for nonorthogonal problems," *Technometrics*, 12, pp.55-67

Konishi S, Kitagawa G (1996) "Generalized information criteria in model selection," *Biometrika*, **83**, No. 4, pp. 875-890

Kramer MA (1991) "Nonlinear Principal Component Analysis Using Autoassociative Neural Networks," *AIChE Journal*, Vol. 37, No. 2, pp. 233-243

Li M, Vitanyi P (1997) *An introduction to Kolmogorov complexity and its applications,* Springer-Verlag New York, Inc.

Linhart H, Zucchini W (1986) *Model Selection,* Wiley Series in Probability and Mathematical Statistics, John Wiley & Sons, Inc.

MacKay D (1992) "A practical Bayesian framework for backprop networks," *Neural Computation*, 4: pp. 448-472

Mallows CL (1973) "Some comments on C_P," *Technometrics*, **15**, No. 4, pp. 661-675

Miller AJ (1990) *Subset selection in regression,* Monographs on Statistics and Applied Probability, Chapman and Hall

Morozov VA (1966) "On the solution of functional equations by the method of regularization," *Soviet Math. Dokl.*, **7**, pp. 414-417

Qin SJ, McAvoy TJ (1992) "Nonlinear PLS Modeling Using Neural Networks," *Computers in Chemical Engineering*, **16**, No. 4, pp. 379-391

Rasmussen B, Hines JW, Uhrig RE (2000a) "Nonlinear Partial Least Squares Modeling for Instrument Surveillance and Calibration Verification," published in the proceedings of the Maintenance and Reliability Conference (MARCON 2000), Knoxville, TN, May 7-10

Rasmussen B, Hines JW, Uhrig RE (2000b) "A Novel Approach to Process Modeling for Instrument Surveillance and Calibration Verification," The Third American Nuclear Society International Topical Meeting on Nuclear Plant Instrumentation and Control and Human-Machine Interface Technologies, Washington DC, November 13-17, 2000

Reed RD, Marks RJ (1999) *Neural Smithing*, MIT Press, Cambridge MA, Chapter 7

Rissanen J (1989) *Stochastic complexity in statistical inquiry,* World Scientific Publishing Co. Pte. Ltd.

Schwaz R (1978) "Estimating the dimension of a model," *Ann. Statist.* **6**, 461-464

Uhrig RE (1991) "Potential Application of Neural Networks to the Operation of Nuclear Power Plants," *Nuclear Safety*, **32**, No. 1. pp 68-79

Upadhyaya BR (1985) "Sensor Failure Detection and Estimation," *Nuclear Safety*

Upadhyaya BR, Holbert K (1989) "Development and Testing of an Integrated Signal Validation System for Nuclear Power Plants," DOE Contract DE-AC02-86NE37959, 1989

Upadhyaya BR, Eryurek E (1992) "Application of Neural Networks for Sensor Validation and Plant Monitoring," *Nuclear Technology*, **97**, 170-176

Urmanov A, Gribok A, Hines JW, Rasmussen B (2001) "Application of Information Complexity in Principal Component Regression Modeling of the Venturi Meter Drift," published in the proceedings of the *Maintenance and Reliability Conference (MARCON 2000)*, Knoxville, TN, May 6-9

Urmanov AM, Gribok AV, Hines JW, Uhrig RE (2000) "Complexity-penalized model selection for feedwater inferential measurements in nuclear power plants," ANS International Topical Meeting on Nuclear Plant Instrumentation, Controls, and Human-Machine Interface Technologies (NPIC&HMIT 2000), Washington, DC

van Emden MH (1971) "An analysis of complexity," *Mathematical Centre Tracts*, **35**, Amsterdam

Vapnik VN (1995) *The nature of statistical learning theory,* Springer-Verlag New York, Inc.

Wallace CS, Freeman PR (1987) "Estimation and inference by compact coding," *J.R.Statst. Soc.* B, **49**, pp. 240-252

Zavaljevski N, Gross KC, Wegerich SW (1999) "Regularization Methods for the Multivariate State Estimation Technique (MSET)," Proc. of the Int. Conf. on Mathematics and Computations, Reactor Physics and Environmental Analysis in Nuclear Applications, September 27-30 1999, Madrid Spain, pp.720-729

21 Application of Neuro-Fuzzy Logic for Early Detection and Diagnostics in Gas Plants and Combustion Chambers at ENEA

Massimo Sepielli[†], P.Carmelo Incalcaterra[†], Massimo Presaghi[†], Paolo F. Fantoni[‡], and Davide Roverso[‡]

[†] ENEA (Ente Nazionale per Energia, Innovazione e Ambiente)
P.O. Box 00060, Rome - Italy

[‡] Institutt for energiteknikk, OECD Halden Reactor Project
P.O. Box 173, N-1751 Halden – Norway

sepielli@casaccia.enea.it

1 Introduction

This chapter presents the work done in collaboration between ENEA and the OECD Halden Reactor Project (HRP), which resulted in the application of HRP neuro-fuzzy tools- PEANO [1, 2, 3, 4] and ALADDIN [5, 6] to problems of gas plant control, either related to measurement calibration or to process instability.

The former part of the chapter describes results and experiences obtained using the PEANO toolbox for validation of critical process sensors at ENEA's ICARO co-generation plant. PEANO's neuro-fuzzy techniques are able to track in real-time the expected behaviour of a complex process, both in steady state and transient conditions. Fuzzy and possibilistic clustering algorithms are used to classify the operating region where the validation process has to be performed. The possibilistic approach used (rather than probabilistic), allows a "don't know" classification, which results in a fast detection of unforeseen plant conditions or outliers. Specialized Artificial Neural Networks (ANN) are used for the validation process, one for each fuzzy cluster in which the operating map has been divided. They work concurrently on each signal pattern presented to the system and the overall contribution is weighted according to the membership function of the pattern in each cluster.

The latter part of the chapter describes the work with the transient identification tool ALADDIN, which has been tested as one of the ANN techniques to be applied to the pattern recognition of flame instability statuses in industrial combustion chambers, in the framework of an Italian national research program on fossil fuel combustion. In particular, acoustic waves have been investigated as input signals for ALADDIN to recognize the flame behavior in the combustion process.

2 Neural Network and Fuzzy Logic Application to Signal Validation

The operation of industrial plants is based on the readings of a set of sensors. Their reliable operation is essential as the output of sensors provides the only objective information from the process. The task of signal validation is to confirm whether the sensors are functioning properly. Faulted or miscalibrated instrumentation channels might lead to the following problems:

- Erroneous identification and diagnosis of abnormal events, which results in possible human errors by the operators in control room.
- Process uncontrollability and instability, when sensors are connected to control and automation systems, resulting in emergency shutdown of the entire process.
- Reduced plant performance and efficiency.

It must be pointed out that these three points might have safety implications in some cases, but they have always negative economical consequences, due to forced shutdowns in the first two scenarios and losses of efficiency in the third one.

Signal validation must be robust to handle multiple sensor faults as well. This requirement is important especially in case of common cause failures in instrumentation channels (for example, common sensing lines in pressure transmitters), which would result in a completely wrong understanding of the plant state.

The signal validation tool PEANO is based on a fuzzy classifier and a set of Artificial Neural Networks (ANNs). The classifier is based on fuzzy and possibilistic clustering techniques and it is trained to identify the incoming signal pattern (a snapshot of process signals) as a member of one of the possible categories (clusters) in which the operating space has been divided. Each cluster is associated with one ANN previously trained only with data belonging to this cluster. During the operation, the classifier provides an automatic switching mechanism to allow the best-tuned ANN to be used. The maximum membership grade of the sample in the particular cluster and the maximum signal mismatch in the neural network module are fed into a Mamdani type fuzzy model to estimate the reliability level of the validation.

The use of neuro-fuzzy models for signal validation has some advantages. The most important are:

- It is not necessary to develop physical models of the monitored process.

- They are less sensitive to measurement noise when compared to physical models.
- They adapt well to non-linear processes (as most real processes are).

2.1 Fuzzy Classification

Let $\bar{x} = [x_1, x_2, \ldots, x_N]^T$ be a vector in $\Re^N$ representing an input dataset. The N components are correlated process signals that constitute a snapshot of the monitored process at a given time. Given $X = (\bar{x}_1, \bar{x}_2, \ldots, \bar{x}_p)$ the $N \times P$ matrix of P patterns covering the $\Re^N$ operating region, the basic idea is to split this region in Q fuzzy clusters and derive a mapping function that assigns each pattern $\bar{x}_k | k = 1 \ldots P$ to each cluster $C_k | k = 1 \ldots Q$ at *some degree*. In pattern recognition the Q clusters are identified by prototype patterns, which in the case of spherical or ellipsoidal clusters are also called *centroids*.

If the Euclidean distance is used to evaluate the closeness of patterns to the centroids, which is the right choice for spherical clusters, the result is the popular *Fuzzy C-means* algorithm. In this application, however, clusters with different shapes and sizes are expected, so that Euclidean distances would not work well. To take care of the not uniform distribution of the patterns, the GK algorithm, from Gustafson and Kessel [7] has been adopted here.

However, in fuzzy clustering, patterns that are not close to any of the identified cluster prototypes, are classified and assigned to the *relatively* most probable cluster, because of the implicit hypothesis that *all* the patterns belong to the obtained partition. There can be uncertainty (or fuzziness) on *where* the incoming pattern should be assigned, but no uncertainty on *if* it should be assigned somewhere. When this probabilistic assumption is applied to signal validation applications, a number of problems may arise:

- Lack of robustness against noisy data.
- It is not possible to say "I do not know," even when this would be the best answer. An incoming pattern might be given a high grade of membership in a cluster, even if it is far away from all the centroids, only because it is *relatively* closer to one specific cluster.

Relaxation of the probabilistic assumption leads to a *possibilistic* approach, which results in a possible solution of the two above-mentioned limitations. A possibilistic classifier initially learns a dataset X of pattern samples. During this process, the model increases its robustness to noisy data and many patterns in X could be discarded as not representative of any developing cluster. When new patterns are examined, the possibilistic model evaluates in which cluster or clusters the incoming pattern could be *possibly* assigned, if any.

The step-by-step procedure used to develop the fuzzy and possibilistic classifiers can be summarized as follows:

- Given a set of samples X, compute an initial set of cluster centroids using the ISODATA [8] algorithm, which has been chosen because it automatically optimizes the number of required clusters.
- Initialize the elements of the partition matrix with crisp values (0 or 1), using ISODATA. Then run the *GK* algorithm, which produces the fuzzy classifier.
- Use the updated partition matrix, from the previous step, to start the iterative process as described by Krishnapuram and Keller [9], to arrive to a possibilistic partition.

2.2 The Artificial Neural Networks Module

Sample data in dataset X are collected in Q training datasets, to be used for training Q supervised neural networks. Each pattern in X is assigned to one or more training sets according to the fuzzy partition, as long as its possibilistic index is above a threshold value h in one or more identified clusters, with $h = (0.5,1)$. The role of the threshold parameter h is twofold:

- patterns not adequately represented in any cluster are discarded, so that they have no influence on the network weights calculation.
- patterns possibly represented in many clusters (responsible of the above mentioned boundary problem) are used in the training set of many corresponding networks.

The network architecture used in this work is a five-layer (three hidden layers), autoassociative feedforward structure trained with the scaled conjugate gradient descent algorithm. The hidden layers use hyperbolic tangent transfer functions, while the output layer is linear. This architecture has been tested to be more robust to process noise and sensor faults.

The input to the ANN's is not limited to the current pattern. To capture the process dynamics, a number of past values of the time series can also be used, together with the current ones, so that the total number of input nodes in each ANN is $N \times R$, where N is the number of signals and R the number of past values used.

2.3 The Reliability Assessment Module

The possibilistic cluster membership has an important role in the final decision whether the network output can be considered reliable or not. A high membership grade in one or two clusters increases our confidence that the data sample is contained in the training volume of one or two neural networks, so that they will be able to recall the output with a low estimation error. On the other side, a low membership value in *all* the clusters is a clear warning that no knowledge was given to the system about that process state. Note that using fuzzy clustering techniques, it would not be possible to have neither *low* values in all the clusters, nor *high* values in more than one.

The reliability function is realized through a fuzzy model, where the input is the maximum membership grade of the sample and the maximum signal mismatch in the neural network module, while the output is the reliability membership grade in three fuzzy sets assessing at what extent the reliability factor can be considered *high, medium* or *low*.

2.4 The ICARO Plant

An application is currently in progress, in a co-operation Halden Project and Italian National Agency for Energy, New Technologies and Environment (ENEA), for the validation of 12 process sensors of a real natural gas-fuelled cogeneration plant. The training data were collected under normal operating conditions, varying from 25% to 100% of nominal power. The layout of the plant and the position of the sensors are displayed in Fig. 1.

The ENEA experimental plant facility of co-generated heat and power (Impianto di Cogenerazione ICARO) is located at the ENEA Casaccia Center, northwest of Rome, Italy.

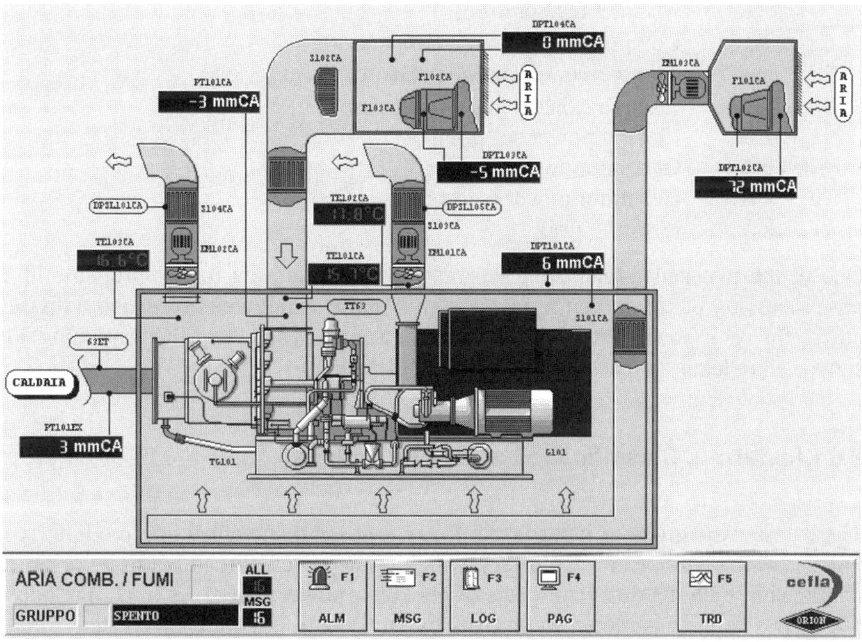

Fig. 1. ICARO cogeneration plant and sensor position

The plant is a CHP (Co-generated Heat and Power) serving roughly 7 MWt district heating and 2 Mwe electric power to the 2000 researches and employees usually working at the Center. Buildings containing offices, workshops, and laboratories cover a site area of 88000m^2.

ICARO is a natural gas turbine-generator system, which operates in connection with an intermediate exhaust gas– pressurized water heat exchanger, which in turn provides hot pressurized water to the district heating. The maximum power is 2 MWe and 7 MWt, and the district heating water temperature does not exceed 134°C.

2.5 Data Preparation

The ICARO test contains different plant power steady-state conditions, coming from two operational days. Data are gathered at a scanning rate of 60". The application has been designed to monitor 12 temperature, pressure, speed, valve opening and power sensors. The list of the selected signals is as follows:

N.	M.U.	Variable
1	rpm	Controller speed
2	%	Fuel valve position
3	bar	Gas pressure
4	°C	Gas flow
5	°C	Air inlet temperature
6	°C	Compressor discharge temperature
7	bar	Compressor discharge pressure
8	°C	Turbine exhaust gas temperature
9	kW	Electric power set-point
10(*)	kW	Generator power (sensor 1)
10	kW	Generator power (sensor 2)
11	kW	Power to grid

One of the two generator power sensors (*) was found out to be unreliable in a pre-processing phase, in that it gave a signal that was not well correlated with the others for each power setting. It was then discarded, and only the remaining 11 signals were taken into account for the ANN training.

2.6 Clustering, Classification and ANN Training

The dataset was given in input to the fuzzy – possibilistic clustering module (see Fig. 2), which defined seven clusters, which, as expected, turned out to be in close relationship with the different levels of generator power. Then, seven autoassociative ANNs were constructed and trained by the system, one for each cluster.

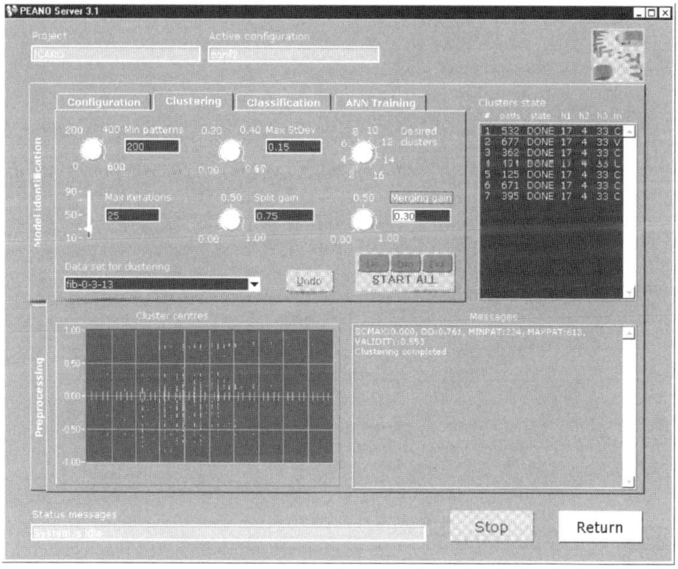

Fig. 2. ICARO signal clustering in PEANO

Patterns were assigned to the respective ANN for training, based on a fuzzy classification step. Since the clusters are fuzzy, patterns in the fuzzy boundary region between two clusters get assigned to the training datasets of the two corresponding ANNs, therefore ensuring a smooth transition during the on-line signal validation.

2.7 Results

In order to test the obtained signal validation model, several sensor faults were generated, and the system was able to detect these faults at a very early stage. As an example, we show here a generator power sensor fault, which is the most important plant signal, simulated with a ramp of 0.01% per sample (i.e., per minute). Results of this test are displayed in Figs 3, 4. The test will be repeated with filtering to avoid spikes on the input signal.

In Fig.3 the measured, actual (process), and estimated values are shown. It can be seen that the system is able to follow very closely the actual process value. In Fig.4 the mismatch between the measured value and the value estimated is shown. The error band in the mismatch plot is calculated by the model according to the expected error of prediction for each particular cluster, calculated in advance during the training. In this case the error band ranges from 20 to 40 kW. When the mismatch exceeds this band, the validation system generates an alarm.

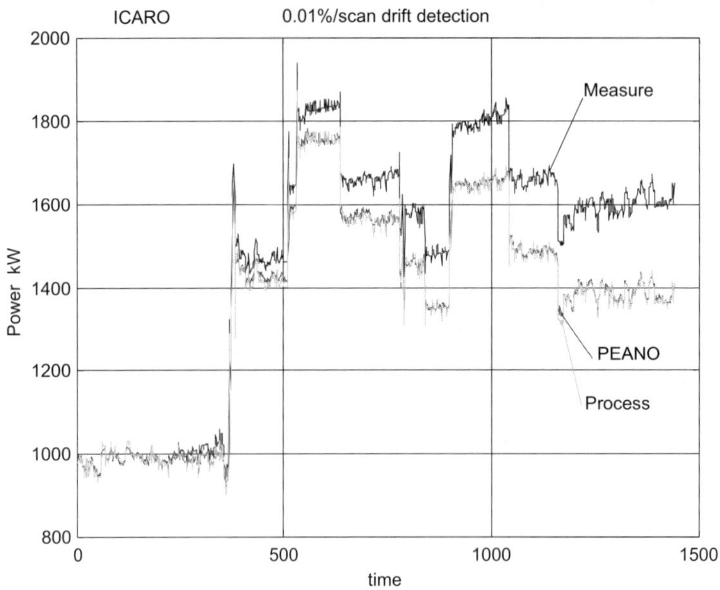

Fig. 3. Measured, actual and PEANO estimated values

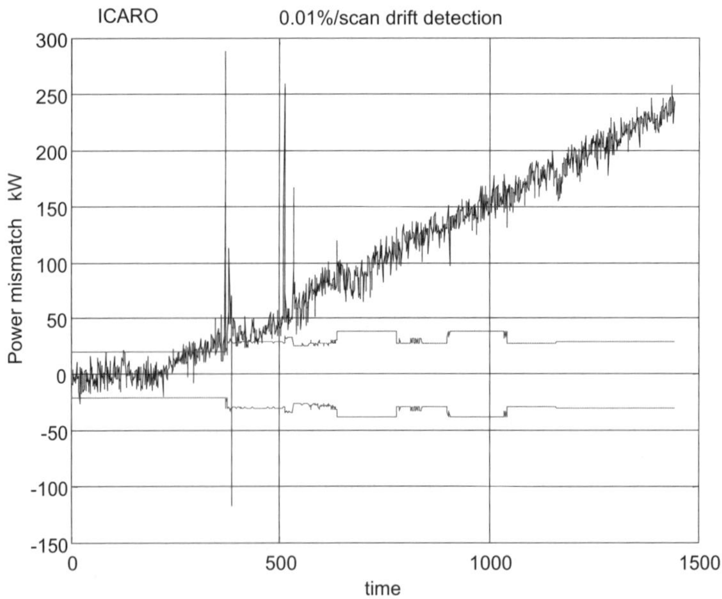

Fig. 4. Mismatch between measured and estimated values

During the *normal operating scenarios,* which the model was trained for, all faults were recognized *correctly*, with the reliability level of *high* or *medium*. An alarm was normally triggered after few samples, as soon as the mismatch between the measured signal and the estimated values exceeded the variable error band.

The model was capable of recognizing the faulty sensors, without affecting the others, out of the 11 monitored. The system is demonstrating to have acquired well the correlation existing among the monitored signals and is ready for providing day-by-day valuable information to the ICARO operators about the calibration state of those sensors.

3 Neural Network Pattern Recognition for Early Detection of Flame Instability

ENEA is involved in a joint program aimed at developing technologies and advanced processes for combustion of fossil fuel. Included in the program, there are specific activities for using artificial intelligence methods such as neural networks, fuzzy logic, and non-linear (chaotic) dynamics [10], in order to perform integrated diagnostics for the characterisation of the combustion stability, energy efficiency and pollutant production.

The ALADDIN event classification system developed at IFE, based on advanced numerical techniques, combining Ensembles of Recurrent Neural Network Classifiers with Wavelet Preprocessing, has been proposed as one technique for recognition of flame instability statuses, due to its capability to perform the prompt detection and recognition of occasional short periods of dynamic nature.

This is the case of combustion in premixed burners, where anomalies occur in some operation regimes. In premixed burners, in fact, fuel and air are premixed before their entering combustion chamber. Differently from other burners, the premixed ones operate at lower temperatures, thus emitting less quantity of pollutants, but being affected by unpredictable operation criticities, that cause a lowering in efficiency, environmental pollution and dangerous structural stresses [11]. The burner can be envisioned as an acoustic cavity, containing both "flame fluid" and "exhaust fluid." The different mechanical impedances and the dimensions of such fluids, determine the resonance frequency of the cavity. Acoustic effects of the cavity move the noise band and center it on the resonance value. Consequently, by monitoring the latter parameter, it is possible to detect its cause: *the variation in flame's activity*.

Figures 5a and 5b represent a schematic view of two-dimensional aspects of the flame due to the up mentioned phenomenon. When dimensional variation in the flame occurs with no variation in the quantities of air and fuel entering the burner, a *flame instability* [11] is experienced. Such instabilities are due to a series of causes that are originated by the reciprocal interaction of many parameters; for instance, pressure fluctuations generated by acoustic noise, act on combustion chemistry and determine chaotic evolutions in the phenomenon.

The photographic series of Fig. 6 (a-f) shows a typical evolution of the flame in instability conditions. The most important information that is possible to gain from noise is the thermal power variation and the dimensional variation of the flame,

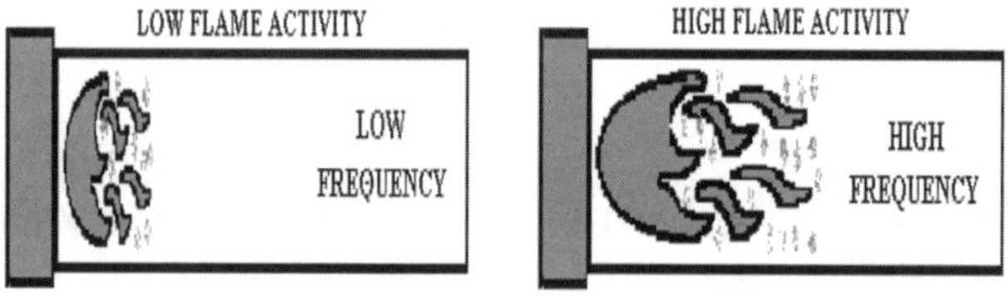

Fig. 5. a) Low flame activity - b) High flame activity

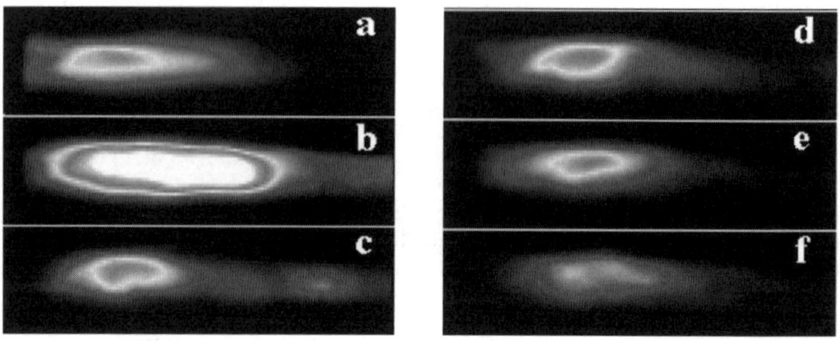

Fig. 6. Typical evolution of flame instability

being the former correlated to noise amplitude and the latter to its dominant frequency.

Acoustic waves and image processing are used to monitor the combustion process and the flame behaviour. Processed signals are originated by two different types of sensors: acoustic- electric transducer (microphone) and pressure gauge connected to combustion chamber.

Both the signals allowed to catch the combustion instability phenomenon, also considering that the microphone signal contained some noise coming from external environment. Fig.7 represents a three-dimensional diagram (time, frequency, amplitude) obtained by the *wavelet transform* technique. In this figure many frequency excursions are noticed, while only once it appears an event associated to high instability. This shows how not always a high variation in the thermal power of the flame is associated to a change in its dimensions.

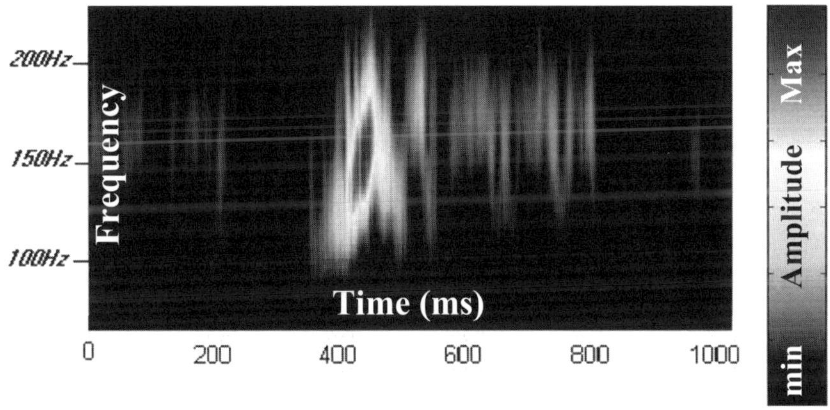

Fig. 7. Three-dimensional diagram (time, frequency, amplitude) obtained by the *wavelet transform*

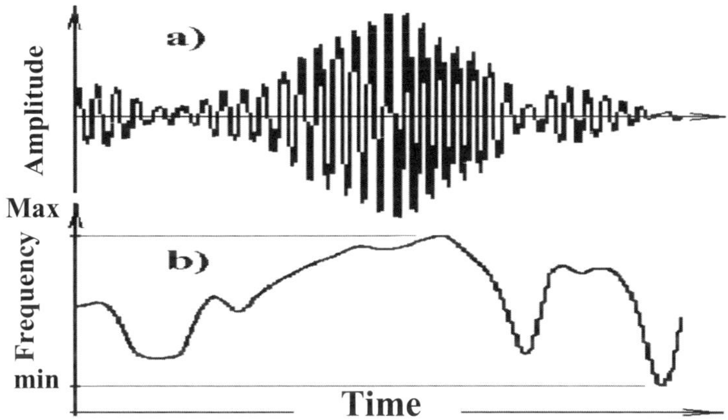

Fig. 8. a) Combustion noise – b) Frequency deviation

Similar results have been obtained when processing the signal by means of a frequency discriminator.

One example is reported in Fig.8, where diagram a) reproduces combustion noise, while b) represents the frequency deviation of the noise itself. It can be evidenced how b) has a trend similar to the envelopment of a) and presents noticeable variations also at times in which a) assumes reduced values.

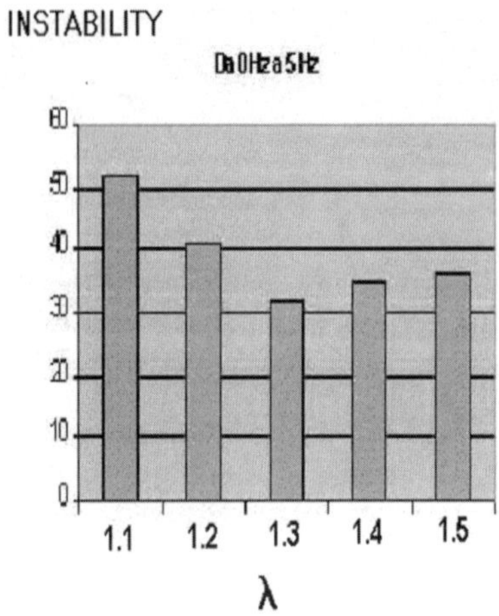

Fig. 9. Instability conditions calculated as pressure energy release versus air-gas flow ratio (λ) in the 0-5 Hz frequency interval

In this way, signal b) has the peculiarity to detect the flame's instability ranges, this in advance to the statistical occurrence of strong instabilities. In. Fig.9, a bar graph is shown generated by five different operation conditions of the burner, respectively having the following values of **λ:** 1.1; 1.2; 1.3; 1.4; 1.5 (**λ** meaning the ratio between introduced air quantity and stechiometric one). Values of **λ** of 1.1 and 1.5 are the operation limits of the investigated burner. The highest bar is referred to the most unstable condition of the flame; the third condition appears to be the most stable one, then the bars start to increase again their height. The bars represent the energy of the signals shown in Fig. 7, b), that have been previously filtered as to eliminate the components that are not correlated to the phenomenon. The asymmetry in the histogram is due to the different instability types (mixtures close to the stechiometric value and mixtures with air excess). Every other device of the front panel is used to set the parameters of the analysis or to control their different phases.

3.1 Preliminary Results

The objective was to devise ways to employ (standard) neural networks trained to recognize the lambda value of 1 second long segments (2000 patterns at 2kHz) of the original noise data. Data was provided for 64 seconds for each lambda value, from 1.1 to 1.5, corresponding to a total of 320 one-second segments. An example of one-second segments of the noise data for each of the five lambda values is shown in Fig. 10.

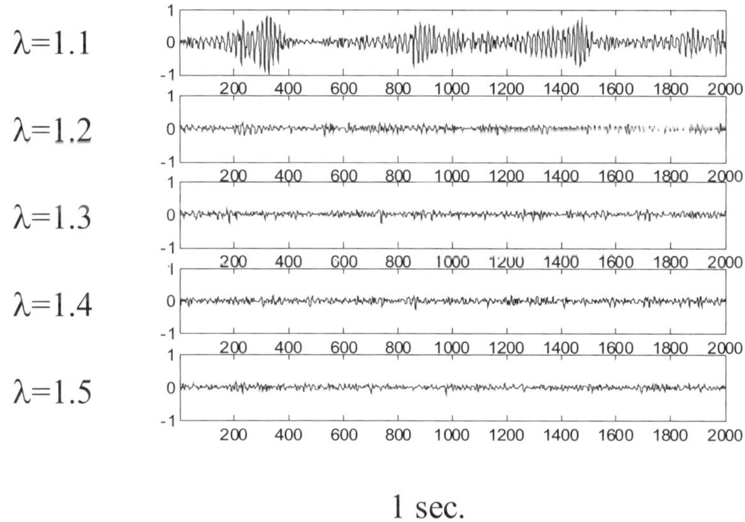

Fig. 10. Original noise data segments

In order to apply neural network models, discriminating features needed to be extracted from this data. A first analysis in the frequency domain, by using the Fourier transform and analyzing the resulting power spectral density functions, did not lead to consistent results that could be used successfully. A second approach used instead the wavelet transform to produce scalograms of the same data, as shown in Fig.11.

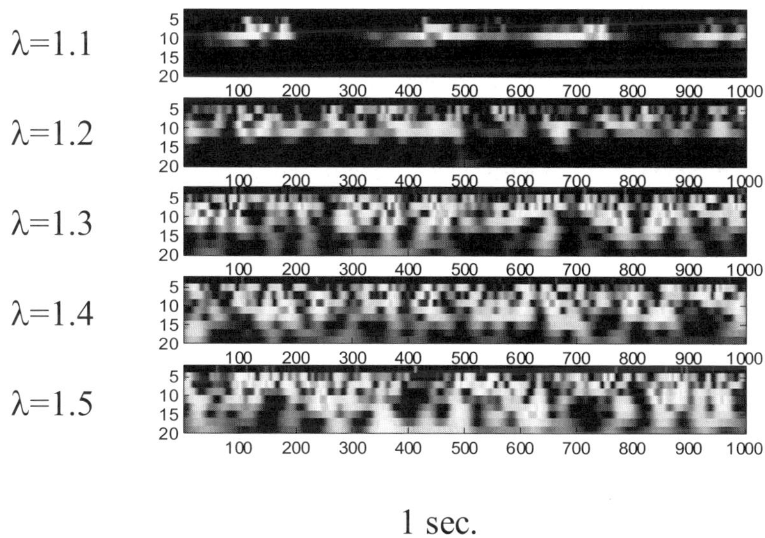

Fig. 11. Wavelet scalograms of noise data segments (nine scales are shown)

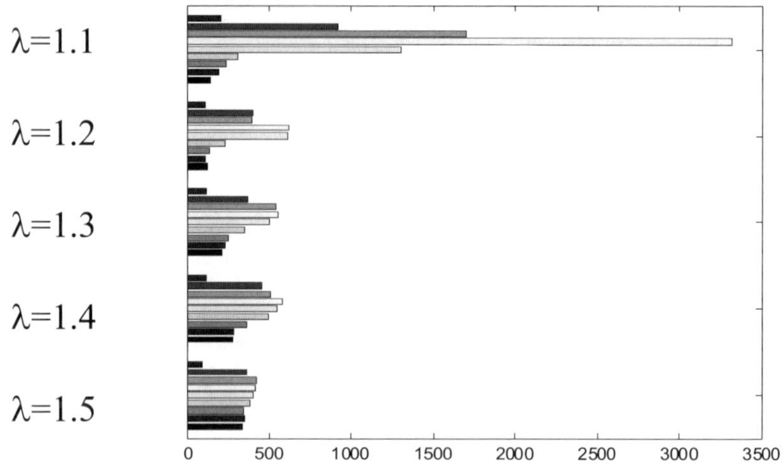

Fig. 12. Aggregated values of the wavelet scalograms

The scalogram values at nine selected scales were then aggregated in nine single values, which could then be directly used as inputs to a neural network classifier. The aggregated values of the scalograms of Fig. 11 are shown in Fig. 12.

Neural network models could then be trained over 80% of segments and tested over the remaining 20% (randomly drawn). By making an average on ten training experiments, the following levels of accuracy were reached:

- 98.7% of training cases are correctly classified
- 88.7% of test cases are correctly classified.

The difference between training and test cases accuracy can be attributed to data scarcity. With a larger amount of data, it would be possible to improve the quality of the learning results. This could be achieved both by using longer data segments as training examples (i.e., longer than 1 second), and by providing a larger number of segments for each instability class

4 General Conclusions and Future Developments

On-line monitor of critical process instrumentation is an important issue, both for safety and economical considerations. Artificial Intelligence methods can be very efficient and robust in this context. This chapter has described the experience and results achieved using a Neuro-Fuzzy system, PEANO, to a cogeneration plant supplying electric energy and heating to the ENEA research center in Rome. The next step is to train PEANO with historical data covering a six-month period of operation to make the system response more reliable and robust, and finally to install the system at the plant and perform real-time tests to assess the capability and limitations of PEANO in this environment.

The problem of flame instability identification seems after these experiments more related to a "status" identification than a "transient" identification, which is the job performed by the ALADDIN tool. This is not a problem in itself, but the dynamic features of ALADDIN, i.e., its recurrent neural networks, are underused. Simpler models seem to be sufficient to recognize the flame statuses. The focal point is the identification of optimal features from the point of view of the discriminating power. Some practical experiments are envisaged and will be done on this respect in the near future.

References

1. Fantoni PF (1996) Neuro-Fuzzy models applied to full range signal validation in NPP. In: Proceedings of Nucl. Plant Instrumentation, Control and Human-Machine Interface Technologies, NPIC&HMIT'96, The Pennsylvania State Univ., USA.
2. Fantoni PF, Mazzola A (1996) Multiple-Failure Signal Validation in Nuclear Power Plants using Artificial Neural Networks. Nuclear Technology.
3. Fantoni PF, Renders JM (1998) On-Line Performance Estimation and Condition Monitoring Using Neuro-Fuzzy Techniques. In: Proccedings of the Workshop on On-Line Fault Detection and Supervision in the Chemical Process Industries, Lyon, France.
4. Fantoni PF, Mazzola A (1996) Accuracy Estimate of Artificial Neural Networks Based Models for Industrial Applications. Braunschweig B, Bremdal B (eds) Artificial Intelligence in the Petroleum Industry.
5. Roverso D (2000) Soft Computing Tools for Transient Classification. Information Sciences 127(3-4):137-156.
6. Roverso D (2000) Multivariate Temporal Classification by Windowed Wavelet Decomposition and Recurrent Neural Networks. In: Proceedings of the 3rd ANS International Topical Meeting on Nuclear Plant Instrumentation, Control and Human-Machine Interface Technologies, NPIC&HMIT 2000, Washington DC.
7. Gustafson DE, Kessel WC (1979) Fuzzy Clustering with a fuzzy covariance matrix. In. Proc. IEEE CDC, San Diego, CA.
8. Tou JT, Gonzalez RC (1974) Pattern Recognition Principles, Addison-Wesley Publishing Company, Reading MA.
9. Krishnapuram R, Keller J (1993) A possibilistic approach to clustering. IEEE Transactions on Fuzzy Systems 1(2):98-100.
10. Annunziato M, Coppola G, Presaghi M, Presaghi R, Puglisi G, Romanello F (2001) Acoustic Analysis Aimed at Characterizing Combustion Instability in Premixed Burners. ENEA RT.
11. McManus KR, Poinsot T, Candel S M (1993) A Review of Active Control of Combustion Instabilities. Progr. Energy Combustion Science 19:1-29.

22 ALADDIN: Event Recognition & Fault Diagnosis for Process & Machine Condition Monitoring

Davide Roverso

Institutt for energiteknikk, OECD Halden Reactor Project
POBox 173, N-1751 Halden, Norway.

roverso@computer.org

1 Introduction

Many industrial processes are characterized by long periods of steady-state operation, intercalated by occasional shorter periods of a more dynamic nature in correspondence of either normal events, such as minor disturbances, planned interruptions or transitions to different operation states, or abnormal events, such as major disturbances, actuator failures, instrumentation failures, etc. This second class of events represents a challenge, and possibly a threat, to the smooth, safe, and economical operation of the monitored process. The prompt detection and recognition of such an event is of the essence for the performance of the most effective and informed response to the challenge.

The most common way of performing event detection and recognition in an industrial plant is to rely on experienced operators, which, by observing the current values of important process variables, as well as their recent history on trend displays, plus eventual alarms generated by the process monitoring system, can usually quickly and reliably diagnose the current event. However, when process variables change with different rates, or are affected by varying lags, it is very difficult for a human operator to track and recognize the current situation. Also, when the changes to the process variables caused by the occurring event are subtle or very slow, and do not cause the alarm system to generate alert signals, the abnormal situation can be easily overlooked by an operator. In both cases, a computerized operator support system (COSS) able to detect and classify these process changes would be of great value.

The task of dynamic event recognition can be approached from different perspectives, such as symptom-based deductive inference (i.e., expert systems), state classification (i.e., *static* pattern recognition), or transient classification (i.e., *dynamic* pattern recognition). The latter is the approach followed in the work here described. The basic assumption behind the transient classification approach is that an event or fault will generate, in time, unique changes in the monitored process. The discrimination of such changes can then in principle lead back to the originating event via an inverse mapping process.

This chapter describes a set of techniques, developed as part of the internationally sponsored OECD Halden Reactor Project, which can be employed to design such a transient classification COSS. The ALADDIN methodology integrates these techniques into a unified model, which has undergone several preliminary tests, and which is now used as the basis for the implementation of a complete system dedicated to the recognition of control system anomalies in nuclear power plants (CARS - Control Anomaly Recognition System, a project sponsored by EPRI, the Electric Power Research Institute, Palo Alto, CA).

In the remainder of this chapter, we first present a brief overview of previous work, focusing in particular on applications of neural networks to transient classification in nuclear power plants. In the three central sections of the chapter we describe, respectively, the recurrent neural network technology at the core of ALADDIN, the use of the wavelet transform to extract compact features from time-varying signals, and a task decomposition algorithm for tackling the scaling issues of large diagnosis tasks. The last section concludes the chapter with a summary and a discussion of current open issues.

2 Previous Approaches to Event Recognition with Neural Networks

Artificial Neural Networks (ANNs) are particularly suited to deal with the problem of event recognition in dynamic processes for several reasons (for a general reference on neural networks see [1]). First of all ANNs can approximate any well-behaved function with an arbitrary accuracy, which is an essential advantage on methods based on linear regression when the problem at hand presents essential non-linearities. The biggest advantage of ANNs, however, manifests itself when dealing with hard problems, e.g., in the case of significantly overlapping patterns, high background noise, dynamically changing environments, and absence of accurate and fast first-principles models.

Focusing first on applications in the nuclear field, we note that possibly the first to demonstrate the feasibility of using ANNs were Bartlett et al. [2]. That work was developed further and enhanced introducing a modular ANN architecture [3]. An important contribution was made by Bartal et al. [4], where they recognized the necessity for a classifier of being able to provide a "don't-know" answer when presented with a transient of a kind not contained in its accumulated knowledge base. An alternative way of dealing with temporal data using an implicit time measure was proposed by Jeong et al. [5]. The same authors later proposed the adaptive template-matching algorithm [6], which allows to describe transients in a two-dimensional continuum of time and severity level. More recent work on transient detection in nuclear power plants include [7], where the authors use a set of specialized neural networks to independently identify each selected transient.

Many attempts have also been recently published which borrow techniques first developed for speech recognition tasks. Among these, we should particularly mention Dynamic Time Warping (see [8] for an application to batch processes, and [9] for an application to knowledge discovery in time-series databases), and Hidden Markov Models [10,11].

3 The Use of Recurrent Neural Networks and their Ensembles

Recurrent ANNs are a class of ANNs that are able to deal with temporal input signals thanks to an internal architecture that is recurrent, i.e., it makes use of feedback connections among the neurons. The Elman recurrent network [12] is a supervised ANN with a *partially* recurrent architecture (i.e., not all neurons have feedback connections) whose main characteristic is an array of neurons that record the internal status of the network. At each time step this internal status is fed-back as an additional input to the network, which then computes the new status and the output vector. An advantage of the recurrent model is that it can discriminate event classes based on differences in their time evolution, and it can recognize events even when presented in slightly different time scales, i.e., it can compensate for faster or slower instances of the same event class.

In [13], our comparative study of alternative neural network designs and models for both state and transient classification was presented. The comparison was performed on a small set of simulated events of the Forsmark 2 Nuclear Power Plant, which is a 969-MWe ABB Boiling Water Reactor in Sweden. The Elman recurrent ANN model came out as the best performing model overall, providing the best accuracy results.

However, recurrent ANNs are known to be hard to train [14], especially when the temporal relationships that are being modelled span relatively long intervals [15]. In our case, training problems would be encountered if the events to be classified were to generate transients with relatively long and similar stabilisation phases, with the most discriminating information coming early in the transients.

A different kind of problem encountered with the recurrent neural network module concerns the stability of the models generated. One observed phenomenon was that different runs of the training procedure led, at times, to models with differing performance on test transients. This is a direct, and expected, consequence of the training procedures that these neural networks employ. Each training session starts with a randomly initialised neural network, and proceeds with a gradient descent algorithm, which tries to minimise the classification error of the network on the training transients. Gradient descent algorithms can get trapped on local minima of the error surface. Different training sessions then, starting from different points on the error surface, can reach different minima, and therefore produce models with differing performance on test transients.

The first kind of problem, i.e., long-term temporal dependencies, can be approached by extracting compact features from the long transient using wavelets, as will be described in more detail later on in this chapter. Introducing ensembles of neural networks approached the second kind of problem, i.e., the problem of local minima.

3.1 Neural Network Ensembles

Several researchers in the field of Machine Learning have investigated and proposed various techniques for combining the predictions of multiple classifiers to produce a single classifier [16,17,18]. The resulting model (referred to as an ensemble) is generally more accurate than any of the original classifiers, tends to be more robust to overfitting phenomena, and avoids the instability problems associated with local minima, which was mentioned earlier.

In an ensemble, each classifier is generally trained separately, and the predicted output of each classifier is then combined to produce the output of the ensemble. However, combining the output of several classifiers is useful only if there is disagreement. Obviously, the combination of identical classifiers produces no gain. In [19] it was demonstrated that an ideal ensemble consists of highly correct classifiers that disagree as much as possible, and that an effective combining scheme is to simply average the predictions of each classifier. As a result, methods for creating ensembles focus on producing classifiers that disagree in their predictions. Neural network techniques that can be employed include methods for training with different topologies, different initial weights, different parameters, and different training sets.

In ALADDIN we adopt the bagging method [17], which tries to generate disagreement among the classifiers by altering the training set each classifier sees during training. Bagging is an ensemble method that creates individuals for its ensemble by training each classifier on a random sampling of the training set, and, in forming the final classification, gives equal weight to each of the classifiers.

Tests of a prototype system based on ensembles of recurrent ANNs have been reported in [20], and were carried out on simulated data of French PWR 900 Nuclear Power Plant transients, corresponding to various occurrences of anomalous rapid load rejection events, i.e., *plant islanding*, including particular malfunctions on plant I&C systems (instrumentation, actuators or closed-loop control systems). CEA (Commissariat à l'Énergie Atomique, France) in cooperation with EDF (Electricité de France) provided the data.

The data provided by CEA consisted of 7 anomalous islandings plus a normal reference case. These transients were used for the training of ALADDIN. Additionally, blind test transients were provided (with amplitudes of the failures differing from the transients used for training) to assess the robustness and sensitivity of the system. The 7 anomalies involved component failures (such as pumps), control function failures, as well as sensor failures.

Overall the system behaved very well, always correctly classifying the transients, even in cases where the test case anomaly size was significantly smaller than in the training case. For completeness, similar tests were performed with the other three neural models described in [13], i.e., the RBF classifier, the Cascade-RBF classifier, and the SOM classifier. The results obtained were in line with the ones from the first case study on the Forsmark 2 data, once more confirming the superior performance of the recurrent ANNs.

4 The Use of Wavelet On-Line Pre-processing - WOLP

In order to tackle long-term temporal dependencies with recurrent ANNs, we first proposed the use of features extracted by wavelet decomposition in [21].

The idea of combining neural networks with multiscale wavelet decomposition has been proposed by a number of authors [22,23,24,25]. These approaches either use wavelets as the neuron's activation functions [24], or in a pre-processing phase by the extraction of features from the time-series data [22,23]. Our approach is of the second kind but differs from earlier work mainly in that it combines wavelet feature extraction with recurrent ANNs, and it shows that promising results can be obtained even with a greatly simplified wavelet feature extraction step. Such simplification is particularly welcome in on-line applications, as well as in high-dimensional applications, i.e., applications where the number of process variables and/or the number of transient classes is considerable.

The scheme here proposed, consists in extracting, over time, three basic features from moving windows over each monitored signal, so that the recurrent ANNs will have three times the number of inputs as before, but with a much lower input rate. These new input signals are the consecutive values of features extracted by Haar wavelet decomposition [26] from a sliding window on the actual signal time-series, and are the following:

1. The mean residual signal taken at he highest, i.e., coarser, scale.
2. The minimum wavelet coefficient over all the scales.
3. The maximum wavelet coefficient over all the scales.

The rationale behind this choice is that (a) captures the general trend of the signal in a compact way, while (b) and (c) capture important discontinuities (e.g., step changes, spikes, etc.) which would otherwise be severely smoothed out by the compression process. The window size is selected so as to correspond to wavelet dyadic decomposition values (i.e., powers of 2), and consecutive windows are chosen with a slight overlap. Since we are using the Haar wavelet on a dyadic window, we do not need to be concerned with edge effects. Because of the ability of continuously applying the wavelet transform on a sliding window, and since we are using the transform as a pre-processing step for the neural network classifier, we have named this technique Wavelet On-Line Pre-processing (WOLP).

The WOLP transformation produces a very compact, but still representative description of the original signal. To better see how WOLP actually works in practice, we show in Fig. 1 an example of application of WOLP to an actual process signal. The upper graph in this figure shows 300 points of the original signal. Also shown in the upper graph are the overlapping windows from which the wavelet features are extracted. In these examples, the windows are 32 patterns long, and the overlap is 5 patterns. This means that a new wavelet transform is applied each time 27 new patterns are acquired from the process. The lower graph shows the output of the WOLP pre-processing. WOLP generates three data streams for each original signal, at a rate equal to the original sampling rate divided by the window slide step size, which in this case is 27.

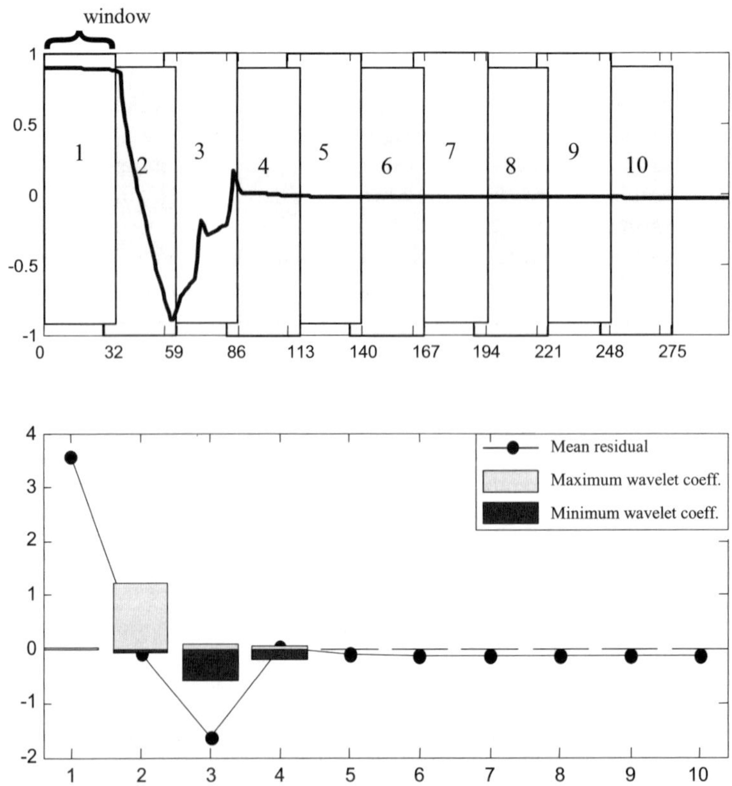

Fig. 1. Example of application of WOLP to a process signal

In the example of Fig. 1, an original data stream of 275 samples generates three data streams of 10 samples each. If we now look in detail at the correspondence between the original signal and the WOLP output, we will see in practice how the WOLP transformation produces a very compact, but still very significant description of the original signal. First, we can see how the *mean residual* feature reflects the general trend of the signal. The mean residual value is very much related to the average signal value within the analysis window, and we could have chosen to use that instead. There are no technical reasons for choosing one over the other. Our choice went to the wavelet mean residual mainly for consistency with the other WOLP outputs.

Shifting our attention to the *maximum* and *minimum wavelet coefficients*, we can notice how the value of the *maximum coefficient* reflects negative trends, step changes, and the negative component of spikes within the analysis window, while the *minimum coefficient* reflects the positive trends, step changes, and positive component of spikes. If we look for example at the third window, we can see how a big increase in the signal value corresponds to a big value of the *minimum coefficient*, and how the small decrease in the signal value (in correspondence with the two small peaks) is matched by a small value of the *maximum coefficient*. When compared with the data corresponding to the second window, we can also notice how the WOLP transformation maintains the correct proportions of the registered changes. The signal decrease in the second window is roughly double the signal increase in the third window. Correspondingly, the *maximum coefficient* in the second window is roughly double the *minimum coefficient* in the third window.

The choice of window size can be used to strike a balance between a high level of transient compression (which greatly improves the performance of the recurrent ANNs) and a resolution still sufficient to discriminate among the event classes.

Successful tests on the PWR 900 data (described earlier in this chapter) were followed by a series of controlled tests conducted on artificially generated multivariate time-series. The aim of these tests was to demonstrate the ability of the system to base its classification decision on a range of discriminating feature, from low frequency to high frequency features, and from early to late developing features.

Two cases of transient behaviour were defined for four distinct variables. By combining these transients in all possible ways, one obtains 16 prototype transients composed of 300 four-dimensional samples, as shown in Fig. 2. These 16 transient classes constitute a very challenging task for a transient classification system like ALADDIN, since for *each* transient class, there are 4 other classes which differ from it by only a single feature, forcing the system to make optimal use of the available discriminating features, which, as seen from Fig. 2, span a range of characteristic behaviours.

Considering these 16 transients as class prototypes, a training set and a test set, each of 800 transients (50 for each of the 16 classes), were generated by random amplitude warping (± 30%), delay (up to 20%), time warping (± 20%), and noise (±1% gaussian). This introduced variation in the signal amplitudes and transient speeds (length) within each event class. The resulting transients varied in length from about 200 to 400 samples. The window size chosen was 64 samples, with a 9-sample overlap between windows, resulting in "compressed" transients of between 4 and 8 12-dimensional[1] "samples" (i.e., windows). The basic recurrent NN architecture used was based on 12 input units, 32 hidden units, and 16 output (classification) units, and was trained for 200.000 epochs. The obtained classification accuracy is shown in Table 1. As it can be seen, the results are very satisfactory, with a 94.1% correct classification rate on the test set, 1.4% misclassification, and 4.5% indecisive classification (i.e., either the system did not classify the transient, or it classified it as belonging to more than one class).

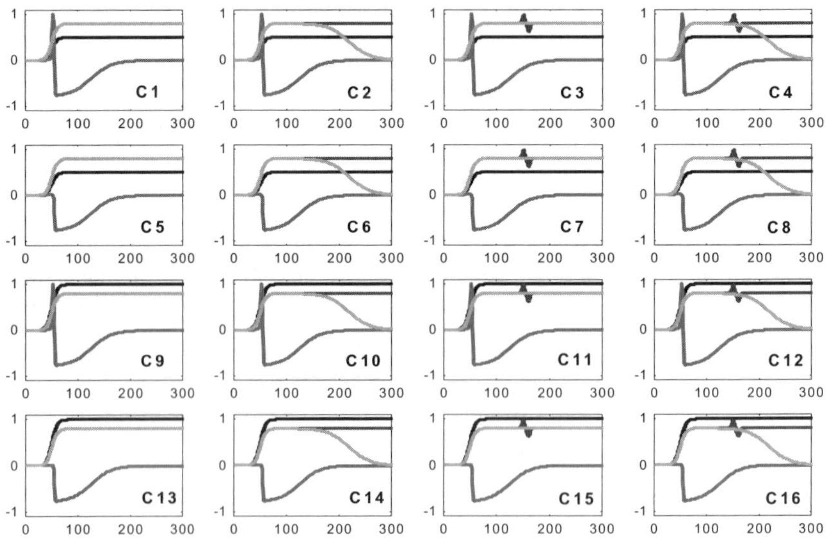

Fig. 2. The 16 class prototypes

Table 1. Classification results

	Correct	Misclassified	Non-classified
Training	95.2%	0.9%	3.9%
Test	94.1%	1.4%	4.5%

A similar test, aimed instead at investigating the discriminatory power of the system when the class differences are in the relative timing of signal features (in this case a small delay in the onset of a peak), was also carried out. The obtained classification accuracy was again above 90%.

[1] 12-dimensional because we extract 3 features from each of the 4 variables.

5 The Use of Autonomous Recursive Task Decomposition - ARTD

Scaling-up of systems and techniques to real-world problems has been, and still is, a primary concern in the field of soft computing[2]. Most industries, or other potential users of soft computing solutions, will inevitably pose the question of whether a particular technique or process will live-up to the expectations when a full-blown application has to be developed.

Issues of scale in a pattern classification task, such as transient classification, are generally related to three basic factors:

1. Amount of data.
2. Input dimensionality.
3. Output dimensionality.

Several techniques have been proposed, and are commonly used, to tackle these issues. When the amount of data to be used for training of the classifier is vast, training time might become extremely long. In this case, most of the proposed solutions involve one form or another of data sampling [27, 28]. Recently, promising techniques for "data squashing" have also been presented [29].

When the input dimensionality is large, i.e., the patterns to be classified are described by a large amount of features, several techniques for dimensionality reduction can be used. These range from feature sub-set selection algorithms [30] which basically eliminate some dimensions (features), to classical Principal Components Analysis (PCA) and other more advanced techniques (e.g., non-linear PCA, Independent Component Analysis (ICA), and data fusion techniques [31]) which instead try to apply a transformation to the original patterns so as to obtain a new representation in a lower-dimensional space.

When the output dimensionality is large, i.e., we have a many-class learning problem, where the patterns to be classified belong to a large number of distinct classes (e.g., fault types,) we have to adopt a task decomposition approach for reducing the size and complexity of the classification models. Several approaches have been proposed in which task decomposition is achieved through the construction of a modular architecture. This can be done in a number of ways:

- *manually*, i.e., explicitly using prior knowledge (see for instance [32];)
- *mechanically*, i.e., in a predetermined fashion, before starting the learning process; for example by decomposing an N-class task into N 2-class (binary) tasks [33] where each classifier discriminates one class from all the others, or into $\binom{N}{2}$ two-class tasks [34], where classes are paired in all possible combinations;

[2] Soft Computing is the branch of Computing Science which includes all Fuzzy Logic, Neural Networks, and Genetic Algorithms research.

- *autonomously*, i.e., the task decomposition is carried out during the learning process. In this category we find methods such as the Hierarchical Mixture of Experts (HME) of Jordan & Jacobs [35], and several other modular approaches. For a comparative study of modular neural classifiers see [36].

The algorithm proposed in this section falls in this third category, and is characterized by a recursive task decomposition process, which bases the decomposition strategy on the performance of partially trained classifiers. We call this new method Autonomous Recursive Task Decomposition (ARTD.)

5.1 The ARTD Algorithm

The Autonomous Recursive Task Decomposition (ARTD) algorithm is a hierarchical decomposition procedure for many-class classification tasks, i.e., tasks involving a large number of distinct pattern classes. ARTD generates a hierarchy of classifiers in a recursive way, by decomposing the task at hand into a set of sub-tasks, and by reapplying the same procedure in turn on each sub-task. The decision to split a task into sub-tasks is based on an analysis of the classification performance of a classifier, which has been only *partially* trained to solve the original task. In the following the ARTD algorithm for the case of neural network classifiers is shown.

The ARTD algorithm generates a tree structure with a neural classifier at each node. All neural classifiers receive the same inputs, but only the networks that receive an activation signal from a network at a higher layer are allowed to compute their classification and propagate the activation downwards. An example of this tree structure is shown in Fig. 3.

The classification tree of Fig. 3 shows an example in which an original 7-class task has been decomposed into two 3-class tasks plus two 2-class tasks. All neural classifiers receive the same original inputs. The outputs of each neural classifier are computed to be the minimum between the actual neural network output and the activation value output from the higher-level network. The values at *leaf* nodes, corresponding to the original classes, constitute the final output of the hierarchical classifier. In the following, an application of ARTD to a number of test cases is discussed.

In order to evaluate ARTD, two separate tests were performed on two different tasks, using different types of classifiers. The first test was performed using ALADDIN, applied to a 32-class transient classification task. The second test was performed using standard feed-forward neural nets (ffNN), applied to the 26-class *Letter Recognition Database* from the UCI Machine Learning Repository [37].

The ARTD Algorithm

Let: $C = \{c_1, \ldots, c_N\}$ be the set of N classes defining the current task

net be an NN classifier; it receives patterns p in input, and has N outputs, one for each class in C

1. Partially train *net* (e.g., for a limited number of epochs and/or limited error goal)
2. Simulate *net* on all the training patterns, and collect the network output patterns, grouped by intended class membership
3. Create a partition S of the set of classes C by clustering the centers of the corresponding sets of output patterns, taking into account the distribution of the patterns in the N-dimensional space of the outputs (e.g., join classes the centers of which are at a distance comparable to the within-class distance of the output patterns). This step generates a set of K superclasses.

$$S = \{s_1, \ldots, s_K\} \;\Big|\; \mathop{\forall}_{j=1}^{K} s_j \subseteq C \wedge \bigcup_{j=1}^{K} S_j \equiv C \wedge K \leq N$$

4. If K=1, go back to Step 1 (unless a maximum number of iterations has been reached, in which case fully train *net* and **return**). If K=N, fully train *net* and **return**. If $1<K<N$, build a new neural network *Snet* with the same inputs as *net*, and K outputs, one for each superclass s_j in S.
5. Fully train *Snet*
6. For each superclass s_j in S containing more than one class c, build a new neural network *Rnet*$_j$ with the same inputs as *net*, and one output for each class c_i in s_j.
7. Apply recursively ARTD to each *Rnet*$_j$
8. **Return**

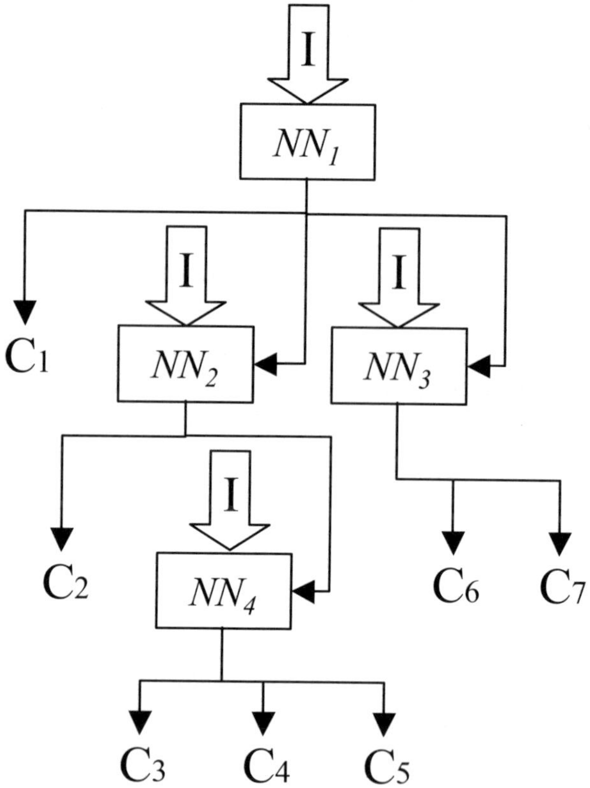

Fig. 3. Sample ARTD classification tree

5.2 The ALADDIN Tests

The ARTD algorithm was developed within the ALADDIN project, primarily to deal with the scaling problems encountered when the number of faults and anomalies to be recognized grows above levels manageable by a single classifier.

The test reported here is based on an artificial dataset composed of a large number of 5-variate time-series (i.e., time dependent trajectories in a 5-dimensional space), belonging to 32 separate classes. Two cases of transient behaviour for the five distinct variables are shown in Fig. 4. By combining these transients in all possible ways one obtains 32 class prototypes. The complete dataset generated consists of 32 examples for each class (16 for training and 16 for testing, for a total of 1024 examples).

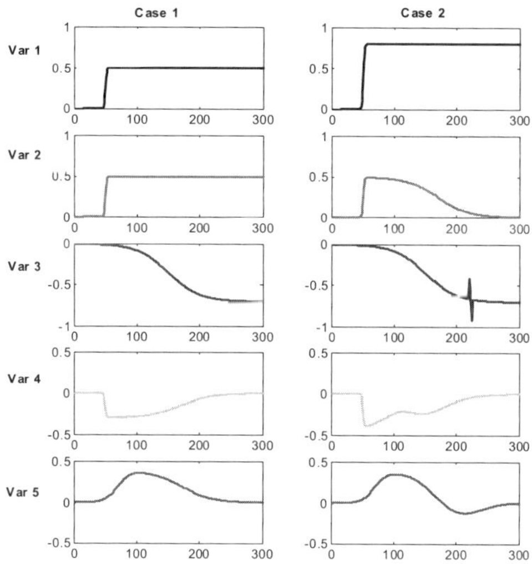

Fig. 4. Basic transient behavior cases

Each example was generated, as in the test data sets described earlier in this chapter, by random amplitude warping, delay, time warping, and noise. Each resulting transient is a 5-variate time-series of lengths varying roughly between 250 and 350 patterns.

In order to test ARTD, two separate versions of ALADDIN were trained on the 32-class task, one without ARTD and one with ARTD. We will refer to these two versions as *flat*-ALADDIN, and *ARTD*-ALADDIN. The resulting recurrent neural network classifiers are shown in Fig. 5 and Fig. 6, respectively.

The 32-class task is a particularly hard task, due to the dynamic nature of the input patterns, and the relatively large number of classes. An analysis of the *ARTD*-ALADDIN result shows that the obtained decompositions closely reflect actual class similarities. It can be noted for example, that most decompositions are binary. This follows from the fact that each of the 5 variables in the time-series could assume one of two slightly different types of behaviour.

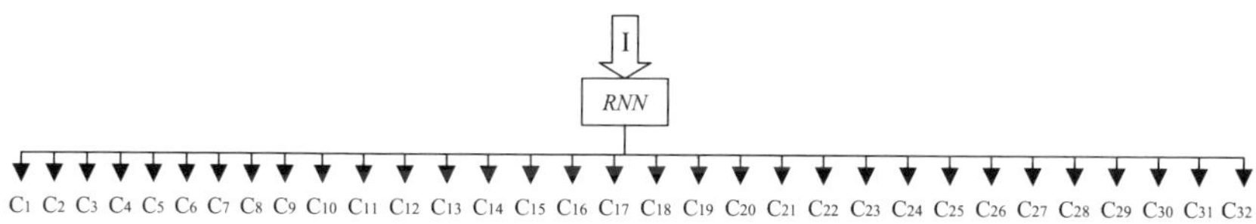

Fig. 5. 32-class *flat*-ALADDIN classifier

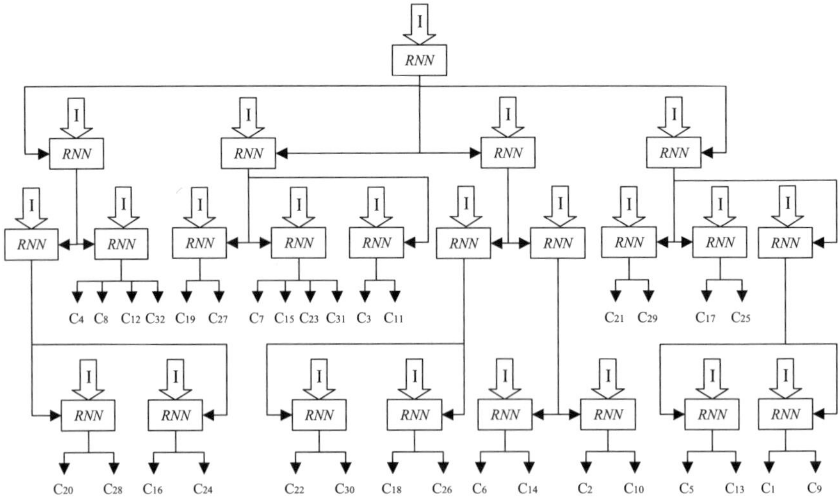

Fig. 6. 32-class *ARTD*-ALADDIN classifier

For example, the neural network at the bottom-right corner of Fig. 6 is specialized in the discrimination of class C_1 from class C_9. The only difference between these two classes is in fact the behaviour of a single variable, with the other four behaving exactly in the same way.

If we then compare training time and classification performance of the two systems, we see that *ARTD*-ALADDIN greatly improves on the original *flat*-ALADDIN. After 61 hours of training, *flat*-ALADDIN could only achieve a 33% correct classification rate[3]. *ARTD*-ALADDIN had already reached a 48% correct classification rate after only 2 hours of training, and reached 85% correct classification after 5 1/2 hours. We can say that in this case ARTD improved ALADDIN's training time at least 30-fold for comparable classification performance. Similar tests performed on the 16-class task described earlier, showed a roughly 4-fold training speed improvement. This seems to substantiate the hypothesis that ARTD's gain is more than linear in the number of classes. This is somewhat expected, since the complexity of a neural network, in terms of number of free parameters (i.e. weights and biases), grows more than linearly in the number of outputs, and so does the complexity of the training algorithm in the number of weights[4].

---

[3] A classification of an example of class *i* is counted as correct when the output *i* is greater than 0.5, and all other outputs values are lower than half of the value of output *i*.

[4] If we fix the architecture of the recurrent neural networks in ALADDIN as having a number of hidden nodes double the number of outputs, we obtain a network with 6464 weights in *flat*-ALADDIN, while we obtain a total sum of 1368 weights in the *ARTD*-ALADDIN networks.

A further advantage in training time is given by the fact that the networks in the *ARTD*-ALADDIN are trained only with the examples corresponding to the classes belonging to that branch of the classification tree.

5.3 The ffNN Test

A separate test was conducted on the 26-class *Letter Recognition Database*, obtained from the UCI Machine Learning Repository [35]. The task here is to discriminate letter categories associated with vectors of 16 integer attributes extracted from raster scan images of the letters. The classifier chosen for the test is a standard three-layer feed-forward neural network (*ffNN*), trained with the back-propagation algorithm.

Also for this task, two separate versions, *flat-ffNN* and *ARTD-ffNN*, were trained on the 26-class letter recognition task. The results obtained, in terms of training time and classification performance, show also in this case a significant gain when the ARTD algorithm is used. After more than 23 hours of training, the *flat-ffNN* reached a 63.9% correct classification rate. *ARTD-ffNN* reached a comparable correct classification rate, 68.4%, after less than 3 hours. The speed gain obtained by the use of ARTD is in this case roughly 8-fold.

As was the case also in the previous tests, the achieved correct classification rates can be of course improved by further training of the classifiers. The objective of these tests, however, was not to compare the maximum correct classification rates achievable by using the various approaches. The objective was instead to show the level of training speed-up that can be achieved by using the task decomposition algorithm ARTD.

5.4 Comments

The ARTD decomposition strategy is based on the analysis of the classification performance (i.e., the outputs) of partially trained classifiers. This differentiates ARTD from previously proposed task decomposition algorithms that base the decomposition strategy on the analysis of the input patterns [38, 39, 40]. This represents a significant advantage especially in cases in which the input patterns are very difficult to cluster, such as in the case of multi-variate time-series. This is the type of task tackled by ALADDIN.

ARTD has been shown to reduce training time significantly, when compared with non-modular classifiers, on two separate test cases. In these examples, improvements in training time ranged from 8-fold in a 26-class static pattern classification task, to more than 30-fold in a 32-class dynamic pattern classification task. This speed-up appears to be roughly exponential in the number of classes of the task at hand, as shown in Fig. 7 below.

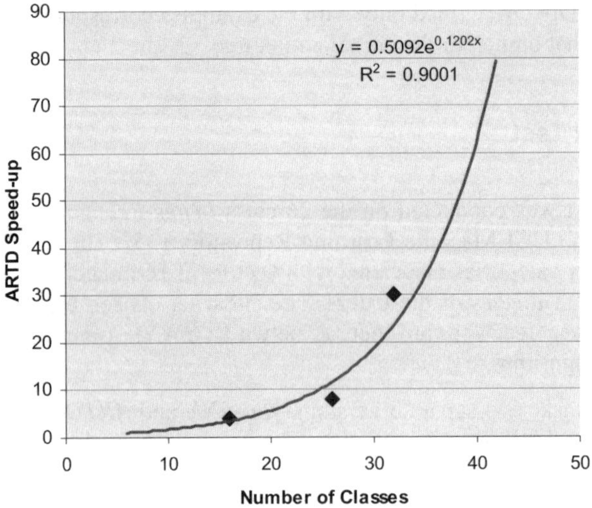

Fig. 7. ARTD Speed-up

The hierarchical classification structure generated by ARTD has other additional advantages over a *flat* architecture. These are discussed in the following.

Partial Diagnosis

First of all, a hierarchical classifier structure provides a series of partial classification results that give additional information on the pattern that is being analysed. For example, an ARTD classifier might fail to completely classify one pattern as belonging to a specific class, i.e. none of the leaves in the classification tree receives a definite activation, but it might still be able to identify to which *set* of classes the pattern is most likely to belong, by giving strong activation at a higher level in the hierarchy.

Incremental Application Development

Another advantage of modular architectures, like the ones generated by ARTD, is the support for incremental application development, i.e. the ability of adding new classes to the existing structure without the need for re-training of the whole classifier.

If we want to add a new class to an existing hierarchical classifier, we give in input to the current classifier a set of examples of the new class and we analyse which network(s) becomes activated. These examples might activate slightly different paths in the current structure. The most appropriate level at which the new class should be added will correspond to the deepest point common to the activation paths of all the examples. The neural network corresponding to this point will then be replaced with an equivalent network with one extra output for the new class, and it will be trained with the examples of the new class and the examples of all the classes belonging to that branch.

Application Maintenance

The advantage of generating a modular architecture is not limited to the incremental application development capabilities that we have just discussed. Application maintenance is another task that is greatly simplified by modularity.

After a full application has been developed, new examples of the currently included classes might become available. The inclusion of these new examples into the "compiled knowledge" of the classifier is a primary maintenance task. Without a modular architecture, this inclusion could be very difficult, and, in the worst case, it could involve re-training of the whole classifier on the augmented database of examples. Modularity clearly facilitates this task. When new examples of a certain class become available, only the modules along the classification path of that class need to be maintained.

Another important aspect of application maintenance involves the direct intervention on the classifier to correct performance problems that might manifest after the completion of the application development. For example, the performance of the system on specific classes might turn out not to be completely satisfactory. This situation calls for a refinement of the classifier, through more extended training, introduction of noise and other distortions to the examples, etc. (the application developer has several techniques available to improve the performance of neural network classifiers). A modular structure is a clear advantage in this situation, allowing for a more focused intervention and, in general, for a much greater control over the application. The application developer might then decide to refine the training of selected modules, or even re-train them with new settings, without having to worry about the problem of unwanted side-effects on other modules of the classifier that are already performing satisfactorily.

The problems of internal interference, typical of "flat" classifiers, are clearly avoided in a modular architecture.

6 Conclusions and Further Work

In this chapter we have described the ALADDIN methodology for dynamic event recognition and fault diagnosis, developed as part of the internationally sponsored OECD Halden Reactor Project. The main focus of this work has been towards the improvement of the practical applicability and scalability of this type of systems to real processes and machinery.

The core of the chapter focussed on the description the three principal techniques that are at the basis of ALADDIN, namely:

- Recurrent neural network ensembles.
- Wavelet On-Line Pre-processing (WOLP).
- Autonomous Recursive Task Decomposition (ARTD).

The utility of each of these techniques, in terms of increased applicability and scalability, was demonstrated by a series of experiments.

Further tests are under way to confirm the encouraging results reported in this chapter. It would not be surprising to find even greater gains in the use of ARTD on bigger and harder tasks than those here described, to the point where a task would simply be intractable for a flat classifier.

Real world applications of systems like ALADDIN in a complex industrial process, where the number of possible faults and anomalies to be considered can reach the hundreds, we believe would not be practically possible without the support of task decomposition techniques like ARTD. The inclusion in ALADDIN of input dimensionality reduction techniques, such as feature selection and non-linear PCA, will be our next step towards the construction of a truly scalable system.

References

1. Hassoun (1995) MH Fundamentals of Artificial Neural Networks. The MIT Press, Cambridge.
2. Bartlett EB, Uhrig RE (1992) Nuclear Power Plant Status Diagnostics Using an Artificial Neural Network. J Nuclear Technology 97:272-281.
3. Basu A, Bartlett EB (1995) Detecting Faults in a Nuclear Power Plant by Using a Dynamic Node Architecture Artificial Neural Network. J Nuclear Science and Engineering 116.
4. Bartal Y, Lin J, Uhrig RE (1995) Nuclear Power Plant Transient Diagnostics Using Artificial Neural Networks that Allow "Don't-Know" Classifications. J Nuclear Technology 110.
5. Jeong E, Furuta K, Kondo S (1995) Identification of Transient in Nuclear Power Plant Using Neural Network with Implicit Time Measure. In: Proceedings of the International Topical Meeting on Computer-Based Human Support Systems: Technology, Methods, and Future. The American Nuclear Society Inc., La Grange Park, IL, pp. 467-474.
6. Jeong E, Furuta K, Kondo S (1996) Identification of Transient in Nuclear Power Plant Using Adaptive Template Matching with Neural Network. In: Proceedings of the International Topical Meeting on Nuclear Plant Instrumentation, Control, and Human-Machine Interface Technologies, NPIC&HMIT'96. The American Nuclear Society Inc., pp. 243-250.
7. Attieh IK, Gribok AV, Hines JW, Uhrig RE (2000) Transient Detection Module to Enhance Nuclear Reactor's Operational Safety. In: Proceedings of the 3rd ANS International Topical Meeting on Nuclear Plant Instrumentation, Control and Human-

national Topical Meeting on Nuclear Plant Instrumentation, Control and Human-Machine Interface Technologies, NPIC&HMIT 2000, Washington DC.

8. Kassidas A, MacGregor JF, Taylor PA (1998) Synchronization of Batch Trajectories Using Dynamic Time Warping. J AIChE 44(4):864-875.
9. Keogh EJ, Pazzani MJ (1999) Scaling up dynamic time warping to massive datasets In: Proceedings of the 3rd European Conference on Principles and Practice of Knowledge Discovery in Databases, Prague.
10. Kundu A, Chen GC, Persons CE (1994) Transient Sonar Signal Classification Using Hidden Markov Models and Neural Nets. IEEE Journal of Oceanic Engineering 19(1):87-99.
11. Kwon KC, Kim JH (1999) Accident identification in nuclear power plants using hidden Markov models. J Engineering Applications of Artificial Intelligence 12:491-501.
12. Elman JL (1990) Finding structure in time. J Cognitive Science 14:179-211.
13. Roverso D (2000) Soft Computing Tools for Transient Classification. J Information Sciences 127:137-156.
14. Mozer MC (1992) Induction of multiscale temporal structure. In: Moody, Hanson and Lippman (eds) Advances in Neural Information Processing Systems 4:349-391. Morgan Kaufmann.
15. Bengio Y, Simard P, Frasconi P (1994) Learning Long-Term Dependencies with Gradient Descent is Difficult. IEEE Transactions on Neural Networks 5(2):157-166.
16. Wolpert D (1992) Stacked Generalization. J Neural Networks 5:241-259.
17. Breiman L (1996) Bagging Predictors. J Machine Learning 24(2):123-140.
18. Freund Y, Shapire RE (1996) Experiments with a New Boosting Algorithm. In: Machine Learning, Proceeding of the 13th International Conference (ICML '96), pp. 148-156. Morgan Kaufmann.
19. Krogh A, Vedelsby J (1995) Neural network ensembles, cross validation, and active learning. In: D. Touretzky and T. Leen (eds) Advances in Neural Information Processing Systems, Vol. 7, pp. 231-238. MIT Press, Cambridge, MA.
20. Roverso D (2000) Neural Ensembles for Event Identification. In Proceedings of Safeprocess'2000, The 4th IFAC Symposium on Fault Detection, Supervision and Safety for Technical Processes, Budapest, HU, pp. 478-483.
21. Roverso D (2000) Multivariate Temporal Classification by Windowed Wavelet Decomposition and Recurrent Neural Networks. In: Proceedings of the 3rd ANS International Topical Meeting on Nuclear Plant Instrumentation, Control and Human-Machine Interface Technologies, NPIC&HMIT 2000, Washington DC.
22. Chen BH, Wang XZ, Yang SH, McGreavy C (1999)Application of wavelets and neural networks to diagnostic system development, 1, feature extraction. Computers and Chemical Engineering 23:899-906.
23. Dechenes CJ (1995) Fuzzy Kohonen Network for the Classification of Transients Using the Wavelet Transform for Feature Extraction. Information Sciences 87:247-266.
24. Koulouris A, Bakshi BR, Stephanopoulos G (1996) Empirical Lerning Through Neural Networks: The Wave-net Solution. In: G. Stephanopoulos and C. Han (eds) Intelligent Systems in Process Engineering. Academic Press, pp. 437-484.
25. Wang XZ, Chen BH, Yang SH, McGreavy C (1999) Application of wavelets and neural networks to diagnostic system development, 2, an integrated framework and its application. Computers and Chemical Engineering 23:945-954.
26. Strang G, Nguyen T (1996) Wavelets and Filter Banks. Wellesley-Cambridge Press.

27. Cochran WG 1977 Sampling Techniques (3rd Ed.), John Wiley & Sons.
28. Lohr S (1999) Sampling: Design and Analysis, Duxbury Press, Pacific Grove, CA.
29. Du Mouchel W, Volinsky C, Johnson T, Cortes C, Pregibon D (1999) Squashing files flatter. In: Proceedings of the 5th International Conference on Knowledge Discovery and Data Mining (KDD'99).
30. Blum AL, Langley P (1997) Selection of Relevant Features and Examples in Machine Learning. Artificial Intelligence 97(1-2):245-271.
31. Hall DL, Llinas J (1997) An introduction to multisensor data fusion. Proceedings of the IEEE, 85(1):6-23.
32. Jenkins R, Yuhas B (1993) A simplified neural-network solution through problem decomposition: The case of the truck backer-upper. IEEE Trans. on Neural Networks 4:718-722.
33. Anand R, Mehrotra KG, Mohan CK, Ranka S (1995) Efficient classification for multiclass problems using modular neural networks. IEEE Trans. on Neural Network, 6:117-124.
34. Lu BL, Ito M (1999) Task Decomposition and Module Combination Based on Class Relations: A Modular Neural Network for Pattern Classification. IEEE Trans. on Neural Networks 10(5):1244-1256.
35. Jordan MI, Jacobs RA (1994) Hierarchical mixtures of experts and the EM algorithm. Neural Computation 6:181-214.
36. Auda G, Kamel M (1998) Modular Neural Network Classifiers: A Comparative Study. Journal of Intelligent and Robotic Systems 21:117-129.
37. Blake CL, Merz CJ (1998) UCI Repository of machine learning databases [http://www.ics.uci.edu/~mlearn/MLRepository.html]. Irvine, CA: University of California, Department of Information and Computer Science.
38. Bartfai G (1994) Hierarchical clustering with art neural networks. In: World Congress on Computational Intelligence, Vol. 2, pp. 940-944, Florida, USA.
39. De Bollivier M, Gallinari P, Thiria S (1991) Cooperation of neural nets and task decomposition. In: Proceedings of the International Joint Conference on Neural Networks, Vol. 2, pp. 845-849, Seattle, USA.
40. Corwin E, Greni S, Logar A, Whitehead K (1994) A multi-stage neural network classifier. In: Proceedings of World congress on Neural Networks, Vol. 3, pp. 198-203, San Diego, USA.

23 PEANO and On-Line Monitoring Techniques for Calibration Reduction of Process Instrumentation in Power Plants

Paolo F. Fantoni

Institutt for energiteknikk, OECD Halden Reactor Project
POBox 173, N-1751 Halden, Norway.

Paolo.Fantoni@hrp.no

1 Introduction

On-line monitoring evaluates instrument channel performance by assessing its consistency with other plant indications. Industry and EPRI [1] experience in the USA at several plants has shown this overall approach to be very effective in identifying instrument channels that are exhibiting degrading or inconsistent performance characteristics.

On-line monitoring of instrument channels provides information about the condition of the monitored channels through accurate, more frequent monitoring of each channel's performance over time. This type of performance monitoring is a methodology that offers an alternate approach to traditional time-directed calibration. On-line monitoring of these channels can provide an assessment of instrument performance and provide a basis for determining when adjustments are necessary. Elimination or reduction of unnecessary field calibrations can reduce associated labor costs, reduce personnel radiation exposure and reduce the potential for miscalibration.

PEANO [2,3,4,5,6,7,8,9] is a system for on-line calibration monitoring developed in the years 1995-2001 at the Institutt for energiteknikk (IFE), Norway, which makes use of Artificial Intelligence techniques and Fuzzy Logic for its purpose. The system has been tested successfully in Europe in off-line tests with EDF (France), Tecnatom (Spain) and ENEA (Italy), and in USA in co-operation with EPRI. PEANO is currently installed and used for on-line monitoring at the HBWR reactor in Halden.

Artificial Neural Networks (ANN) and Fuzzy Logic can be combined to exploit learning and generalization capability of the first technique with the approximate reasoning embedded in the second approach. Real-time process signal validation is an application field where the use of this technique can improve the diagnosis of faulty sensors and the identification of outliers in a robust and reliable way.

PEANO implements a fuzzy-possibilistic clustering algorithm to classify the operating region in which the validation process has to be performed. The possibilistic approach (rather than probabilistic) allows a "don't know" classification that results in a fast detection of unforeseen plant conditions or outliers.

In a previous prototype of PEANO, the classifier, based on the classic ISODATA [10] algorithm, identified the incoming signal pattern (a set of reactor process signals) as a member of one of nine clusters covering the entire universe of discourse represented by the possible combinations of steady-state and transient values of the input set in the *n*-dimensional input world. Each cluster was associated with one ANN that was previously trained only with data belonging to that cluster, for the input set validation process.

During the operation, while the input point moved in an *n*-dimensional world (because of process state changes or transients), the classifier provided an automatic switching mechanism to allow the best tuned ANN to do the job. The results were satisfactory, but the following two drawbacks were identified:

1. the boundary problem. That model did not consider properly the transition area among clusters, where the input pattern could be considered a member of two or even three clusters at the same time. As a consequence of that, the switching mechanism changes the working ANN abruptly, when the input pattern moves in ambiguous areas, resulting in unexpected changes in performance.
2. a crisp classifier always tries to find a reasonable membership cluster, even when the input pattern is far away from all the identified clusters. In turn, the activated ANN always gives a response, so that there is no way to have a reliability measure for the output. This is a very well-known problem of every black-box approach (and there is nothing more than ANN modeling that can be called black-box), that could justify the very few industrial ANN applications currently existing in the world, in the nuclear field.

The two problems above were solved by introducing a *possibilistic fuzzy clustering technique* coupled to a set of concurrent specialized ANN and a fuzzy model (Mamdani type) for reliability estimation. Fig. 1 shows a schematic diagram of the overall Signal Validation system.

There are two main advantages in using this architecture: the accuracy and generalization capability is increased compared to the case of a single network working in the entire operating region and the ability to identify abnormal conditions, where the system is not capable of operating with a satisfactory accuracy, is improved.

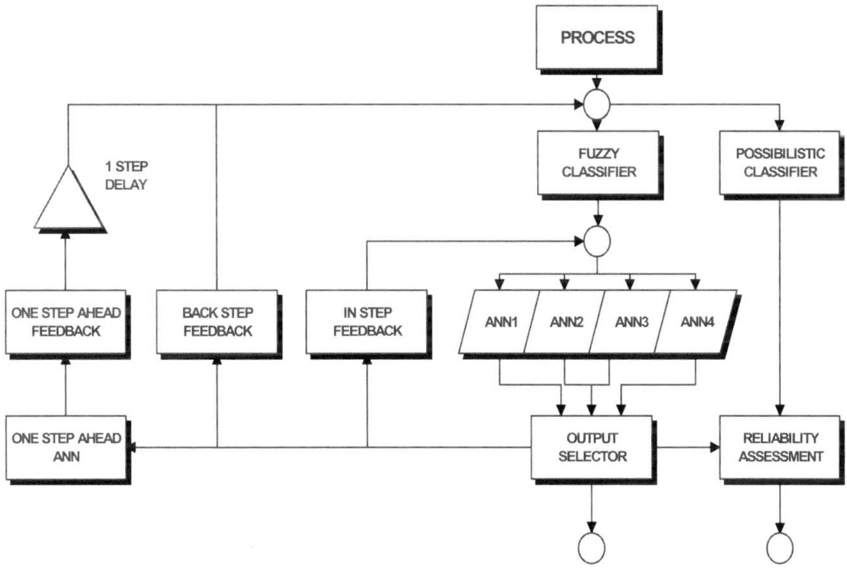

Fig. 1. Functional Diagram of PEANO

2 Fuzzy Classification

Let $\bar{x} = [x_1, x_2, \ldots, x_N]^T$ a vector in $\Re^N$ representing an input dataset. The N components are correlated process signals that constitute a snapshot of the monitored process at a given time. Given $X = \left(\bar{x}_1, \bar{x}_2, \ldots, \bar{x}_p\right)$ the $N \times P$ matrix of P patterns covering the $\Re^N$ operating region, the basic idea is to split this region into Q fuzzy clusters and derive a mapping function that assigns each pattern $\bar{x}_k | k = 1 \ldots P$ to each cluster $C_k | k = 1 \ldots Q$ at *some degree*. This transformation is expressed by the following equation:

$$\bar{x}_k \Rightarrow \{u_{1k}, u_{2k}, \ldots u_{Qk}\} \tag{1}$$

$$u_{ik} \in [0,1], i = 1 \ldots Q, k = 1 \ldots P \tag{2}$$

where u_{ik} is the membership grade of the pattern $\bar{x}_k$ in cluster C_i. In pattern recognition the Q clusters are identified by prototype patterns, which in the case of spherical or ellipsoidal clusters are also called centroids, so that the representation of a fuzzy classifier for a given X $(N \times P)$ matrix dataset with Q clusters is completely defined by:

$$B = (\bar{\beta}_1, \bar{\beta}_2, \ldots, \bar{\beta}_Q) \tag{3}$$

$$U = \left(\bar{u}_1, \bar{u}_2, \ldots, \bar{u}_Q\right)^T \tag{4}$$

$$\bar{\beta}_m = \left[x_1^{Cm}, x_2^{Cm}, \ldots, x_N^{Cm}\right]^T \tag{5}$$

$$\bar{u}_m = \left[u_{m1}, u_{m2}, \ldots, u_{mP}\right]^T \tag{6}$$

where B is the $N \times Q$ matrix of the cluster prototypes and U is the $Q \times P$ matrix of the membership grades of X, also called the *fuzzy C-partition*.

The fuzzy partition problem, as expressed in Bezdek [11], can be solved with the minimization of an objective function, which can be written as:

$$J(B, U, X) = \sum_{i=1}^{Q} \sum_{j=1}^{P} \left(u_{ij}\right)^m \Delta^2\left(\bar{x}_j, \bar{\beta}_i\right) \tag{7}$$

where $m \in [1, \infty]$ is called the fuzzifier parameter and Δ is a function representing the distance between two vectors. When $m = 1$ the classifier is crisp and when $m >> 1$ fuzziness is maximized. $m = 2$ is the recommended value, for most applications.

The choice of the Δ function depends on the expected shape of the clusters. If the Euclidean distance is used, which is the right choice for spherical clusters, the resulting algorithm is the popular *Fuzzy C-means* algorithm. In this application clusters with different shapes and sizes are expected, so that Euclidean distances would not work well. To take care of the not uniform distribution of the patterns in the dataset, the GK algorithm, from Gustafson and Kessel [12], has been used. Here the distance function is expressed as:

$$\Delta_{i,j}^{\ 2} = \left(\det C_i\right)^{\frac{1}{N}} \left(x_j - \beta_i\right)^T C_i^{-1} \left(x_j - \beta_i\right) \tag{8}$$

where $\boldsymbol{C_i}$ is the *fuzzy-covariance* matrix for cluster *i*, defined as:

$$C_i = \frac{1}{\sum_{j=1}^{P} \left(u_{ij}\right)^m} \sum_{j=1}^{P} \left(u_{ij}\right)^m \left(x_j - \beta_i\right)\left(x_j - \beta_i\right)^T \tag{9}$$

In fuzzy clustering U must satisfy the following three conditions:

$$\sum_{i=1}^{Q} u_{ik} = 1, k = 1 \ldots P \tag{10}$$

$$u_{ik} \in [0,1],\ i = 1 \ldots Q,\ k = 1 \ldots P \tag{11}$$

$$0 < \sum_{j=1}^{P} u_{ij} < P, i = 1 \ldots Q \tag{12}$$

Condition (10) reflects the probabilistic requirement that the total probability for an input dataset pattern to belong to *any* cluster is 1. In other words, patterns not reflecting any of the identified cluster prototypes are classified and assigned to the *relatively* most probable cluster, only because of the implicit certainty that *all* the patterns belong to the established partition. There can be uncertainty (or fuzziness) on *where* the incoming pattern should be assigned, but no uncertainty on *if* it should be assigned somewhere. When this methodology is applied to signal validation applications, a number of problems may arise:

- Lack of robustness against noisy data. There is no compensation for the noise in the calculation of B and U.
- It is not able to say, "I do not know", also when this would be the best answer. An incoming pattern might be given a high grade of membership in a cluster, even if it is far away from all the centroids, only because it is *relatively* closer to one specific cluster.

Relaxation of requirement (10) leads to a possibilistic approach that results in a possible solution of the two above-mentioned limitations.

A possibilistic classifier initially learns a dataset X of pattern samples (in other words it calculates B and U). During this process, the model increases its robustness to noisy data and many patterns in X could be discarded as not representative of any developing cluster. When new patterns are examined, the possibilistic model evaluates in which cluster or clusters the incoming pattern could be *possibly* assigned, if any.

Following Krishnapuram and Keller's work [13], minimization of the objective function (7), without the constraint in (10), results in the following equations for U and B:

$$u_{ij} = \frac{1}{1 + \left(\frac{\Delta^2\left(\bar{x}_j, \bar{\beta}_i\right)}{\eta_i} \right)^{\frac{1}{m-1}}} \tag{13}$$

where $m > 1$ and

$$\beta_i = \frac{1}{\sum_{j=1}^{P} \left(u_{ij}\right)^m} \sum_{j=1}^{P} \left(u_{ij}\right)^m \bar{x}_j \tag{14}$$

where η_i is computed by:

$$\eta_i = \left(\det C_i\right)^{\frac{1}{N}} \tag{15}$$

The step-by-step procedure used to develop the fuzzy and possibilistic classifiers can be summarized as follows:

- Given a set of samples *X*, compute an initial set of cluster centroids using the ISODATA algorithm, which has been chosen because it automatically optimizes the number of required clusters.

- Initialize the elements of the partition matrix *U* with crisp values (0 or 1), using ISODATA. Then run the *GK* algorithm, which produces the fuzzy classifier.

- Use the updated matrix *U* and *B* , from the previous step, to start the iterative process as shown in Eqs. (13) and (14) to arrive to a possibilistic partition.

3 The Artificial Neural Network Module

Sample data in dataset *X* are collected in *Q* training datasets, to be used for training *Q* supervised neural networks. Each pattern in *X* is assigned to one or more training set according to the fuzzy partition, as long as its possibilistic index in *U* is above a threshold value *h* in one or more identified clusters, with $h = (0.5, 1)$. The role of the threshold parameter *h* is twofold:

- sample patterns not adequately represented in any cluster are discarded, so that they have no influence on the network weights calculation.
- sample patterns *possibly* represented in many clusters (responsible of the above mentioned boundary problem) are used in the training set of many corresponding networks.

The network architecture used in PEANO consists of five layers (three hidden layers), feedforward structure trained with the conjugated backpropagation algorithm, as shown in Fig. 2. The number of neurons in the bottleneck layer is a critical parameter to reach a good generalization capability. The hidden layers use hyperbolic tangent transfer functions, while the output layer is linear. This architecture has been tested to be more robust to process noise and sensor faults [14].

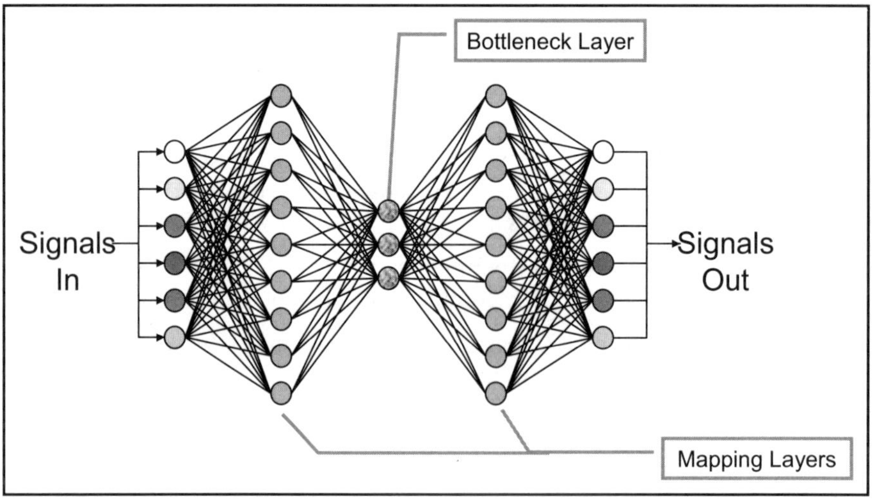

Fig. 2. The 5-layers architecture in PEANO

The input to the ANN's is not limited to the current pattern. To capture the process dynamics, a number of past values of the time series are used, together with the current ones, so that the total number of input nodes in each ANN is $N \times R$, where N is the number of signals and R the number of past values used. PEANO implements moving averages from the current time-step to a number of selectable past values in a Fibonacci sequence. This allows considering both short and long time constant effects in the process dynamics.

The three feedback loops shown in Fig. 1 are used during the recall (on-line validation).

The *in-step feedback*, in case of mismatch in one or more signals, re-evaluates the corrected pattern to get better estimates in the not affected channels. This feedback is triggered only if the signal mismatch is estimated to be greater than two standard deviations, to avoid instabilities.

The *back-step feedback* corrects mismatching signals for the future evaluations, when those values will be used as past values, as mentioned before. This is also triggered by the same threshold value.

The *one-step-ahead* feedback monitors the next coming pattern for possible large deviations. Large deviations have a negative effect on the overall performance, because they lead to false classifications with the result that a non-optimal ANN is triggered for recall, only as a consequence of a large deviation in one or more channels. Large deviations from the expected values are corrected before the classification, resulting in a much more accurate and stable validation. In Fig. 1, a predictive ANN calculates one-step-ahead expected values, but tests have shown that the current values (at time t) can be used as a rough estimation of the values at time t+1.

The recall algorithm used to validate an input pattern applies the concept of *presumption of no sensor fault, if possible.* This concept is based upon the well-tested hypothesis that when a neural network confirms the sensor input (no fault condition) and the relative cluster membership grade is high, the network output is reliable.

This leads to the following recall strategy:

- For each process sample: get the most representative cluster, which is the one with the highest membership grade u_1 in U
- Recall the output using the neural network associated to this cluster
- Calculate the maximum signed deviation, as follows:

$$err_1 = \max\left|s_j - o_j\right| \times \operatorname{sgn}\left(s_j - o_j\right) \quad j = 1,\ldots,N \tag{16}$$

where $(s_j - o_j)$ is the mismatch for the *j*-th signal (s_j is the j-th signal value and o_j is the corresponding validated value).

- If err_1 is very low, accept the result, otherwise recall the pattern using also the network with the second highest grade of membership, u_2

- Calculate the following weighted error:

$$werr = \frac{err_1 \times u_1 + err_2 \times u_2}{u_1 + u_2} \tag{17}$$

where err_2 is the maximum error with the second network. Now if:

$$abs(werr) > abs(err1) \tag{18}$$

Accept the output from the first network, otherwise calculate a weighted output from:

$$wout = \frac{out_1 \times u_1 + out_2 \times u_2}{u_1 + u_2} \tag{19}$$

where out_1 and out_2 are the vector outputs from the two networks.

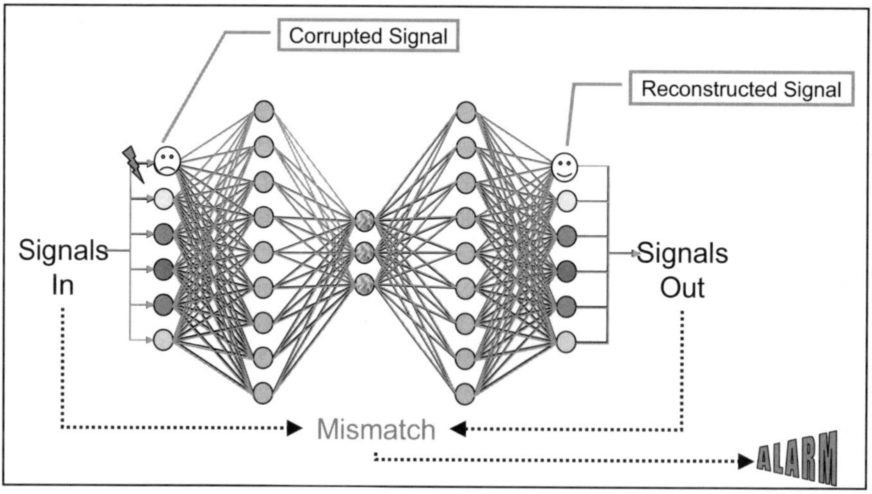

Fig. 3. Mismatch detection in a 5-layer Autoassociative ANN

3.1 The Reliability Assessment Module

In a previous work we tried to solve the reliability problem connected to the use of neural networks, using a Radial Basis Function network associated to a crisp pattern classifier [15]. The algorithm implemented now in PEANO extends the idea, exploiting the unique features of a possibilistic classifier.

The possibilistic cluster membership has an important role in the final decision whether the network output can be considered reliable or not. A high membership grade in one or two clusters increases our confidence that the data sample is contained in the training volume of one or two neural networks, so that they will be able to recall the output with low estimation error. On the other side, a low membership value in *all* the clusters is a clear warning that no network has been trained to recall such a pattern. Note that using fuzzy clustering techniques, it would not be possible to have either *low* values in all the clusters, or *high* values in more than one.

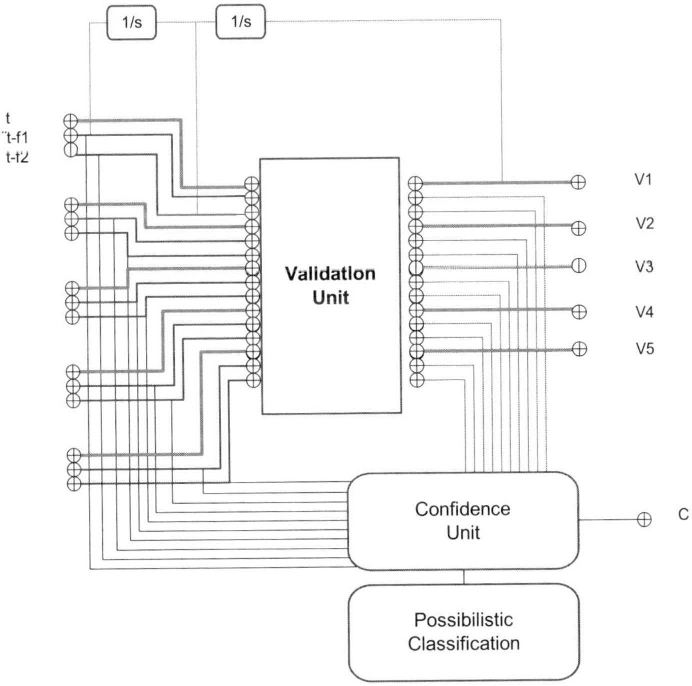

Fig. 4. Reliability assessment using Moving Average Delayed Nodes

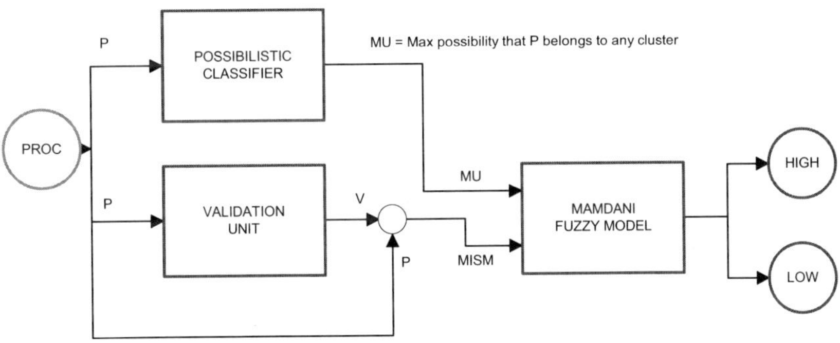

P = Vector of process signals
V = Vector of validated signal values
MU = Max membership value
MISM = P-V

Fig. 5. The Mamdani fuzzy model for reliability assessment

To understand how the reliability assessment function works, we need to discuss for a moment the uncertainties associated with the validation process.

In auto-associative ANNs, see Fig. 3, the inputs and the outputs are identical. During the training phase, when pairs of input-output patterns are presented, the ANN learns the correlation between the process variables, and, once it is fully trained, the outputs represent the best estimates of the instantaneous values of the current input variables. The used information is the difference Δs between inputs (I) and outputs (O), which is called the *mismatch*:

$$\Delta s_j = O_j - I_j\,, \quad 1 \leq j \leq n \tag{20}$$

where n is the number of monitored sensors. When the sensors are working properly, all of these estimates are virtually identical to the input measured process variables and the differences are very close to zero. If any error is introduced in the input vector (sensor fault or drift), a difference between the input value of the faulty sensor and its output is found. Δs might be considered the sum of the network estimation error (E^{NN}) and the measurement error (E^{meas}):

$$\Delta s = E^{NN} + E^{meas} \tag{21}$$

When E^{NN} is minimized with an appropriate training, Δs can be considered to be a parameter representing the measurement error and it is used for the validation of the signal.

However, the value of E^{NN} is known only for the training dataset, where the maximum error is in fact the target of the training process. The network behavior when a new input pattern is processed cannot be anticipated, so that the error associated with the ANN prediction may vary from an order of magnitude less than E^{meas} to several order of magnitude larger than that. In the latter case, since Δs becomes very large, the network conclusion is that one or more sensors are faulty, which is not the correct diagnosis.

E^{NN} is a measure of the network reliability and it can be estimated from Eq. (21) when E^{meas} is known. Since the measurement error for the delayed nodes should always be zero (the past values have already been corrected), the delayed output mismatch (see Fig. 4) can be used to estimate the ANN error.

Fig. 5 shows the complete reliability model, which is realized through a fuzzy model, where the input is the maximum membership grade of the sample and the maximum delayed signal mismatch in the neural network module, while the output is the reliability membership grade in three fuzzy sets assessing at what extent the reliability factor can be considered *high, medium,* or *low*. This fuzzy model applies *Mamdani-type* implication rules, with the following fuzzy rules:

IF *max-grade* is *low* AND *max-mismatch* is not *low*

THEN *rel-grade* is *low*

IF *max-grade* is *high*

THEN *rel-grade* is *high*

IF *max-grade* is *medium*

AND *max-mismatch* is *medium*

THEN *rel-grade* is *medium*

For each data sample presented to the system, the three fuzzy rules are *fired* at different degree, resulting in three different membership values for *rel-grade* in the three fuzzy sets *high, medium,* and *low*. These values give a clear idea about the accuracy of the network output.

4 Real-Time Monitoring with PEANO

PEANO has a Client-Server architecture, as shown in Fig. 6. The server is connected to the process through a TCP/IP communication protocol and the results of the validation activity is transferred to the client programs, also using TCP/IP.

Fig. 7 shows the display of a PEANO client during an on-line validation test.

The error bands in the mismatch plots are calculated by PEANO during the training according to the expected error of prediction for each particular cluster and signal. The error bands should be interpreted as follows:

- **narrow error band**: It is normally set at 2 standard deviations of the expected error. Exceeding this band is considered as the first warning, especially if the situation persists.
- **wide error band**: It is normally set at 3 standard deviations. Exceeding this band is considered a definite alarm condition

Table 1 shows typical error band values (wide band) in percent of the signal range, for a PWR test.

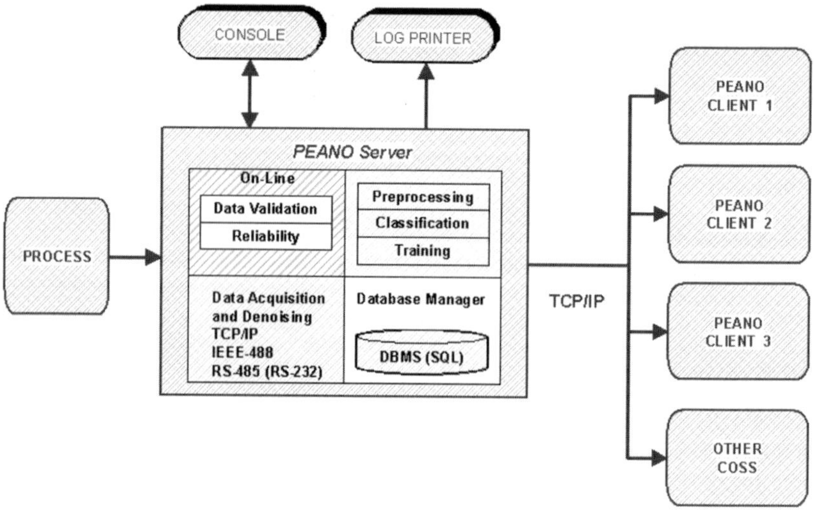

Fig. 6. The PEANO Client-Server architecture

Table 1. Typical PEANO accuracy in real-time tests

Signal	rated	Error band
Reactor power	100 %	0.24 %
Temp. control rods	232 steps	0.3 %
Coolant temperature	306 °C	0.1 °C
Pressurizer pressure	155 bar	0.1 bar
Pressurizer level	64.4 %	0.3 %
Feedwater flow	2088 kg/s	5 kg/s
Steam generator level	45 %	0.012 %
Steam pressure	69 bar	0.12 bar

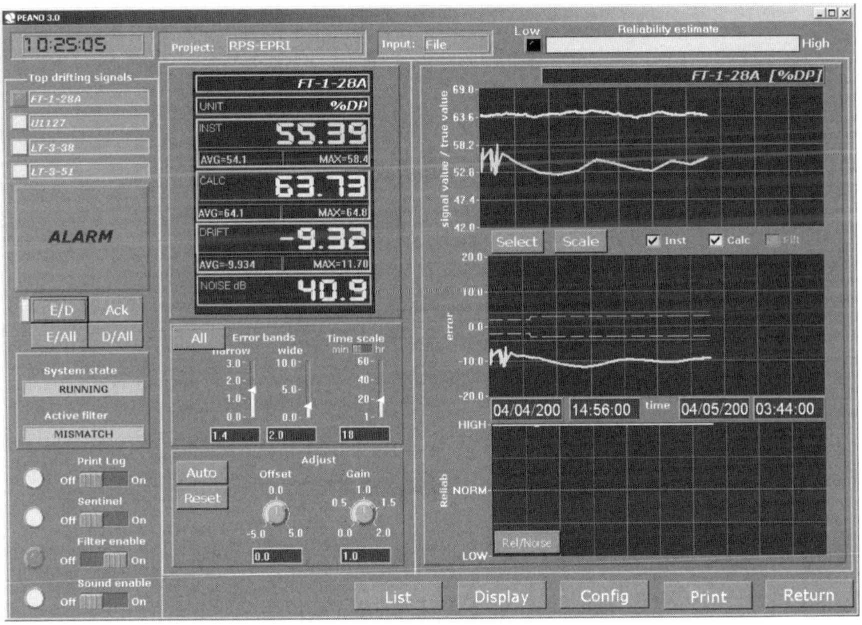

Fig. 7. The PEANO Client Display during monitoring

PEANO includes a real-time digital filter module that allows the interactive design of Infinite Impulse Response (IIR) filters to be used during the monitoring. Filters can be used as input noise suppression or mismatch noise suppression, as indicated in Fig. 8.

Mismatch filters can be efficiently used to minimize spurious alarms due to normal instrument noise or spikes, while input filters can improve significantly the reliability of the validation task.

The filter design module includes Butterworth, Chebyshev and Elliptic filters. It is possible to design and automatically use different filters for each signal, according to the frequency response of the different sensors. It is also possible to design and save in the database different filter configurations and activate one of them during the operation.

Note that, while the mismatch filter can be switched on or off in each client station, the server controls the input filter.

The noise trend diagram and noise level in the client display is not affected by the use of input or mismatch filters: it will always show the original instrument noise below the Nyquist frequency.

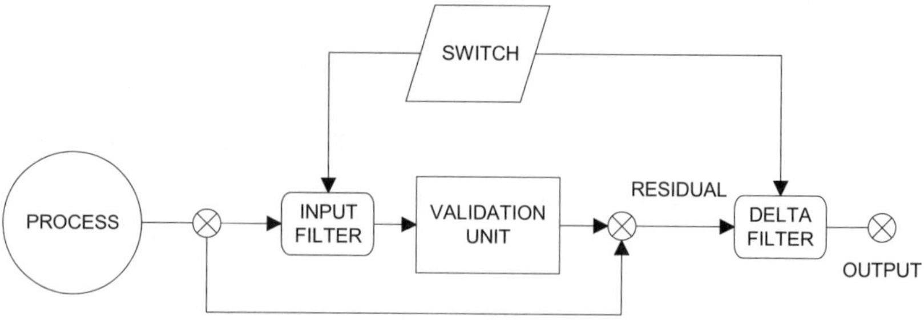

Fig. 8. Position of the noise suppression filters in the validation chain

4.1 Performance Test Results Using PEANO

Fig. 9 to Fig. 13 describe some results achieved with PEANO, using real data from a US PWR nuclear plant (data provided by EPRI). PEANO was used to monitor 55 process signals up to the secondary side of the steam generators. The data was taken from 3 months of operation (March to May 2000) at different operation conditions.

Fig. 9 shows the re-calibration of a steam flow sensor while the plant was in operation. It is visible that after this intervention, the mismatch was well within the 2 standard deviation limits.

Fig. 10 shows the monitoring of a feedwater flow sensor during the power-up ramp. While the mismatch between PEANO and the instrument is always within the limits, it is interesting here to note that the tolerance band width changes with the plant operation conditions, because the accuracy of the system cannot be the same in all the possible conditions.

Fig. 11 is an interesting example of span drift of a steam flow sensor. Span drifts are difficult to detect because they show up only at some location of the instrument range. In this real-life example, the instrument was perfectly inside the calibration range until the power level came close to the rated level (high end of the instrument range). At this point the instrument started to drift and eventually finished outside the allowed tolerance band. At the plant, this drift was discovered only one month later.

Fig. 12 shows the embedded noise-filtering characteristic of systems based on auto associative ANN. The 2 mapping layers together with the bottleneck layer (see Fig. 2) act as a low-pass filter, which is highly desirable when the objective is the early detection of drifts.

Fig. 13 shows the results of a simulated drift in a feedwater sensor. Here it is possible to see how PEANO follows closely the theoretical true value of the signal while the sensor is drifting. The lower diagram (the mismatch diagram) confirms that this drift would have been detected quite early, as soon as the mismatch line passe through the tolerance band.

4.2 Using PEANO for Estimation of Unknown Parameters

As described in sec. 3, the input and the output layers of an autoassociative network are identical. For this reason the ANN architecture used in PEANO and showed in Fig. 2 is suitable for signal validation, but not for signal estimation or prediction.

However, the Neuro-Fuzzy model that PEANO implements, composed by a pattern recognition module followed by a bank of autoassociative ANN, turns PEANO in an efficient system for signal estimation. In PEANO this is called "external signal definition", where one of the input signals is tagged as "external", set to 0 and treated in a different way by the system. The only requirement that exists in order to handle an input as external, is that a sufficient correlation exists among this input and the other. Basically, the algorithm used by PEANO works as follows:

- The system finds a rough value for the external input that *maximizes* the input pattern (composed by the external signal and all the other inputs) fuzzy membership in one of the defined clusters.
- The pattern containing the rough value (a reasonable value) of the external input is passed to the ANN associated to the cluster where the pattern belongs to. At the output of this ANN one has the final value for the external input. In other words, the ANN bank acts as a fine tuner of the pattern recognition module.

In this operation mode, PEANO accomplishes the double task of calculating the unknown parameter and validating (correcting) the known parameters used to calculate it.

An example of this technique is shown in Fig. 14, where 3 concurrent systems are used to estimate the emission of O_2, CO, and NOx from a gas turbine plant (data provided by Foxboro Scada, Italy). Three concurrent systems are needed because the calculated O_2 content must be used as input for the CO calculation and the CO is also input for the NOx estimation. Tests on this system have shown that the estimation of the 3 gases is robust to any drift or fault in any of the known input parameters.

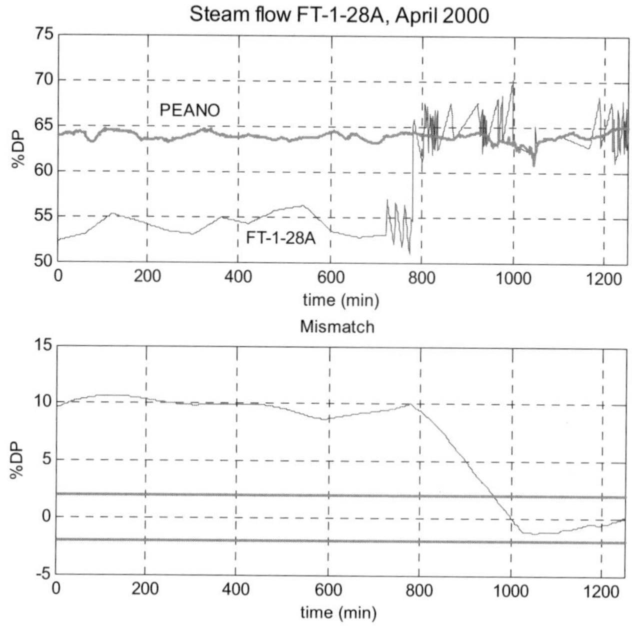

Fig. 9. Re-calibration of a Steam flow channel under PEANO supervision

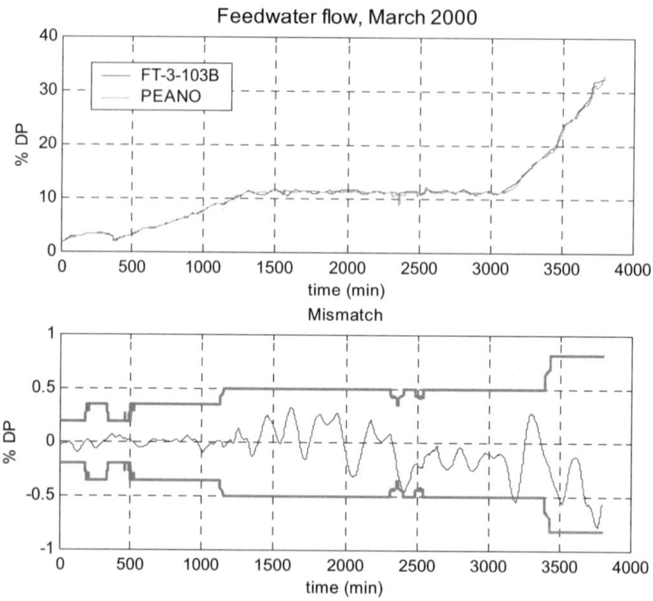

Fig. 10. Feedwater flow monitoring during a power-up ramp

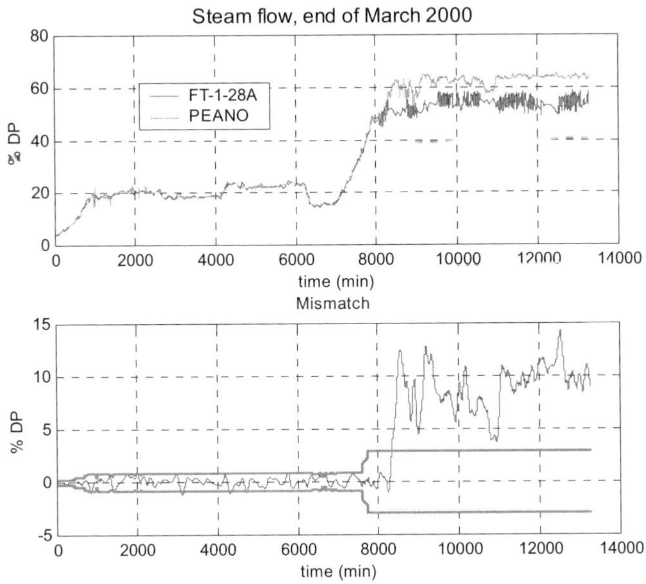

Fig. 11. Span drift detection during a power-up ramp

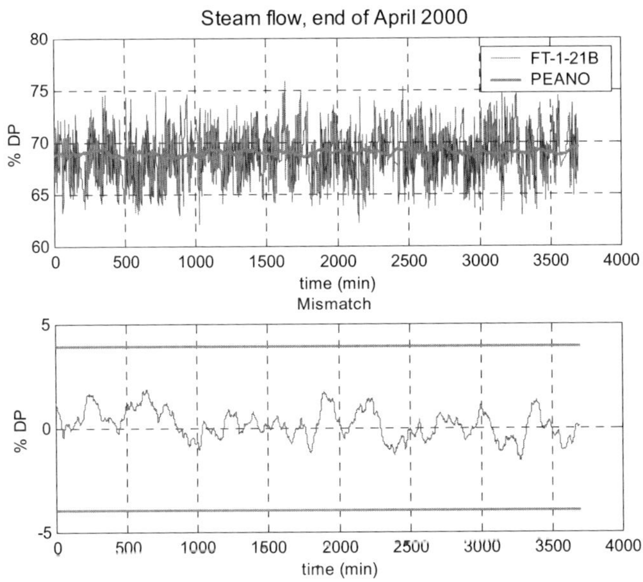

Fig. 12. Noise suppression with PEANO

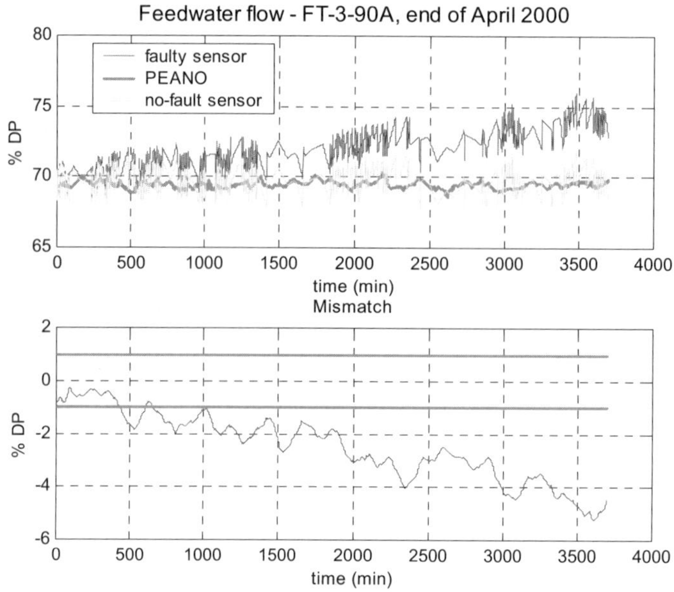

Fig. 13. Simulation of feedwater flow drift at full power

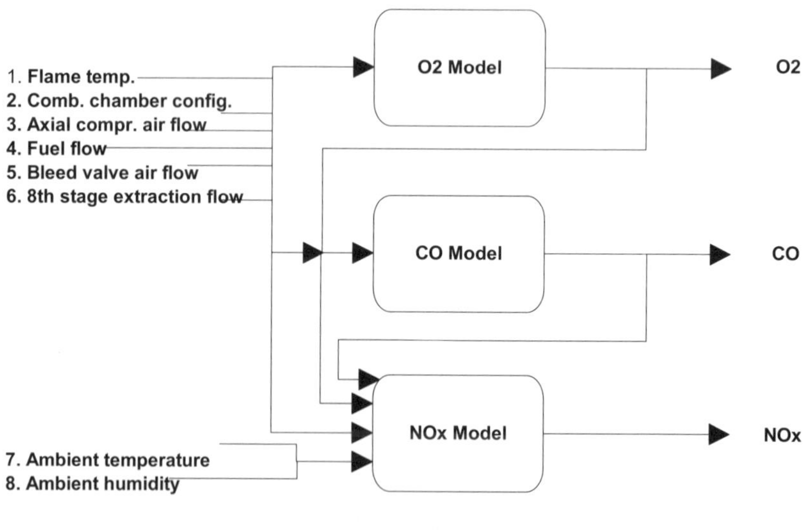

Fig. 14. Concurrent PEANO operation for emission gas prediction

Acknowledgements

PEANO has been developed at the OECD Halden Reactor Project by the author and many scientists from all over the world. A non-exhaustive list of scientists that contributed to this project is:

Attila Racz and Sandor Lipcsei, KFKI, Hungary

Mario Hoffmann, IFE, Norway

Davide Roverso, IFE, Norway

Stefan Figedy, VUJE, Slovac Republic

Alessandro Mazzola, University of Milan, Italy

Mauro Vitor De Oliveira, CNEN, Brasil

Aimar Sorenssøn, IFE, Norway

Hilary Selmer Olsen, IFE, Norway

The author wants to thank also EDF and CEA (France), Tecnatom (Spain), ENEA and Foxboro Scada (Italy), EPRI, Edan Engineering, AMS and The University of Tennessee (USA) for their help in providing the data that was used to test PEANO and for the help given during the development. Special thanks must be given to Dr Robert E. Uhrig (UT), Wesley Hines (UT) and Hashem Hashemian (AMS) for their constant contributions and efforts.

References

1. EPRI (2000), On-Line Monitoring of Instrument Channel Performance, TR-104965-R1 NRC SER.
2. Fantoni P, Mazzola A (1994) Applications of Autoassociative Neural Networks for Signal Validation in Accident Management. In: Proceedings of the IAEA Specialist Meeting on Advanced Information Methods and Artificial Intelligence in Nuclear Power Plant Control Rooms, Halden, Norway.
3. Fantoni p, Mazzola A (1995) Transient and Steady State Signal Validation in Nuclear Power Plants using Autoassociative Neural Networks and Pattern Recognition. In: Proceedings of SMORN VII, Avignon, France.
4. Fantoni P, Mazzola A (1996) A Pattern Recognition-Artificial Neural Networks Based Model for Signal Validation in Nuclear Power Plants. In: Annals of Nuclear Energy, Vol. 23, No.13, pp 1069-1076.

5. Fantoni P, Mazzola A (1996) Multiple Failure Signal Validation in Nuclear Power Plants using Artificial Neural Networks. In: Nuclear Technology, Vol. 113, No. 3, pp 368-374.

6. Fantoni P (1996) Neuro-Fuzzy Models Applied to Full Range Signal Validation in NPP. In: Proceedings of NPIC&HMIT'96, The Pennsylvania State Univ., PA.

7. Fantoni P, Figedy S, Racz A (1998) PEANO, A Toolbox for Real-Time Process Signal Validation and Estimation. HWR-515, OECD Halden Reactor Project (restricted).

8. Fantoni P, Renders J (1998) On-Line Performance Estimation and Condition Monitoring using Neuro-Fuzzy Techniques. In: Proceedings of the Workshop on On-Line Fault Detection and Supervision in the Chemical Process Industries, Lyon, France.

9. Fantoni P (2000) A Neuro-Fuzzy Model Applied to Full Range Signal Validation of PWR Nuclear Power Plant Data. In: International Journal of General Systems, Vol. 29(2), pp 305-320.

10. Tou JT, Gonzalez R C (1974) Pattern Recognition Principles. Addison Wesley Publishing Company, Reading, MA.

11. Bezdek JC (1981) Pattern Recognition with Fuzzy Objective Function Algorithms. Plenum Press, New York, NY.

12. Gustafson DE, Kessel W C (1979) Fuzzy Clustering with a Fuzzy Covariance Matrix. In: Proceedings of IEEE CDC, San Diego, CA.

13. Krishnapuram R, Keller J (1993) A Possibilistic Approach to Clustering. In: IEEE Transactions on Fuzzy Systems, Vol. 1, No. 2, pp 98-100.

14. Bartal Y, Lin J, Uhrig RE (1995) Nuclear Power Plant Transient Diagnostics Using an Artificial Neural Networks that Allow "Don't-Know" Classifications. In: Nuclear Technology, Vol. 110.

15. Fantoni P, Mazzola A (1996) Accuracy Estimate of Artificial Neural Networks Based Models for Industrial Applications. In: Artificial Intelligence in the Petroleum Industry. Symbolic and Computational Applications 2, Chapter 17, B Braunschweig & B Bremdal Editions.

Index